S0-CAH-184

REACTIVE INTERMEDIATES

REACTIVE INTERMEDIATES

A Serial Publication

VOLUME 1

Edited by

MAITLAND JONES, JR.
Princeton University

ROBERT A. MOSS
Rutgers University

A WILEY-INTERSCIENCE PUBLICATION
JOHN WILEY & SONS
New York • Chichester • Brisbane • Toronto

CHEMISTRY
6237-1174

Copyright © 1978 by John Wiley & Sons, Inc.

All rights reserved. Published simultaneously in Canada.

Reproduction or translation of any part of this work beyond that permitted by Sections 107 or 108 of the 1976 United States Copyright Act without the permission of the copyright owner is unlawful. Requests for permission or further information should be addressed to the Permissions Department, John Wiley & Sons, Inc.

ISBN: 0-471-01874-1

Printed in the United States of America

10 9 8 7 6 5 4 3 2 1

CONTRIBUTORS

D. Bethell
Robert Robinson Laboratories
University of Liverpool
Liverpool, England

Peter P. Gaspar
Department of Chemistry
Washington University
Saint Louis, Missouri 63130

K. N. Houk
Department of Chemistry
Louisiana State University
Baton Rouge, Louisiana 70803

Maitland Jones, Jr.
Department of Chemistry
Princeton University
Princeton, New Jersey 08540

Leonard Kaplan
Union Carbide Corporation
Post Office Box 8361
South Charleston, West Virginia 25303

W. J. le Noble
Department of Chemistry
State University of New York
at Stony Brook
Stony Brook, New York 11794

Ronald H. Levin
Department of Chemistry
Harvard Universtiy
Cambridge, Massachusetts 02138

Walter Lwowski
Department of Chemistry
New Mexico State University
Las Cruces, New Mexico 88003

Robert A. Moss
Wright and Rieman Laboratories, Rutgers,
The State University of New Jersey
New Brunswick, New Jersey 08903

10154

QD476
A1R4
v.1
CHEM

PREFACE

Reactive Intermediates: A Serial Publication has several purposes. It is intended to preserve the currency of the Wiley series *Reactive Intermediates in Organic Chemistry* by providing authoritative, critical, and selective analysis of the relevant recent literature for each major type of reactive organic intermediate. To ensure the authoritative and critical nature of the coverage we have attempted to recruit expert authors for each subject area, insisting that each author, on the basis of published monographs or research articles, be clearly entitled to "expert status" in his field.

Reactive Intermediates: A Serial Publication is also intended to stand on its own as a critical survey of the recent literature, to be read for both current awareness and interpretive value by those actively involved or contemplating involvement in reactive intermediate research.

There are already several exhaustive annual surveys of organic chemistry or reaction mechanisms. *Reactive Intermediates* is manifestly not intended to compete with these publications. It is not intended to document the inexorable growth of the literature, but to carefully select and then evaluate those recent contributions that the authors believe require most urgent attention by students and researchers. Its goal is the stimulation of new and quality research.

The various authors were hence exhorted to be selective and critical. To further that end, they were given considerable latitude in matters of format and style. Inevitably, the contention between freedom and format has made for some loss of unity, but we feel it also has also fostered a livelier, more provocative, and, finally, more valuable book.

In one or two cases, individual authors decided that a more traditional, historically based survey could make a more potent contribution to the advance of their field than the more restricted, highly focused presentation that we generally favored. Contributions thus range from the more classical to the supercritical, with most falling between these extremes and close to the mark that we have chosen as normative.

Nominally, literature coverage in this volume of *Reactive Intermediates: A Serial Publication* is focused on 1975-1976. However, at the authors' discretion, particularly important contributions from 1973, 1974, and early 1977 have been included. It is anticipated that a second volume of *Reactive*

Intermediates will appear in 1980, covering the relevant literature for 1977 through 1979.

We hope that this initial volume of *Reactive Intermediates: A Serial Publication* will be received with the same spirit of enthusiasm for chemistry in which it was conceived and written.

Maitland Jones, Jr.
Robert A. Moss

Princeton, New Jersey
New Brunswick, New Jersey
May 1978

*As a convenience to the reader, the authors have marked with an asterisk those references that they consider most significant for detailed discussion, analysis, or rereading.

CONTENTS

REACTIVE INTERMEDIATES

1

ARYNES

RONALD H. LEVIN

Harvard University, Cambridge, Massachusetts 02138

I. INTRODUCTION

There were numerous significant advances in the field of aryne chemistry during 1975-1976. Several new species have been generated and the structure-reactivity profiles of some of the more familiar examples have become more fully characterized. These developments are discussed in the following sections.

II. *ortho*-BENZYNE

A. Isolation

In 1973 Chapman, Mattes, McIntosh, Pacansky, Calder and Orr reported the successful matrix isolation of *o*-benzyne.[1] Irradiation of either benzocyclobutenedione (**1**) or phthaloyl peroxide (**2**) at 8 K in an argon matrix led to the eventual formation of *o*-benzyne. The infrared (IR) spectrum was reported for the range 450 to 1650 cm^{-1}. When the *o*-benzyne was generated in a matrix containing furan and this matrix was subsequently allowed to warm to 50 K, there was a close correspondence between the disappearance of the IR bands at 469, 736, 849, 1038, 1053, 1451, 1607, and 1627 cm^{-1}, and the emergence of bands attributable to the *o*-benzyne-furan Diels-Alder adduct (**3**). This observation, coupled with the fact that the former signals were only produced when an appropriate *o*-benzyne precursor was irradiated, allowed these eight signals to be assigned to *o*-benzyne.

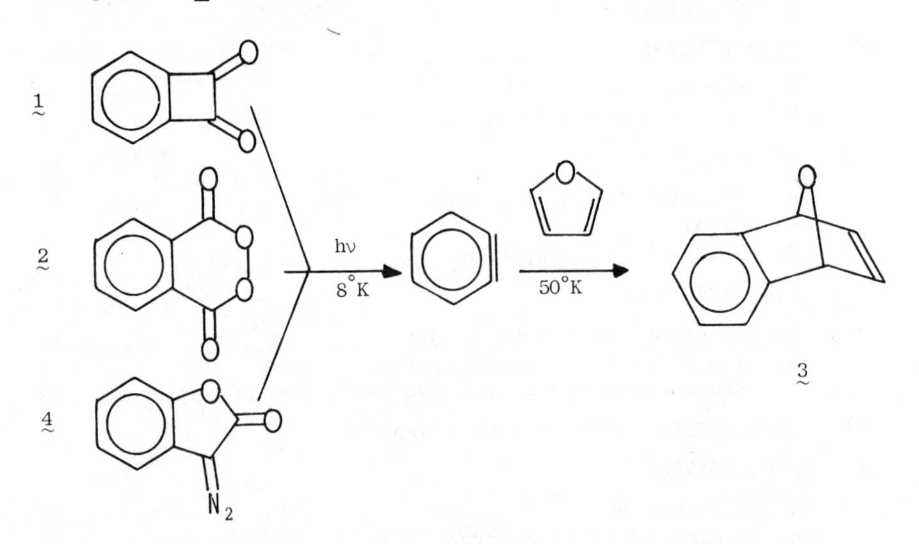

FIGURE 1.1. The low-temperature photolytic generation of *o*-benzyne

Approximately 2 years later Chapman's group reported that 3-diazobenzofuranone (4) was an even better precursor for matrix-isolated o-benzyne.[2] Significantly, the region of IR observation was now extended to 2600 cm^{-1}, and a weak absorption at 2085 cm^{-1}, again attributable to o-benzyne, was detected.

B. Structure

Utilizing these IR data, Berry and Laing carried out a normal coordinate analysis.[3] Their starting assumptions were minimal, namely, that o-benzyne would have C_2 symmetry and $\angle 123 \geq \angle 234$. An initial set of trial force constants was taken

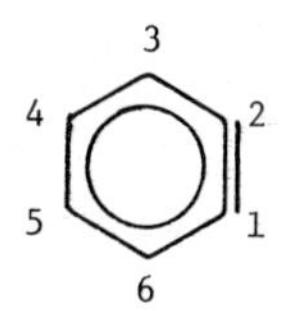

FIGURE 1.2. Numbering used for o-benzyne

from the literature using benzene as a model. However, due to the symmetry differences between o-benzyne and benzene, certain force constants had to be estimated. The input data were continually varied and refined until differences between the predicted and experimentally observed infrared frequencies were 1.1% or less. Berry's analysis actually led to the prediction of 15 IR-active bands within the region experimentally investigated. An explanation of the "missing" bands presumably has its origin in simple dipole-intensity arguments.

In addition to these frequencies, a normal coordinate analysis provides a set of force constants, bond angles, and bond lengths. The latter two items are tabulated in Table 1, column A. Minor changes in the force constants, bond angles, or bond lengths led back to IR frequencies that were quite different from those actually observed. In addition, several structures for o-benzyne suggested in the literature were also tested and again led back to IR frequencies divergent from those experimentally observed. Nonetheless, it must be realized that there is no guarantee that this normal coordinate analysis has provided a unique solution.

Since the calculated frequencies are much more sensitive to changes in the force constants than to variations in bond angles and lengths, these latter quantities are usually obtained through the use of empirical relationships. Badger's rule, for example, has been used to relate the equilibrium bond length

TABLE 1. o-Benzyne Structural Parameters

	A	B	C	D
r_{12}(Å)	1.304[a]	1.349[b]	1.297[c]	1.333[d]
r_{23}(Å)	1.414[a]	1.401[b]	1.385[c]	1.387[d]
r_{34}(Å)	1.384[a]	1.387[b]	1.417[c]	1.419[d]
r_{45}(Å)	1.414[a]	1.405[b]	1.419[c]	1.392[d]
∠123	123.0° [a]	-- [b]	127.4° [c]	126.5° [d]
∠234	118.5° [a]	-- [b]	109.7° [c]	109.5° [d]
∠345	118.5° [a]	-- [b]	122.9° [c]	124.0° [d]

3 4 2 5 1 6

[a] o-Benzyne parameters obtained from Berry's best fit to the IR frequencies.[3]

[b] o-Benzyne parameters obtained from Badger's rule using the best-fit force constants.[3]

[c] o-Benzyne structure obtained from MINDO/3 calculations with configuration interaction included.[4]

[d] Average values from the crystal structure of benzocyclopropene **5**.[5]

(R_e) to the stretching-force constant for that particular bond (K) through a set of predetermined constants (Y,Z); for carbon

$$K(R_e - Y)^3 = Z \quad (1)$$

skeletons, $\underline{Y} = 0.68$. Unfortunately, the limits at which such empirical relationships break down are usually ill-defined. In the present case $\underline{Z}$ was estimated using the stretching-force constant and carbon-carbon bond length from benzene. With these values of $\underline{Y}$ and $\underline{Z}$ and the o-benzyne force constants that yielded the best-fit IR frequencies, the o-benzyne bond lengths presented in Table 1, column B, were obtained. As one would expect, the disparity between columns A and B increases with approach to that area of the molecule where benzene is least appropriate as a model. Furthermore, working back from the Badger's rule structure leads to a set of IR frequencies that match poorly with those experimentally observed.

Returning to the structure that results from the best fit to the IR data, it is of interest to compare it to one of the geometries suggested by recent theoretical calculations, as well as to benzocyclopropene. In regard to the former, we have tabulated the results of Dewar's (MINDO/3 with CI) calculations in column C of Table 1.[4] The only benzocyclopropene whose structure has been reported is derivative **5**;[5] its pertinent structural details are presented in column D of Table 1. The changes from columns D to C are small, but in the directions expected for a further shortening of the C_1-C_2 bond. Berry's

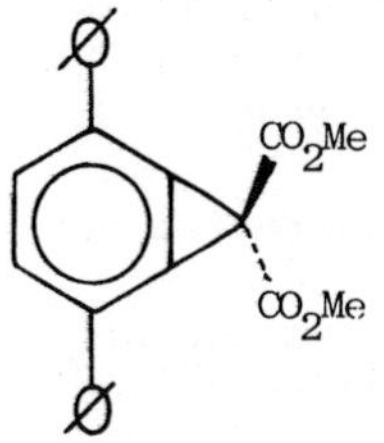

CO_2Me
CO_2Me
Ø
Ø

FIGURE 1.3. Benzocyclopropene **5**

suggested structure is strikingly different (in particular, note ∠234). Essentially, it may be viewed as a minimally deformed benzene ring. However, to achieve this comparatively relaxed geometry, some of the in-plane, dehydro-orbital overlap must be sacrificed. The MINDO/3, on the other hand, has apparently concluded that the resultant increase in the dehydro-orbital overlap more than compensates for the increased distortion energy attendant on molecular deformation.

A direct experimental evaluation of the Berry-Dewar model should be straightforward. Berry has already calculated the IR-active frequencies for o-benzyne-d_4 with his suggested geometry, and the same should be done for Dewar's structure. Preparation and observation of the IR spectrum of authentic perdeuterio o-benzyne should then provide a convincing and critical test.

III. meta-BENZYNE

A. Structure

While o- and p-benzyne have been the subjects of numerous experimental studies, their meta congener has received comparatively little attention. The few existing references to m-benzyne have generally considered the species depicted by structure 6. However, recent calculations by Dewar (MINDO/3

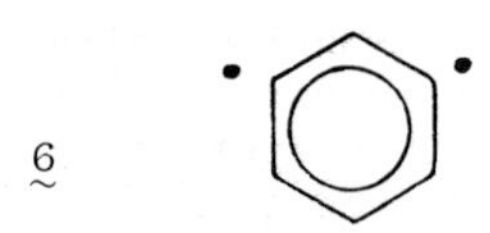

FIGURE 1.4. m-Benzyne (6)

with CI)[4] and Washburn (STO-3G)[6] have suggested that m-benzyne is best considered as a species with a singlet ground state, a stability comparable to that of o-benzyne, and perhaps most significantly, as a species with a considerable 1,5-bonding interaction. The essential structural features from these two calculations are presented in Table 2. In both cases, the $C_1 - C_5$ internuclear separation is predicted to be considerably reduced from the 2.416 Å found in benzene.

B. Generation and Properties

Considering that bicyclohexatriene 7 might provide a better description of the local energy minimum than 6, Washburn

FIGURE 1.5. Bicyclo[3.1.0]hexa-1,3,5-triene (7)

attempted its generation.[6] When exo,exo-2,6-dibromobicyclo-[3.1.0]hex-3-ene (8) was treated with potassium t-butoxide in tetrahydrofuran containing dimethylamine at -75° C, 6-dimethylaminofulvene (9) was rapidly produced in 90% yield. Three pathways were considered to explain this result. Elimination of HBr from the starting dibromo compound may occur in three ways to produce 10 through 12. Subsequent elimination of a second molecule of HBr could then produce 7, which in turn

TABLE 2. Calculated Structural Data for m-Benzyne

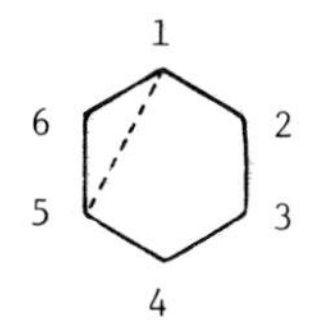

	Reference 4	Reference 6
r_{12}(Å)	1.389	1.391
r_{23}(Å)	1.411	1.389
r_{16}(Å)	1.373	1.364
r_{15}(Å)	2.120	1.500
$\angle 123$	117.1°	107.7°
$\angle 234$	115.2°	111.4°
$\angle 165$	101.1°	66.7°
$\angle 612$	134.8°	163.3°

could be captured by dimethylamine to yield **13**. A [1,5]-sigmatropic shift, known to be facile in the bicyclo[3.2.0]hepta-1,3-diene homolog, could then lead to the observed product, **9**. The possible intermediacy of 6-bromofulvene along the reaction coordinate leading from **8** to **9** was eliminated from consideration by showing that it was relatively stable under conditions that rapidly transformed a mixture of **8** and potassium t-butoxide into 6-t-butoxyfulvene. Arguments were also brought to bear against the intervention of **11** and **12** by noting that elimination from the model compound, **14**, occurred slowly under conditions more forceful than those required to effect the **8** → **9** interconversion.

Mechanisms in addition to those discussed by Washburn may also be developed. One such route was recently advanced by

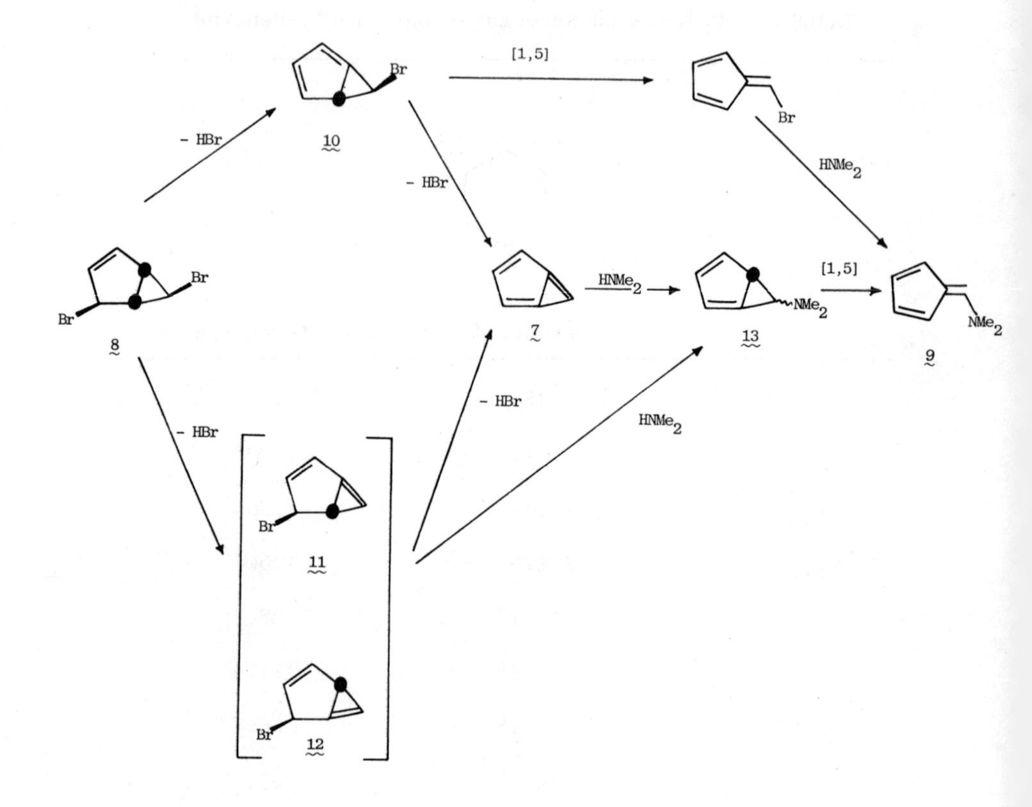

FIGURE 1.6. Possible mechanisms for conversion of 8 into 9

Amaro and Grohmann.[7] In this scheme a nucleophile attacks 10 with concomitant ring opening to 15; subsequent loss of bromide

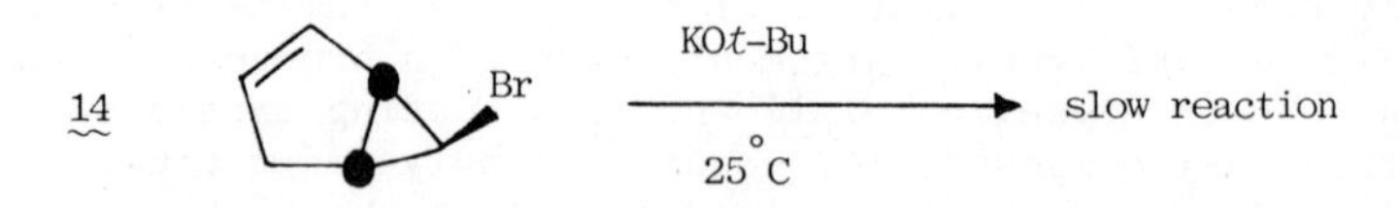

FIGURE 1.7. Elimination from 14 is slow

would produce the observed fulvene product. However, a panoply of deuterium, carbon-13, and halogen labeling studies recently reported by Washburn and Zahler vitiate such possibilities.[8]

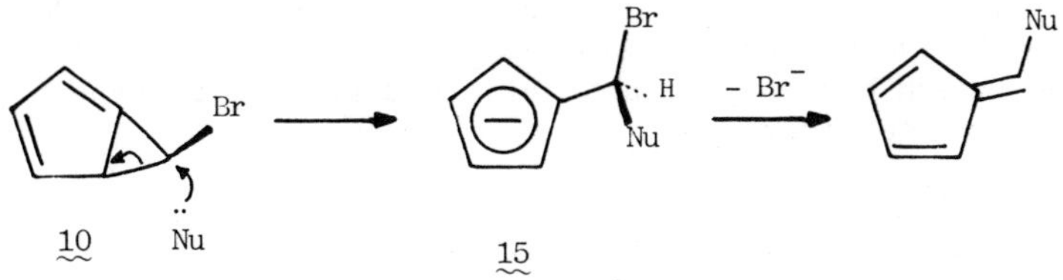

FIGURE 1.8. Nucleophilic attack on 10

For these experimental findings to remain consistent with the intermediacy of bicyclohexatriene 7, the intricate mechanism illustrated in Figure 1.9 is required. The complex scheme invokes reversible cyclopentadienyl anion formation to explain the partial loss and randomization of some of the deuterium labels. This anion may expel bromide to produce 7, which then suffers irreversible <u>trans</u>-1,4-addition of the elements of Nu-H. To account for these regio- and stereochemical observations, there was recourse to specific solvent cage effects and theoretical predictions of high electron density at C-2 and C-4 in 7. Further transformation into 13 again requires a highly specific shift of H_{syn} to reproduce the observed isotope distribution. This occurrence is also explained on the basis of PMO theory and computer calculations.

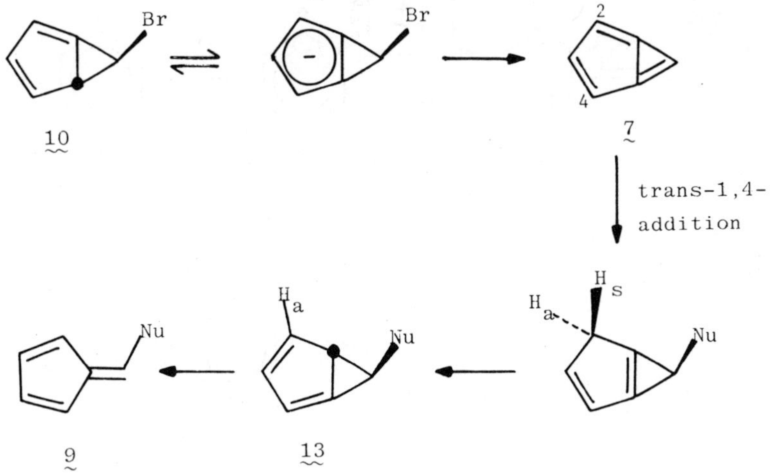

FIGURE 1.9. Mechanism of fulvene formation from 10

These findings may well be in accord with the intermediacy of **7**, but isotopic loss and scrambling hamper a definitive interpretation. Experiments designed to intercept **7** via cycloaddition (perhaps with diphenylisobenzofuran, *vide infra*) could prove most informative. Although further effort to establish conclusively the intermediacy of **7** seems most appropriate, Dewar and Washburn have supplied an alternative to *m*-benzyne **6** worthy of careful consideration.

IV. *para*-BENZYNE

A. Structural Possibilities

In 1972 Jones and Bergman reported that gas-phase or solution pyrolysis of *cis*-1,2-diethynylethylene yielded benzenoid products.[9] These authors advanced *p*-benzyne (**16**), formally the

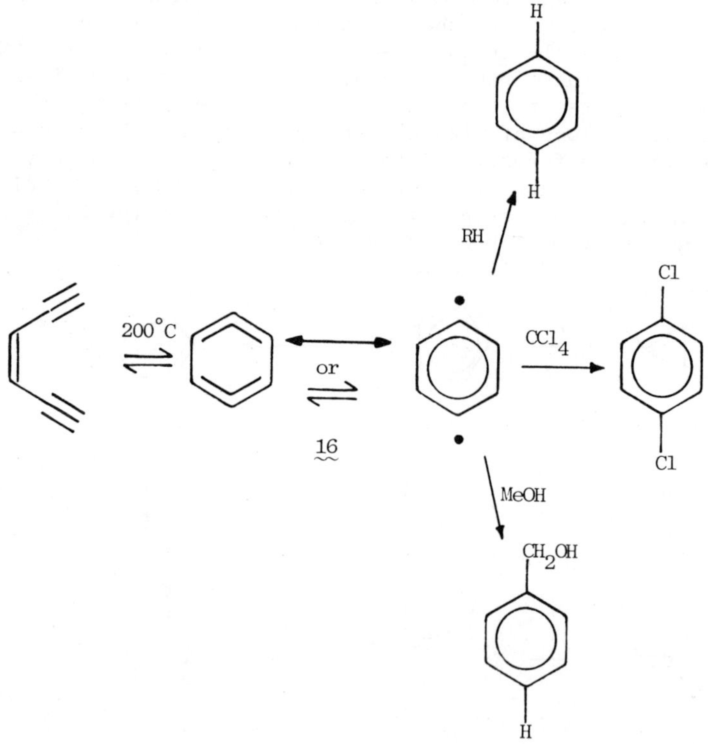

FIGURE 1.10. Generation and reactions of *p*-benzyne **16**

product of a Cope rearrangement from the ene-diyne, as a likely intermediate.

Shortly thereafter Dewar, using MINDO/3, found that diyl 16 should be more stable than the alternate p-benzyne structure, butalene (17).[10] Significantly, both structures were found to be local minima.

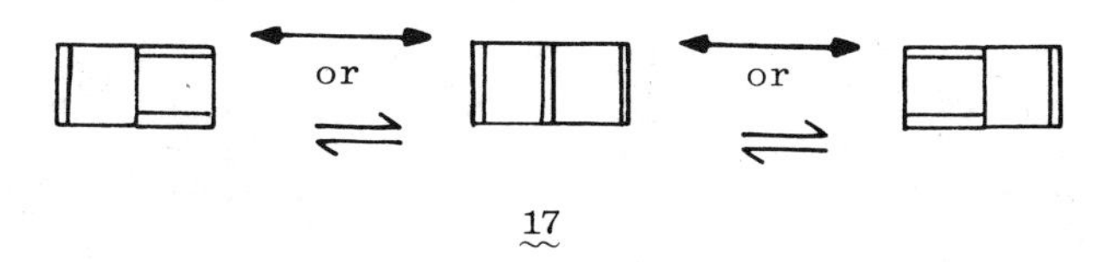

FIGURE 1.11. Butalene

B. Butalene

During 1975 Breslow's group reported on the generation and trapping of this p-dehydrobenzene isomer.[11] These workers found that treatment of 1-chlorobicyclo[2.2.0]hexadiene (18) with $LiNMe_2$ in $HNMe_2$-THF at 0^0 C produced N,N-dimethylaniline in quantitative yield. Repetition of this experiment in the presence of $DNMe_2$-THF gave aniline with the label distribution pictured in Figure 1.12. At -35^0 C, para-deuteration still amounted to 76%, but ortho- and meta-deuteration were substantially reduced.

Cl H

18

$LiNMe_2$ → $DNMe_2$ THF

NMe_2 .48D .48D .15D .15D .76D

6.3% d_0
25.0% d_1
38.0% d_2
22.0% d_3

FIGURE 1.12. Deuterium distribution in N,N-dimethylaniline produced from 18

If the reaction was effected at -35^0 C in the presence of both $DNMe_2$ and diphenylisobenzofuran (DPIBF), adduct 19, with deuterium located as shown (within experimental error), is obtained in 10% to 15% yield. As illustrated in Figure 13, this result could be explained by two different mechanisms, one requiring DPIBF to trap 17, and the other obliging DPIBF to intercept Dewar-aniline 20. This latter alternative seems unlikely, however, since Dewar benzene 18 did not react with DPIBF.

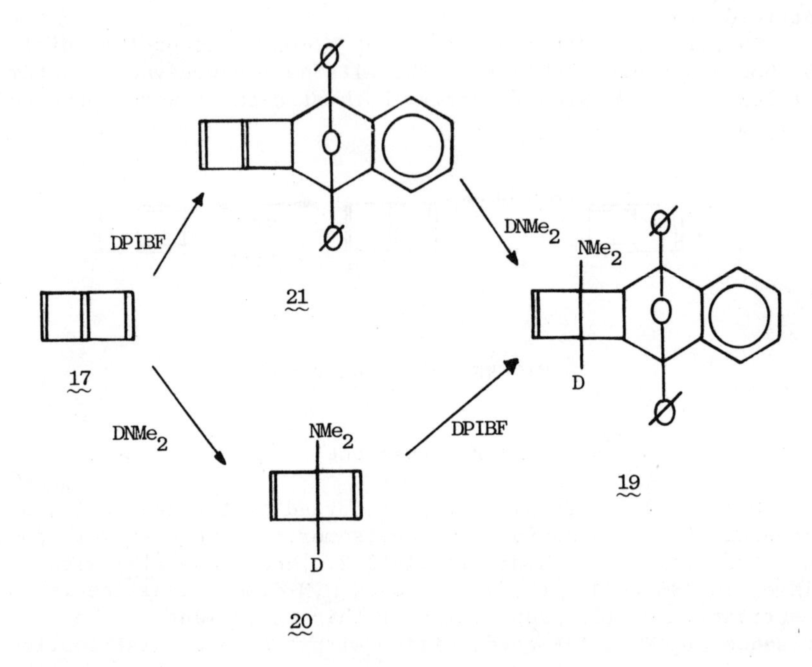

FIGURE 1.13. Possible mechanisms for conversion of 17 into 19

In addition to the experiments described in the preceding paragraphs, the appropriate controls were run to demonstrate that chlorobenzene is not converted to N,N-dimethylaniline, nor does this aniline incorporate deuterium under the reaction conditions. All of these observations may be accommodated if butalene is a reaction intermediate. Diels-Alder trapping with DPIBF would be expected to produce cyclobutadiene 21, which could then add $DNMe_2$ across the central double bond to afford the least strained bicyclo[2.2.0]hexane derivative 19. In the absence of DPIBF, the deuteroamine apparently adds to butalene. As depicted in Figure 1.14, several modes of addition are possible. For some of these adducts to rearrange to benzene derivatives, hydrogen shifts are required. Indeed, in a footnote, Breslow reports the detection of an unstable intermediate involved in the latter stages of the reaction.[11] Should these migrations be base-catalyzed, a ready explanation for the formation of d_2 and d_3 species, as well as for ortho- and meta-deuterioanilines, would be at hand. Furthermore, the relative propensities toward 1,2, 2,3, and 1,4 addition could be temperature-dependent, accounting for the marked change in deuterium distribution with temperature.

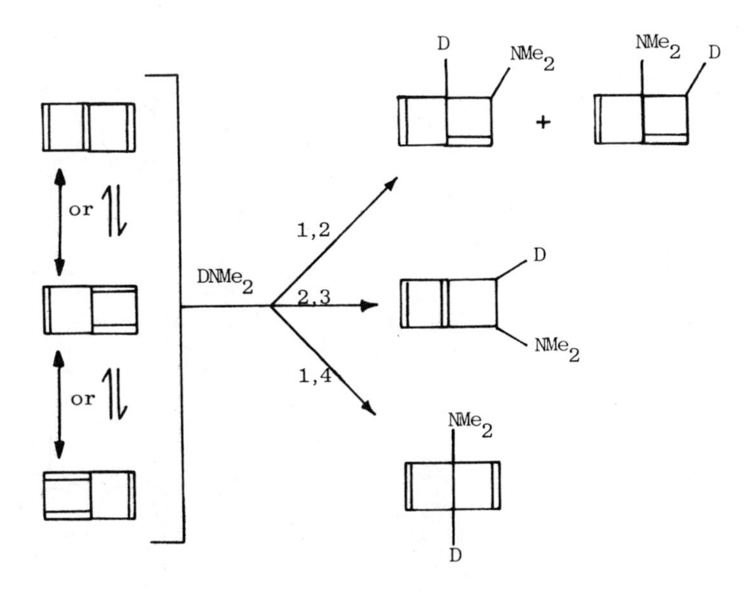

FIGURE 1.14. Possible modes of deuteroamine addition to butalene

These results serve to establish butalene as a viable intermediate. The differing chemistries observed for 16 and 17 confirm Dewar's suggestion that two distinct isomers of *para*-benzyne are capable of existence.[4b]

C. 9,10-Dehydroanthracene

More recently, Chapman, Chang, et al. have used the matrix-isolation technique in an attempt to observe and characterize 9,10-dehydroanthracene (22).[12] Photolysis of 23 at 77 K in a hydrocarbon glass containing 5% carbon tetrachloride produced both anthracene and 9,10 dichloroanthracene on warming. These results suggest the intermediacy of 22, the dibenzo analog of diyl 16.

Photolysis at even lower temperatures allowed the ultraviolet (UV) spectrum of a reactive intermediate to be recorded. Attempts to observe an ESR signal were unsuccessful. This latter result tentatively removes carbenic structures, which in this system would presumably exist as ground-state triplets, and radicals from consideration. Somewhat tenuous arguments can also be made against dibenzobutalene 24 and 1,5-didehydro-3,4-benz[10]annulene 25, in that a different UV spectrum might be expected. Diyl 22 is consistent with the observed spectrum, since a basic anthracene pattern in combination with several

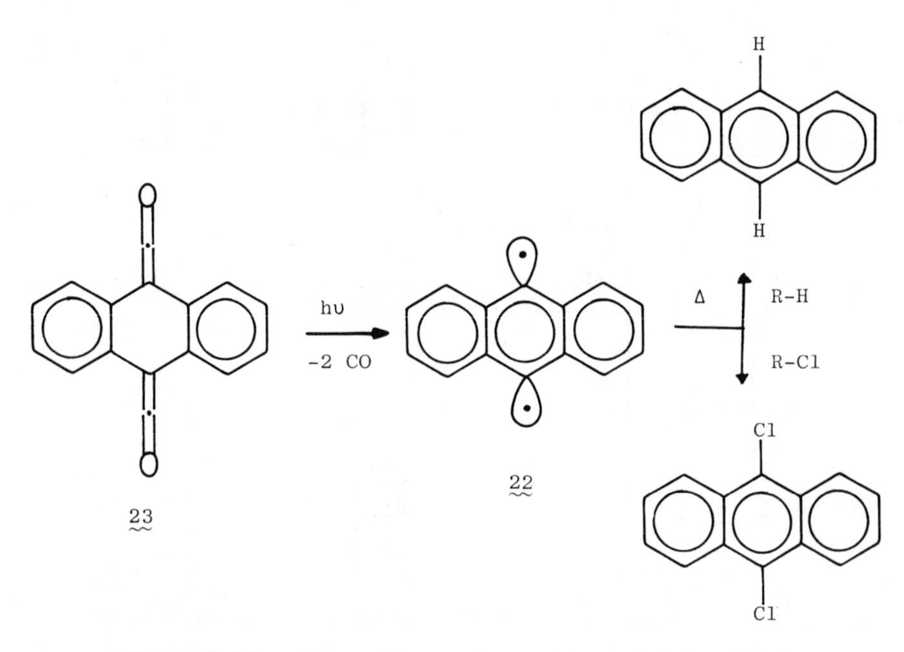

FIGURE 1.15. Generation and trapping of 9,10-dehydroanthracene (22)

long wavelength absorptions[13] is observed. Although the intermediacy of 22 seems established, further characterization of the observed transient appears warranted.

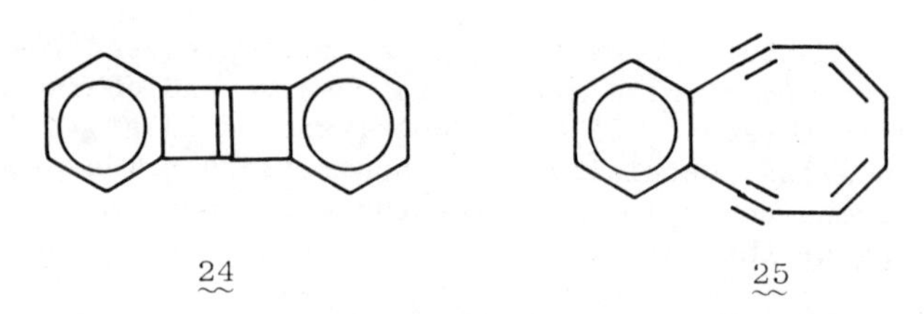

FIGURE 1.16. Dibenzobutalene (24) and 1,5-didehydro-3,4-benz[10]annulene (25)

V. 1,8-NAPHTHYNE

There have been two recent reports on the chemistry of 1,8-dehydronaphthalene (26).[14,15] One article treated the reactions of 26 with acyclic dienes,[14] and the other explored the additions of 26 to cyclic polyenes.[15]

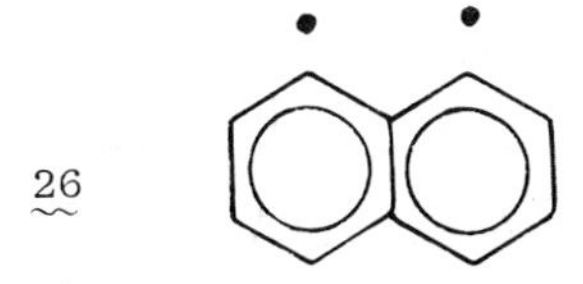

FIGURE 1.17. 1,8-Naphthyne (26)

A. Reactions with Acyclic Dienes

The data from the former study are presented in Table 3, and several interesting trends can be discerned. The formation of the 1,2-adducts, presumably a concerted process,[16] is sensitive to steric effects (cf. D/E and L/M). Noticeably, however, one of the most highly substituted 1,2-adducts, Q, is formed in highest relative yield. Since the 1,2- and 1,3-addition modes must be competitive, the relative increase in Q is probably reflecting a retardation in the rate of formation of 1,3-adduct R. The 1,3-adducts are thought to arise via biradical 27. The rate of formation of such biradicals should be sensitive to both the degree of substitution at the terminus attacked and the degree of substitution on the resultant allyl moiety. Increased terminal substitution should impede initial biradical formation. This interpretation provides a ready explanation for the Q/R ratio. Reducing the degree of terminal substitution (cf. N/R) results in an increase in the relative amount of 1,3-adduct as expected. Turning to the effect of substitution on the allyl radical fragment, substitution at C_1 and/or C_3 of the residue should be stabilizing. Indeed, a comparison of the relative yields of B, F, and N, where only the degree of methylation on the allyl moiety is varied, shows an increase in the expected order.

The abstraction products are also thought to arise from biradical 27. If the two radical centers cannot assume the appropriate orientation, 1,3-addition is precluded and radical abstraction should dominate. Judging from models, methylation at C_1 or C_2 of the allyl fragment should decrease the population of the 27-*syn* conformations, while methylation at C_3 should have little or no effect. As anticipated, the relative amount of abstraction product increases along the series P, C, H, K.

TABLE 3. Products from the Addition of 1,8-Naphthyne to Acyclic Dienes[14]

Diene	Total Yield	1,2-Adducts[a]	1,3-Adducts[a]	Abstraction[a]
	10%	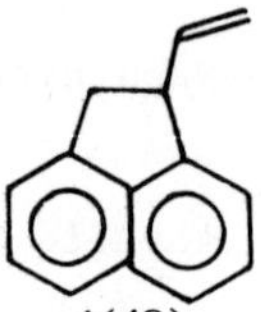 A(43)	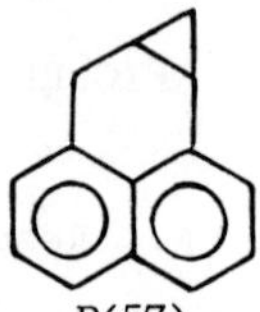 B(57)	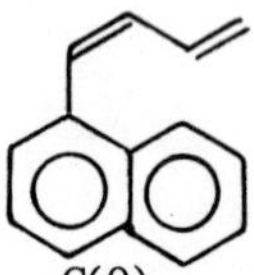 C(0)
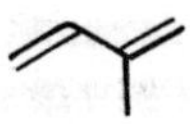	-	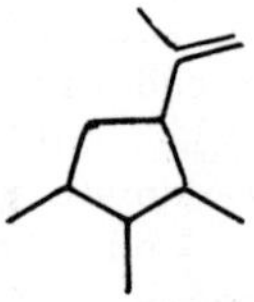 D(18) 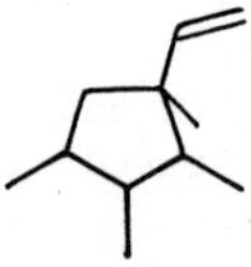 E(10)	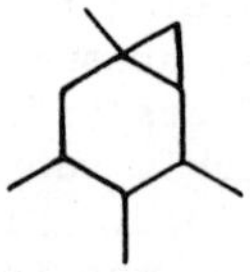 F(68) 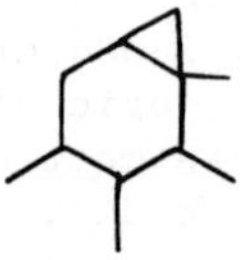 G(0)	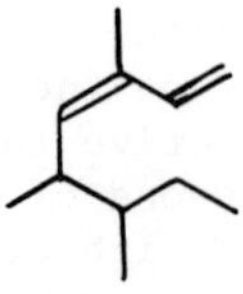 H(4)
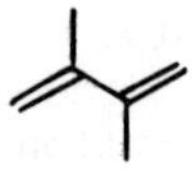	-	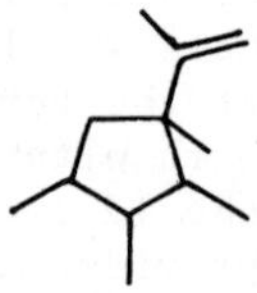 I(46)	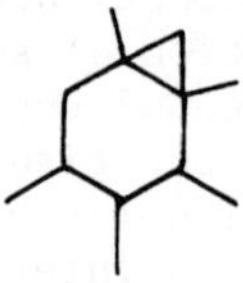 J(26)	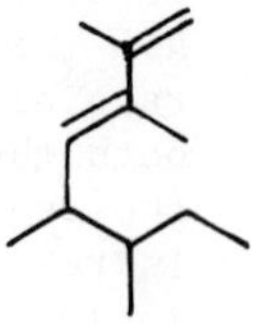 K(28)
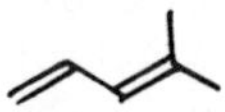	-	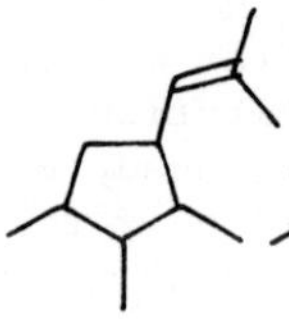 L(23) 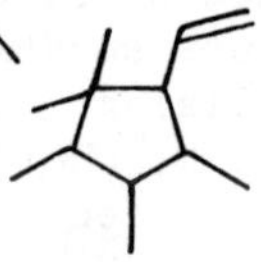 M(4)	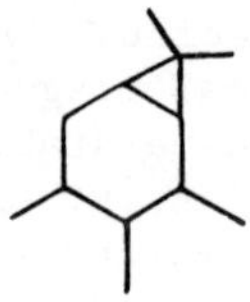 N(72) 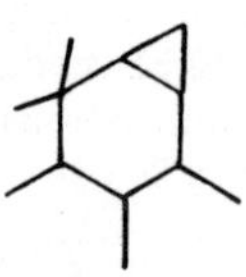 O(0)	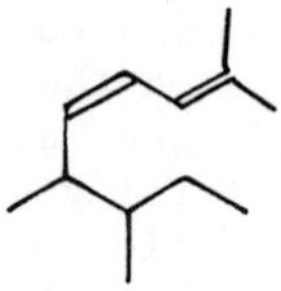 P(0)
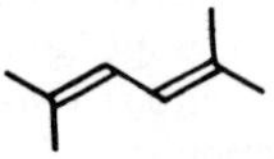	-	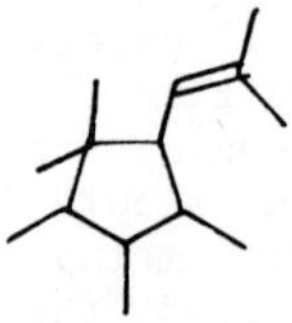 Q(74)	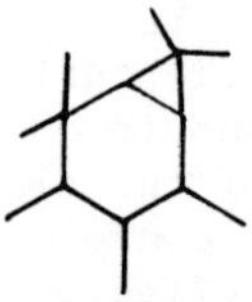 R(19)	—

[a] The relative product yields are given in parentheses.

B. Reactions with Cyclic Polyenes

The other study involving cyclic polyenes[15] is often sketchy in experimental detail, but several observations merit attention. The systems investigated were cycloheptatriene, cyclopentadiene, 6,6-dimethylfulvene, furan, and norbornadiene.

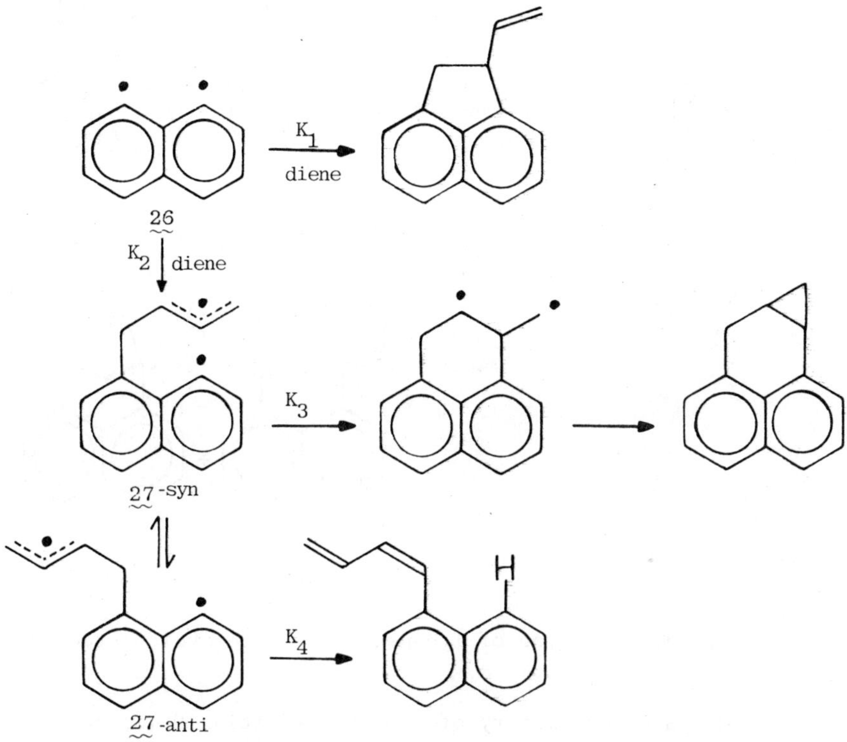

FIGURE 1.18. Reactions of 1,8-naphthyne with acyclic dienes

Abstraction and/or 1,2-addition were generally found to predominate, and only in the case of cyclopentadiene was a 1,3-adduct obtained. Curiously, the 1,4-adduct, 31, was absent. In an earlier investigation, 31 was shown to have been present in 2% overall yield and to have been a primary product.[19] By the same token, 30 was not detected in the earlier[19] investigation.

1,8-Dehydronaphthalene is generated by the low-temperature lead tetraacetate oxidation of 32. Clearly, the various metallo-coordinated species present could play havoc with the product distribution as well as the spin state of 26.

Experiments directed toward the generation of 26 from alternate sources would be quite valuable.

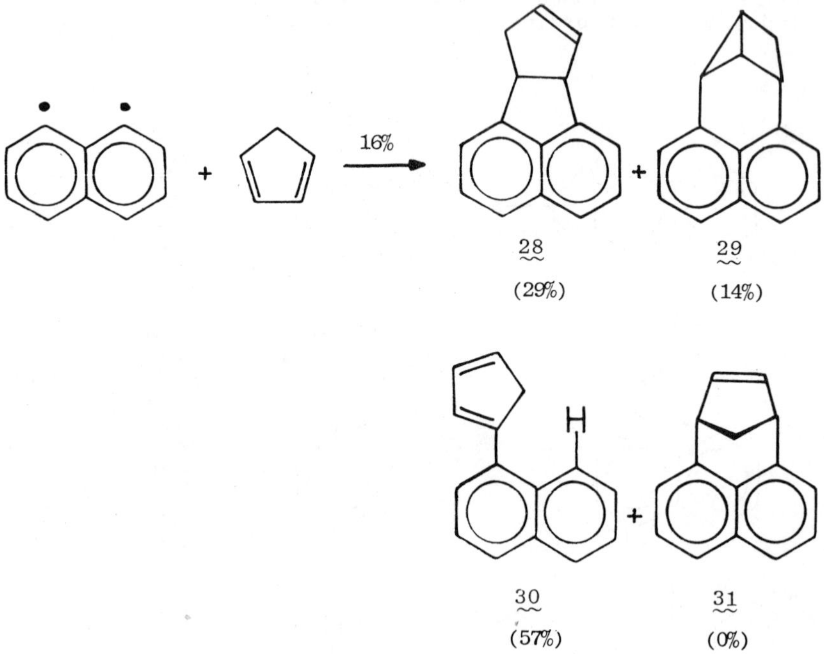

FIGURE 1.19. Reaction of 1,8-naphthyne with cyclopentadiene

C. Stereochemistry of the Cycloaddition Processes

The interesting cycloaddition of 26 to the stereoisomeric hexa-2,4-dienes has also been studied by Meinwald's group.[20] It was found that the 1,2-cycloadducts were formed stereospecifically, whereas the same mixture of 1,3-adducts was formed from both *cis,cis*- and *trans,trans*-hexa-2,4-diene. These results provide further support for the concerted and stepwise nature of the 1,2- and 1,3-cycloaddition modes of 1,8-dehydronaphthalene, respectively.

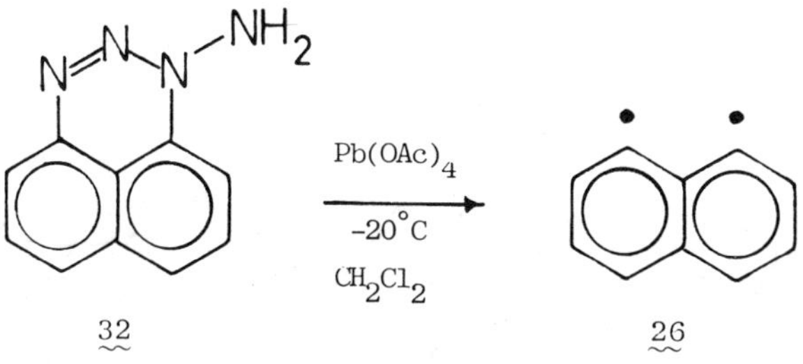

NH_2

$\xrightarrow[\text{-20°C}]{Pb(OAc)_4}$

CH_2Cl_2

32   26

FIGURE 1.20. Generation of 1,8-naphthyne

VI. NORBORNYNE: MECHANISM OF FORMATION

Although this chapter is entitled "Arynes," the coverage has been extended to include other cyclic acetylenes of interest. In this category falls the remarkable report by Gassman and Valcho on the generation of norbornyne (33).[21]

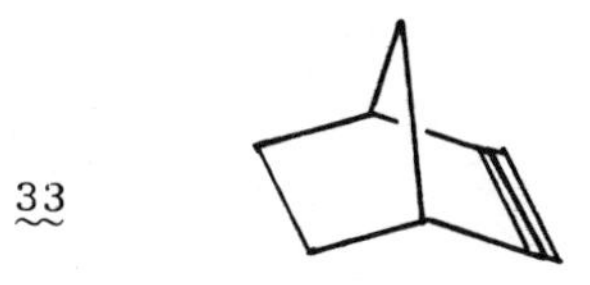

FIGURE 1.21. Norbornyne (33)

The reactions of organolithium reagents with 2-chlorobicyclo[2.2.1]hept-2-ene (34) have a rich and varied chemistry as illustrated in Figure 22. Suffice it to say that 35 is produced via a direct coupling reaction,[22] whereas 36 through 38 have carbenic origins.[21,23]

Focusing now on 39, three critical experiments aimed at determining its origin were performed.[21] When 34 was treated with n-butyllithium and the reaction quenched with D_2O, 38 and 39 were formed in a 1.0 : 1.6 ratio, the latter being labeled to the extent of 87% at C-3. This is consistent with the production of 40 in the penultimate step leading to at least 87% of the 39 formed. If instead 34-d_1, labeled as shown in Figure 1.24, was used as the substrate, then the 38:39 ratio was dramatically altered to 16:1 and a large fraction of the 39 produced was undeuterated. The striking shift in the product ratio following isotopic substitution at C-3 suggests that the

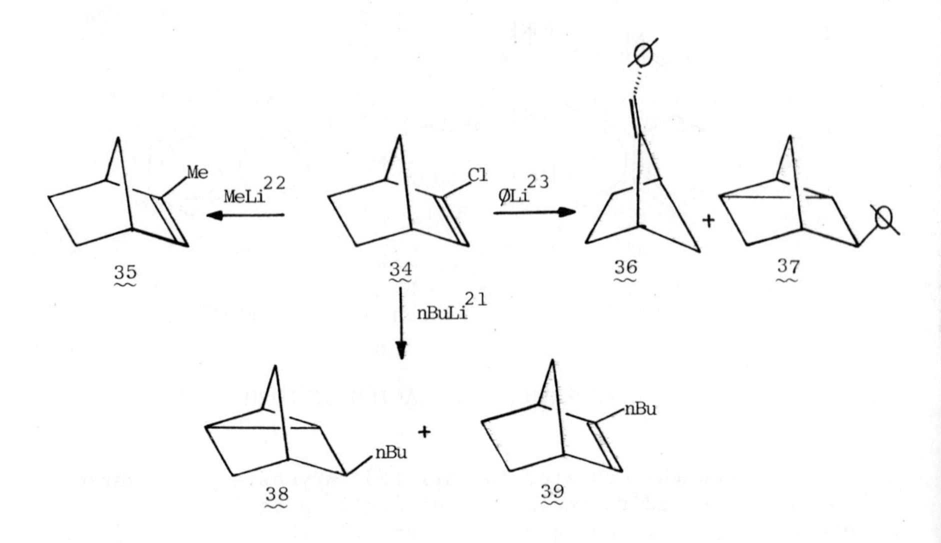

FIGURE 1.22. Reactions of 2-chloronorbornene (34) with organolithium reagents

FIGURE 1.23. Quenching of 40 with D_2O

rate-determining step in the formation of 39 involves proton (deuteron) abstraction to yield 41. The retention of 19% (0.15 : 0.81) deuterium at C-3 parallels the formation of 13% 39-d_0 in the earlier D_2O quench experiment and again suggests that two mechanisms (85% : 15%) may be involved in the production of 39.

In a final experiment, optically active (+)-34 was treated with n-butyllithium in the normal fashion to produce 39 with the enantiomeric composition illustrated in Figure 1.26. This result again supports the dual-mechanism hypothesis with the bulk of 39 being derived from an achiral precursor and a minor amount (within experimental error of the 13 to 19% estimates

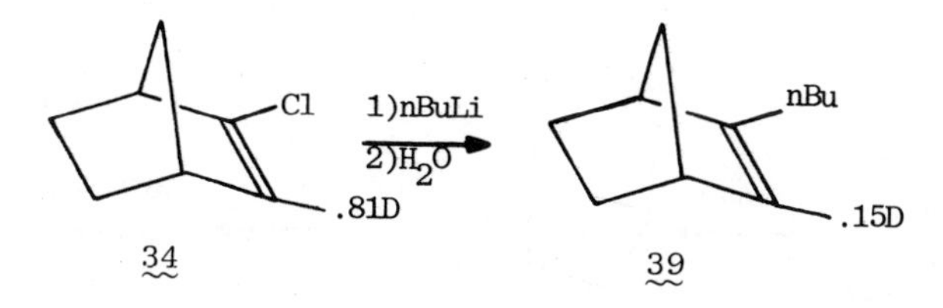

FIGURE 1.24. Reaction of deuterated 2-chloronorbornene with alkyllithium

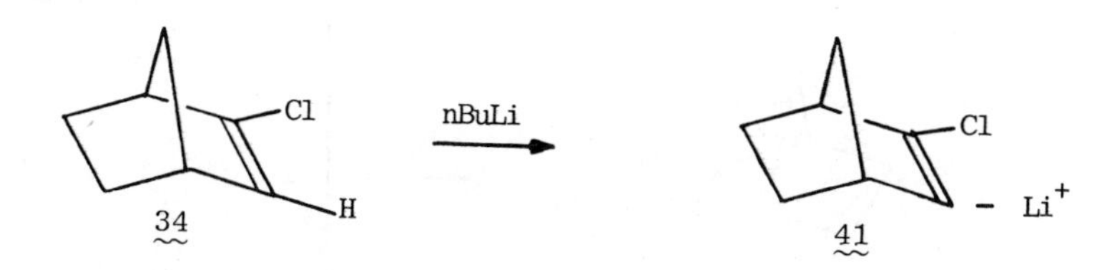

FIGURE 1.25. The rate-determining step in conversion of 34 into 39

derived in the preceding paragraphs being formed by a direct coupling route (cf. 35) with retention of configuration.[22]

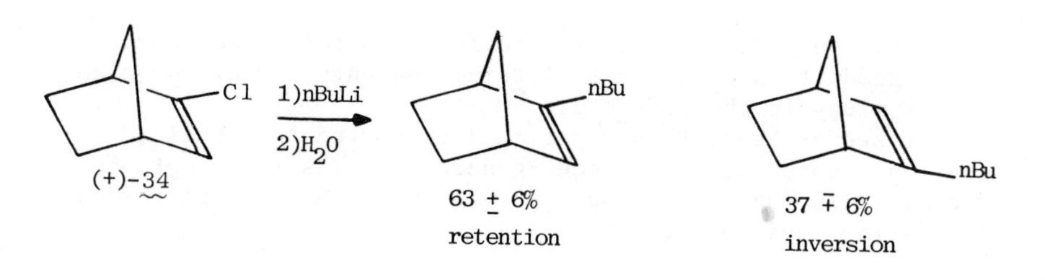

FIGURE 1.26. Loss of optical activity in conversion of 34 into 39

The accumulated data strongly suggest that 52%[24] of the 2-chloronorbornene molecules react via the scheme depicted in Figure 1.27. The most straightforward route from 41 to a symmetric intermediate would involve the precedented

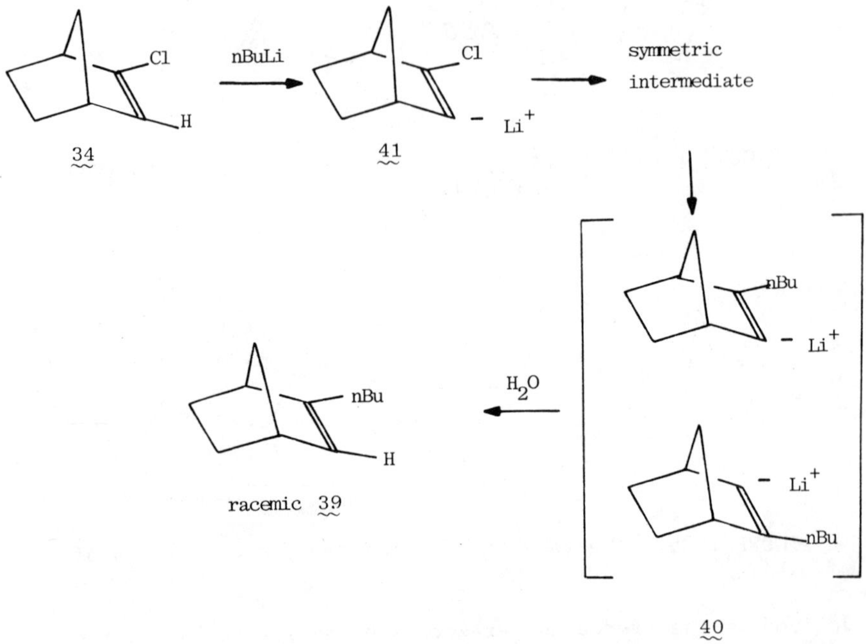

Cl nBuLi Cl symmetric intermediate
H - Li+
34 41
nBu
- Li+
H_2O
nBu
H
- Li+
racemic 39
nBu
40

FIGURE 1.27. Mechanistic scheme for conversion of 34 into 39

elimination of lithium halide[25] to form, in this case, norbornyne. Readdition of an organolithium reagent to yield 40 also finds analogy in the literature.[25]

It seems clear, then, that this contorted and strained acetylene, norbornyne, has been generated. Its interception by other trapping agents could provide an array of bizarre and interesting substances.

VII. MISCELLANEOUS

A. 2,3-Thiophyne

Numerous problems have surrounded the area of hetaryne chemistry over the years. Recently, however, a report on the flash vacuum pyrolysis of the anhydride of thiophene-2,3-dicarboxylic acid has presented evidence to strongly support the intermediacy of 2,3-thiophyne (42).[26] Most notable is the interception of 42 by unactivated dienes such as 2,3-dimethylbutadiene and furan, in a 4+2 fashion.

FIGURE 1.28. Cyclobuta[1,2-d]benzyne

B. Cyclobuta[1,2-d]benzyne

The interesting molecule cyclobuta[1,2-d]benzyne (43) has also been advanced as a reaction intermediate.[27] In addition to providing an intriguing dimer, it was argued that this aryne should be stabilized by the ring fusion; some preliminary evidence supporting this position was advanced. The stabilization arguments were based on the structural differences between benzene and benzocyclobutene as applied to computationally derived structures for o-benzyne. As discussed earlier, the o-benzyne geometry resulting from Berry's normal coordinate analysis[3] differs substantially, especially in regard to bond angles, from the most recent theoretically deduced structure. It will, therefore, be of interest to see if the more definitive experiments proposed by Hillard and Vollhardt[27a] bear out their earlier predictions of increased stability for 43.

C. Polymer-Bound o-Benzyne

An interesting report on the use of polymer supports to effectively stabilize arynes has appeared.[28] The authors reasoned that if the average aryne-aryne separation could be increased above what it is in normal solution, the rate of bimolecular dimerization leading to biphenylenes should be reduced. If attachment of the aryne precursor to a resin would achieve this goal, then the average lifetime of these polymer-bound arynes should be increased. This approach makes use of the simple relationship between concentration and bimolecular rates, but it achieves the desired reduction in concentration via substrate immobilization. Hence it was suggested that the term "pseudodilution" be used to describe this phenomenon.

When an appropriately functionalized o-benzyne precursor was affixed to the resin and the aryne subsequently generated, none of the biphenylene-like dimer was detected ($< 0.3\%$). If the same basic precursor was used to generate the aryne without prior attachment to the resin, the dimers were produced in 65% yield. To further demonstrate the validity of this approach, the polymer-bound aryne was generated as above and a time, t, allowed to transpire before a trapping agent was added. The yield of adduct was found to decrease with both time and increasing functionalization of the support. Although the process responsible for the decrease in adduct yield with time was not identified, the aryne half-life was estimated to be in the neighborhood of 50 s. This is to be contrasted with the gas-phase value, which is on the order of a few hundred microseconds.[29]

VIII. REFERENCES

1.* O. L. Chapman, K. Mattes, C. L. McIntosh, J. Pacansky, G. V. Calder, and G. Orr, J. Am. Chem. Soc., 95, 6134 (1973).

2.* O. L. Chapman, C.-C. Chang, J. Kolc, N. R. Rosenquist, and H. Tomioka, J. Am. Chem. Soc., 97, 6586 (1975).

3.* J. W. Laing and R. S. Berry, J. Am. Chem. Soc., 98, 660 (1975).

4.* (a) M. J. S. Dewar, Pure and Applied Chemistry, 44, 767 (1975) (only minor changes were found in his earlier paper); (b) M. J. S. Dewar and W.-K. Li, J. Am. Chem. Soc., 96, 5569 (1974).

5. E. Carstensen-Oeser, B. Müller, and H. Dürr, Angew.Chem., Int. Ed., 11, 422 (1972).

6.* W. N. Washburn, J. Am. Chem. Soc., 97, 1615 (1975).

7. A. H. Amaro and K. Grohmann, J. Am. Chem. Soc., 97, 5946 (1975).

8.* W. N. Washburn and R. Zahler, J. Am. Chem. Soc., 98, 7827, 7828 (1976).

9. R. R. Jones and R. G. Bergman, J. Am. Chem. Soc., 94, 660 (1972).

10. See reference 4(b) above. The calculated energy difference between 16 and 17 varied between 19 and 36 kcal/m depending on the electronic state of 16 and the use or neglect of CI.

11.* R. Breslow, J. Napierski, and T. C. Clarke, J. Am. Chem. Soc., 97, 6275 (1975).

12.* O. L. Chapman, C.-C. Chang, and J. Kolc, J. Am. Chem. Soc., 98, 5703 (1976).

13. Such absorptions typify systems with a radicaloid center situated perpendicular to a π system. See, for example, I. Moritani, S.-I. Murahashi, H. Ashitaka, K. Kimura, and H. Tsubomura, J. Am. Chem. Soc., 90, 5918 (1968).

14.* J. Meinwald, L. V. Dunkerton, and G. W. Gruber, J. Am. Chem. Soc., 97, 681 (1975).

15.* M. Kato, S. Takaoka, and T. Miwa, Bull. Chem. Soc. Jap., 48, 932 (1975).

16. The cycloaddition of 1,8-dehydronaphthalene to simple olefins has been found to proceed stereospecifically (see, however, the discussion on addition to dimethylfumarate and maleate).[17] Hoffmann and co-workers suggest that the HOMO of structure 26 is antisymmetric, thereby permitting such 1,2-additions to proceed in a concerted manner.

17. C. W. Rees and R. C. Storr, J. Chem. Soc. C, 765 (1969).

18. R. Hoffmann, A. Imamura, and W. J. Hehre, J. Am. Chem. Soc., 90, 1499 (1968).

19. J. Meinwald and G. W. Gruber, J. Am. Chem. Soc., 93, 3802 (1971).

20. L. V. Dunkerton, Ph.D. thesis, Cornell University; Diss. Abstracts, 35, 5327-B (1975).

21.* P. G. Gassman and J. J. Valcho, J. Am. Chem. Soc., 97, 4768 (1975).

22. (a) P. G. Gassman and T. J. Atkins, Tetrahedron Lett., 3035 (1975); (b) P. G. Gassman, J. P. Andrews, Jr., and D. S. Patton, J. Chem. Soc., Chem. Commun., 437 (1969).

23. P. G. Gassman and T. J. Atkins, J. Am. Chem. Soc,, 92, 5810 (1970).

24. Figured at (1.6/2.6)(0.85).

25. (a) L. K. Montgomery and L. E. Applegate, J. Am. Chem. Soc., 89, 2952 (1967); (b) L. K. Montgomery, A. O. Clouse, A. M. Crelier, and L. E. Applegate, J. Am. Chem. Soc., 89, 3453 (1967); (c) L. K. Montgomery, F. Scardiglia, and J. D. Roberts, J. Am. Chem. Soc., 87, 1917 (1965); (d) L. K. Montgomery and J. D. Roberts, J. Am. Chem. Soc., 82, 4750 (1960); (e) F. Scardiglia and J. D. Roberts, Tetrahedron, 1, 343 (1957); (f) G. Wittig and G. Harborth, Chem. Ber., 77, 306 (1944); (g) see also G. Wittig, J. Weinlich, and E. R. Wilson, Chem. Ber., 98, 458 (1965); (h) G. Wittig and P. Fritz, Angew. Chem., Int. Ed., 5, 846 (1966); (i) A. T. Bottini, F. P. Corson, R. Fitzgerald, and K. A. Frost, III, Tetrahedron, 28, 4883 (1972); (j) G. Wittig and J. Heyn, Justus Liebigs Ann. Chem., 756, 1 (1972); (k) G. Kobrich, Angew. Chem., Int. Ed., 11, 473 (1972).

26.* M. G. Reinecke and J. G. Newsom, J. Am. Chem. Soc., 98, 3021 (1976).

27.* (a) R. L. Hillard, III and K. P. C. Vollhardt, J. Am. Chem. Soc., 98, 3579 (1976); (b) R. L. Funk and K. P. C. Vollhardt, Angew. Chem., Int. Ed., 15, 53 (1976).

28.* P. Jayalekshmy and S. Mazur, J. Am. Chem. Soc., 98, 6710 (1976).

29. M. E. Schafer and R. S. Berry, J. Am. Chem. Soc., 87, 4497 (1965), and references cited therein.

2

CARBANIONS

W. J. le NOBLE

Department of Chemistry
State University of New York at Stony Brook
Stony Brook, New York 11794

I. INTRODUCTION

Much pleasure has surely been derived in recent years by those organic chemists who fear that their science is drying up, from two developments that seem to say that our understanding of carbanions, the most pervasive species in all of organic chemistry, has important kinks in it. One of the new directions is based on experimental information concerning gas-phase acidities and the other, on the insight that the behavior of presumed carbanions is, in fact, often that of ion pairs. It becomes clear that much of the "truth" about carbanions to be found in current textbooks, however logical and convenient from a pedagogical point of view, is incorrect. Who would have suspected, even 5 years ago, that the well-known, well-studied, and well-understood effects of polar substituents on the strength of simple acids in solution would mysteriously disappear or even be reversed in the gas phase, or that the nucleophilicity of anions would be so drastically dependent on the proximity of counterions or solvent molecules!

Whereas this suggests that carbanion chemistry may have been on the wrong track, it is not at all obvious where the newly arriving information is going to take us. My objective in writing this chapter has not been to speculate about this, but rather to describe in some detail what facts have become available, and which older ideas have thereby been invalidated. In addition, a number of tentative responses, further results, and new techniques are described, which, it may be hoped, will instigate those colleagues who have not yet given much thought to these matters. There's room here; there are wide open spaces here, in fact, for new ideas, and for new research programs, large or small.

Many of my remarks are concerned with anions in general, but, because any new information concerning stable anions must ultimately also affect our knowledge of carbanion intermediates, one trusts I may be forgiven for these liberties, taken with the instructions of the editors.

II. ACIDS AND ACID STRENGTHS

A. Traditional Methods and Results

The ubiquity of the equilibrium, Eq. (1),

$$HA \rightleftharpoons H^+ + A^- \qquad (1)$$

its simplicity, and the incredible range of values of the equilibrium constant (80 powers of 10) have easily made it the most-studied chemical reaction. In the early stages, chemists studied those acids for which the tendency to donate a proton to water could be measured by means of suitable indicators; with the invention of the glass electrode, it became possible to obtain a direct measure of the activity of H^+ and hence to determine the true thermodynamic value of the pK_a in water to perhaps ± 0.02 units. Because many organic acids have only limited solubilities in water and because pairing becomes a problem in pure alcoholic solvents, it became customary to use 50% aqueous ethanol as the solvent, at 25°C and 1 atm. A wide variety of acids was thus examined; trends were noted, and these were gathered and summarized by Brown, McDaniel, et al. in their classic review of 1955.[1] These trends may be briefly recapitulated, as follows.

Polar substituents have large electrostatic effects; thus a nearby carboxylate center substantially raises the pK_a of a carboxylic acid, whereas an ammonium group lowers it. Neutral electronegative atoms increase the acid strength; thus the pK_a values of acetic acid and of fluoro-, chloro-, bromo-, and iodoacetic acid are 4.76, 2.66, 2.86, 2.86, and 3.12, respectively. Internal hydrogen bonds may affect the acid strength in either direction, depending on whether they operate more strongly on the acid or on the anion. The second ionization constant of oxalic acid is relatively small, for example (pK_a = 4.19); the acid strength of 2,6-dihydroxybenzoic acid is high (pK_a = 2.30) [see Eqs. (2) and (3)].

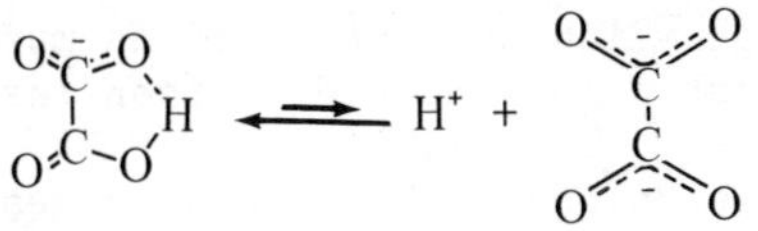

$$\rightleftharpoons H^+ + \qquad (2)$$

$$\rightleftharpoons H^+ + \qquad (3)$$

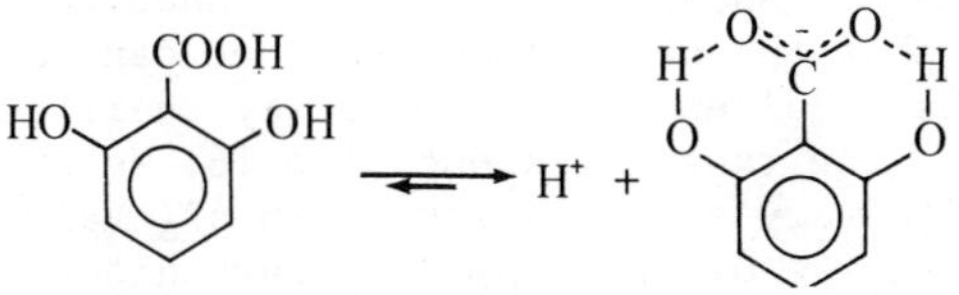

Resonance can be an important factor, raising the acid strength of phenols, for example, when conjugation with a *p*-substituted electron-withdrawing group is possible [Eq. (4)].

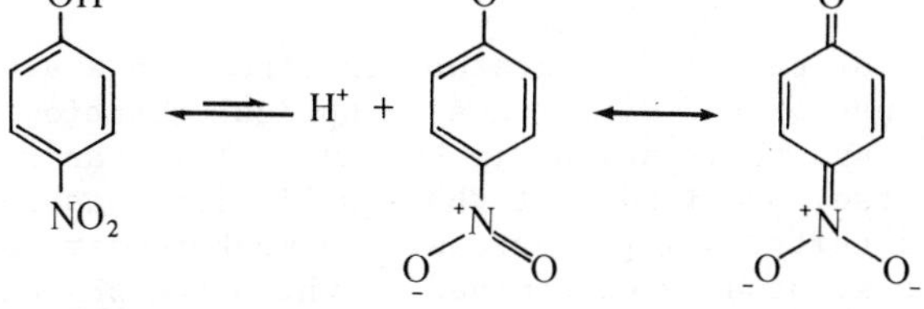

$$\rightleftharpoons H^+ + \qquad \longleftrightarrow \qquad (4)$$

Steric effects are important if they can either prevent achievement of the planar conformations necessary to make resonance effective (steric inhibition of resonance) or if they interfere with the conformation required for effective hydrogen bonding. Finally, it should be noted that hybridization has an important effect; increases in *s*-character in the bond holding the proton produce corresponding increases in acidity.[2]

The theories advanced to explain the effect of electronegative atoms have been especially influential in the views that organic chemists now hold of substituent effects in all reactions. Bjerrum, Smallwood, et al. all have described variations of a simple electrostatic model,

$$\overset{\delta-}{X}-\overset{\delta+}{C}H_2-C(=O)O^- \quad H^+$$

in which the C-X dipole interacts with the ionizable proton.[3] The changes in electrical work as a function of the magnitude of the dipole can be calculated, and this procedure predicts differences in pK_a that are in qualitative agreement with those observed. The principal uncertainties concern conformations, and the range of possible values for the dielectric constant of the medium through which the interactions must operate (80 for water, and at most 5 for the hydrocarbon residue).

B. Extensions to Wider pK_a Ranges

Both water and ethanol can themselves function as either acid or base, so that the acid strengths that can be measured in mixtures of these solvents are limited by their autoprotolysis constants to the range of about 0 to 16. We are not concerned here with very strong acids, with pK_a values below this range, because few of these produce carbanions.

The first systematic extension into the area of weaker acids was made by Conant,[4] who made use of visible spectroscopy to determine the position of Eq. (5)

$$\text{HIn} + \text{NaIn}' \rightleftharpoons \text{NaIn} + \text{HIn}' \tag{5}$$

for a number of pairs of indicators in ether. In a similar vein, McEwen made use of the difference in optical rotation of many chiral acids and their anions in benzene.[5] This can readily be done, of course, provided that the equilibrium constant is not too far from unity. Using progressively weaker pairs of acids, a series was eventually determined in which the pK_a was extended all the way from methanol (pK_a = 16) to cumene (pK_a = 37); of course, the absolute error in pK_a kept pace as the series was extended. A latter-day referee would raise questions about ion-pairing effects, but the Conant-McEwen series is remarkably accurate, even by present standards. Transmetallation equilibria have also been studied by Streitwieser and Hammons,[6] Applequist and O'Brien,[7] and Dessy, Kitching, et al.[8] Each of these studies produced somewhat different results; perhaps Cram's attempt to place all of them on a common scale is now most useful.[9]

As extensions to ever weaker acids are pursued, one of the more difficult obstacles is the slowness of the equilibrations. Accordingly, a kinetic approach was devised by Streitwieser, Caldwell, et al.[10] Assuming, on the basis of the Brønsted relationship, that more ionizable protons should also exchange more rapidly [an assumption not held universally[11]] pK_a values were assigned to a large variety of hydrogen-containing compounds on the basis of their rates of tritium exchange with lithium cyclohexylamide. A wide range of tritium activities and temperatures was employed, and much was learned about variables such as the shape of carbanions and substituent effects (see further in the text that follows).

Still another approach, recently described by Breslow and Chu,[12] concerns those hydrocarbons, RH, for which the cation R^+ is relatively stable. Although this is a limitation, when the method is applicable, it gives thermodynamic pK_a values and it can be applied even when the proton in question is less ionizable than others in the molecule. The approach is based on the sequence:

$$RH \rightleftharpoons ROH \xrightleftharpoons{H^+, -H_2O} R^+ \xrightleftharpoons{e^-} R^\bullet \xrightleftharpoons{e^-} R^- \qquad (6)$$

The free energy involved in the first step does not vary much from one hydrocarbon to another; it can be calculated, for example, from lists of bond energies or heats of formation, and it is the source of relatively little error. The free-energy change of the second step can be obtained from the pK_{R^+} value, the pH at which the alcohol and the carbonium ion are present in equal concentration at equilibrium. The last two steps are appraised by means of reduction potentials. The radical in the sequence is subject to dimerization so that a rapid method of reduction is required; cyclic voltammetry is used. The method gives results generally comparable to others and has permitted measurement of some of the weakest acids yet known; thus 1,2,3-tri-*t*-butylcyclopropene has a pK_a of 65. In a recent extension[13] the need for relatively stable cations was circumvented by the use of known bond energies to appraise the equilibrium, Eq. (7),

$$RH \rightleftharpoons \dot{R} + \dot{H} \qquad (7)$$

in conjunction with a polarographic measurement of the conversion of $R^\bullet$ to R^-; the corresponding conversion of $H^\bullet$ to H^+ is evaluated by means of a known standard acid. In this way the p$\underline{K}_a$ for the tertiary proton of isobutane could be calculated from the potential of the second polarographic wave of t-butyl iodide, [Eq. (8)].

$$t\text{-BuI} \overset{e^-}{\rightleftharpoons} I^- + t\text{-Bu}^\bullet \overset{e^-}{\rightleftharpoons} t\text{-Bu}^- \tag{8}$$

and the known Me_3C-H bond energy. The resulting p$\underline{K}_a$ was about 70 to 72, in agreement with the commonly held notion that, among alkyl anions, the tertiary ones are the least stable.

Related to these studies of weakly acidic hydrocarbons has been the discovery and employment of the so-called superbases, which are not only able to convert even very weak acids to their anions, but do so reasonably rapidly and cleanly in synthetically useful concentrations. Potassium 3-aminopropylamide is an example.[14] It can be prepared from the parent bisamino compound and potassium hydride, with excess amine serving as the solvent. As one demonstration of the power of this base, it effects the isomerization of interior to terminal acetylenes within a period of seconds at 0° [Eq. (9)]. This reaction requires temperatures far in excess of 100°, and takes many hours if the more traditional base sodium amide is used.

$$\text{(interior acetylene)} \xrightarrow{K^+\ ^-NH(CH_2)_3NH_2} \text{(terminal acetylide)}^-\ K^+ \tag{9}$$

A similar function can be performed by certain neutral bisamines in reactions in which an acid must be thoroughly removed as it is formed. 1,8-Bis(dimethylamino)naphthalene is one of these "proton sponges";[15] it is sufficiently hindered to be totally unreactive toward ethyl iodide, yet the p$\underline{K}_a$ of its conjugate acid is 12.43. Highly hindered bases such as lithium 2,2,6,6-tetramethylpiperidide ("H$^+$arpoons") have found similar uses.[16]

The extension of information to even weaker acids has, in recent years, permitted quantitative assays of new concepts. Homoaromaticity is prominent among these. Thus it has been shown that the rate of base promoted allylic D-exchange in **1** is 50,000 times faster than that in **2**; by way of contrast, the analogous ratio for the diphenyl derivatives **3** and **4** is only 3.3. The delocalization of negative charge into the phenyl rings evidently diminishes the opportunity for homoaromatic delocalization as in 5.[17] Undoubtedly, such interpretations will be on even firmer ground once pK_a values are available (thus see the caveat on p. 37).

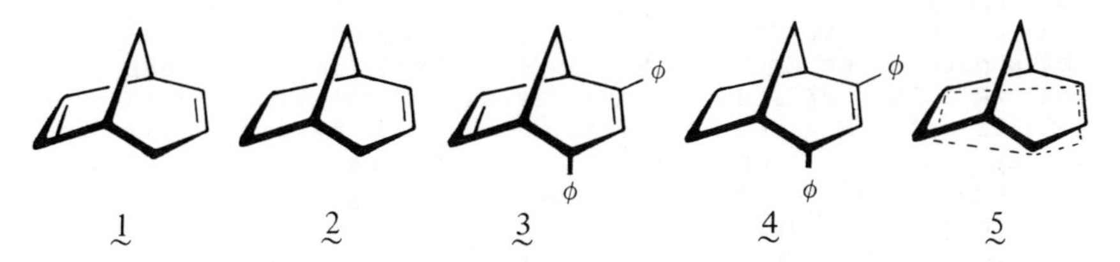

Similarly, the concept of bicycloaromaticity has been tested in this way. Grutzner has reported[18] that anion 6 deprotonates in THF 10 to 100 times faster than 7; he concluded that the latter anion is subject to extra (laticyclic) stabilization due to the

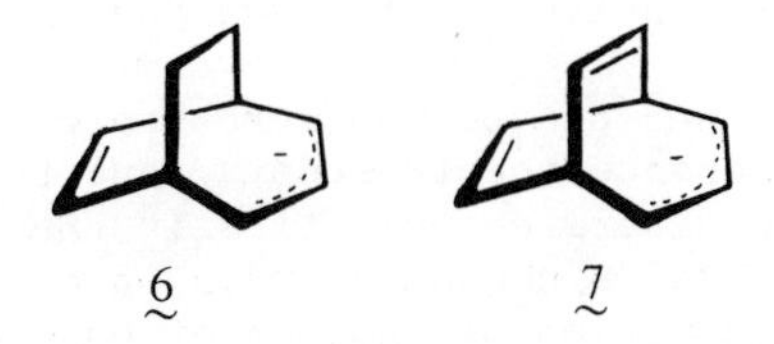

third π-bridge. Such stabilization is predicted on theoretical grounds when the three π-bridges do not all contain either (4n) or (4n + 2) π-electrons.[19] The same conclusion was drawn by Goldstein and Natowsky,[20] who found detritiation of the corresponding, labeled hydrocarbons by t-butoxide to be fastest for the triene.

C. Extensions to Wider Solvent Ranges

One obvious aspect of "the" pK_a scale is that it is dependent on the medium in which the ionization takes place; neither the proton nor the anions are unencumbered in any solvent. The hydrogen bond provides a strong anion-solvent interaction, hence considerable change may be expected when the traditional, protic media, such as water or aqueous alcohol, are replaced by aprotic ones. This is particularly true for those anions in which charge delocalization is minimal. Thus simple alcohols such as methanol have pK_a values in water of about 16; in dimethylsulfoxide (DMSO) this quantity is close to 30.[21] This change is very likely due to the fact that the hydrogen-bonding interactions of the ions with water must be much stronger than those of the undissociated acids, whereas none of these species interact strongly with DMSO.

Even in aprotic solvents, large variations are conceivable

due to ion-pairing variations; an example is phenylacetylene. Conant and Wheland found its pK_a to be 15.8 in ether;[4] Streitwieser and Reuben reported it to be 20.5 in cyclohexylamine,[22] and Bordwell assigned a value of 26.5 in DMSO.[23] These large variations in K_a make sense if it is realized that ion pairs tend to remain associated in low dielectric constant media such as ether but dissociate in DMSO (see text that follows). Further studies have revealed that even the order of acidities for a given set of acids need not be the same in such widely varying solvents as these; the degree of ion pairing and thus the variation in pK_a with solvent, is greatest for anions with localized charges.

The relative freeness of ions in DMSO has recently attracted much attention in relation to the determination of pK_a scales in that solvent. Bordwell[23] and Ritchie[21] have recently contributed to this area. Bordwell has primarily employed indicators, whereas Ritchie has made much use of the glass electrode.

D. Extensions to the Gas Phase

In 1968 Brauman and Blair, who were exploring the new tool of ion cyclotron-resonance spectroscopy, reported that they had found it possible to determine acidities in the gas phase.[24] Although such measurements cannot be made on a routine basis, and although the information may not be of much direct practical value, the data are likely to have a profound effect on fundamental knowledge; after all, if the equilibrium constants can be determined in the absence of all solvent interactions, the latter can, in principle, be quantitatively assayed. Since solvation energies are known to be quite large (e.g., typically a 100 times larger than the energies associated with substituent effects), it does not seem surprising in retrospect that startling observations were made.

As an example, it was reported that toluene is more acidic than *t*-butanol. It is likely that the reason for this is that the formation of a delocalized species such as the benzyl anion, in the gas phase, is not as expensive energetically as is the formation of methoxide ion. In solution the localized negative charge of methoxide ion presumably derives much more stabilization from solvation (the relative weakness of the simple alcohols in DMSO was noted earlier here), leading to an apparent reversal of acidities.

One of the problems with gas-phase ions is that their structure cannot be readily ascertained; in fact, they cannot easily be recovered. Although this may not always pose a problem, it is perhaps well to remember that when the benzyl cation (often considered a model of cation stability in solution) is generated in the gas phase it rearranges to the tropylium ion with great ease.[25]

McMahon and Kebarle[26] have recently begun to determine gas-phase pK_a values for an extensive series of compounds, and further anomalies have come to light. Thus acetonitrile has been found to be more acidic than methanol, and malononitrile has been found to be more acidic than chloroacetic acid. Even within a series of closely matched analogs, unexpected results are the order of the day. It was reported, for example, that among the haloacetic acids, acid strength increases in the order, F < Cl < Br, opposite to the sequence that applies in solution. The latter result has traditionally been ascribed to inductive effects and, as noted above, order-of-magnitude calculations based on fundamental physical properties had given encouraging results. Now it turns out that this differential inductive effect either does not exist or operates in a sensible manner only while it is greatly overshadowed by solvation effects! Since the inductive effect is routinely invoked by physical organic chemists in their explanation of a great variety of observations, we need to worry about how much credence to lend to many of these rationalizations.

The obvious need now is for precise thermodynamic data concerning the solvation of ions. A beginning has been made in several laboratories; as an example, we may mention a study by Haberfield and Rakshit,[27] who measured the enthalpies of hydration of the haloacetic acids. These values can be combined with the aqueous and gas-phase ionization enthalpies to calculate ΔH_{hydr} of the anions, using the cycle shown in Eq. (10).

$$\begin{array}{ccc}
XCH_2COOH(g) & \xrightarrow{\Delta H_{ion'n(g)}} & XCH_2COO^-(g) + H^+(g) \\
\uparrow \Delta H_{vap} & & \downarrow \Delta H_{hydr} \quad \downarrow \Delta H_{hydr} \\
XCH_2COOH(l) & & \\
\downarrow \Delta H_{hydr} & & XCH_2COO^-(aq) + H^+(aq) \\
XCH_2COOH_{aq} & \xrightarrow{\Delta H_{ion'n(aq)}} &
\end{array} \qquad (10)$$

The result in this instance was indeed found to be that the hydration enthalpies follow a trend opposite in direction and larger in magnitude than the "anomolous" gas-phase acidities and that <u>this</u> factor is responsible for the "normal" and well-known solution acidities.

III. SHAPE AND CHARGE DISTRIBUTION OF CARBANIONS

It is well known that a pyramidal shape characterizes carbanions in which the charge is not delocalized; thus such bridgehead precursors as 1-*H*-undecafluorobicyclo[2.2.1]heptane

(8) have acidities comparable to open-chain analogs such as 2-*H*-heptafluoropropane.[28] Delocalization, however, generally

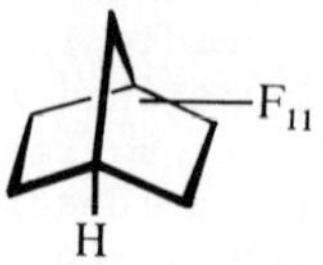

F_{11}

H

8

leads to planar anions; for instance, whereas 9-fluorofluorene is a weaker acid than fluorene, benzal fluoride is much stronger acid than toluene. Presumably, these opposite effects of a fluoro substituent are traceable to a change in anion shape, and it seems reasonable to assume that the fluorenyl anion must be planar.[29] Vinyl anions are nonlinear; thus cyclopropene exchanges its vinylic protons readily.[30]

In recent years it has been shown that the shape of cyclopropyl anions may be affected by substituents. Walborsky, Impastato, et al. found that treatment of optically active 1-methyl-2,2-diphenylcyclopropane with lithium diisopropylamide at -60° gives the corresponding 1-lithiocyclopropane with retained configuration; thus alkylation products are still optically active.[31] If the same experiment is done with 1-cyano-[32] or 1-ethynyl-2,2-diphenylcyclopropane,[33] the alkylation products are racemic; 1-isocyano-2,2-diphenylcyclopropane leads to retained product at -70°, but racemic mixtures are obtained at 0°.[34] There are complications in the interpretation, of course; racemic products do not rule out rapidly converting pyramidal anions, and retention does not rule out planar anions tightly paired with a gegenion, but it is strongly inferred that the 1-methylcyclopropyl anion is pyramidal, that the 1-isocyano analog rapidly inverts, and that the 1-cyano- and 1-ethynyl anions are effectively planar. Carbonion delocalization is undoubtedly responsible for these latter observations.

Another important feature present in carbanions capable of delocalization is the charge density at various positions. The first chemist to assign such densities on a nonchemical basis[35] was Kloosterziel, who used analysis of the proton chemical shifts of the anions of crotonaldehyde and of pentadiene (9 and 10) to show that the negative charge resided principally on the oxygen atom of the former, causing its terminal methylene protons to appear at lower field than in the

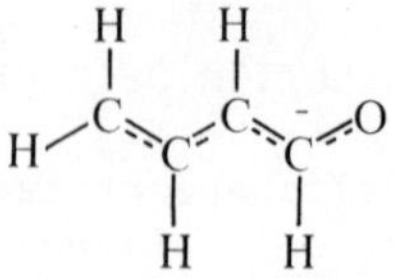

9

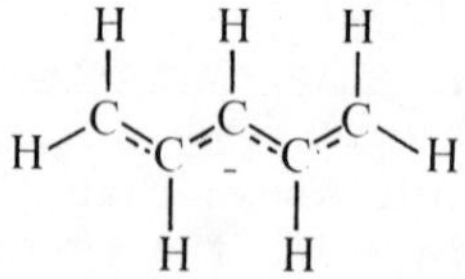

10

analogous hydrocarbon anion. Coupling constants were used by these authors to verify that both anions assume _E_,_E_-("W") conformations; _Z_,_Z_-("U") conformations have been detected by means of nuclear Overhauser effects.[36] Bates, Brenner, et al.[37] used ^{13}C NMR in a similar way. Grutzner[38] has applied these techniques to allylic Grignard reagents; he was able to show, for example, that the monosubstituted reagents are purely in the terminal form, Eq. (11)

$$RCHMgClCH{=}CH_2 \longrightarrow RCH{=}CH{-}CH_2MgCl \qquad (11)$$

and that the _E_- and _Z_-conformations equilibrate rapidly on the NMR time scale, even at -80°.

These advances are permitting a choice between the rival theories used to predict product ratios when such anions undergo reaction; product stabilities, charge control, and least motion have all been advanced to predict product composition. In the reaction of ammonium chloride with 1,3-diarylallylic anions **11**

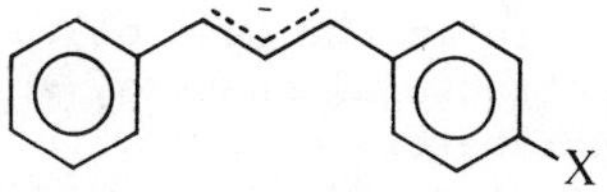

X

11

it has been shown that the site of protonation tends to be that of greatest charge density, so that, at least in this reaction, charge control provides the best prediction.[39]

The relation of chemical shifts and charge density is also implicit from many NMR studies of such new concepts as homoaromaticity. Thus H_6 and H_7 in anion **12** appear at higher field by 2.30 ppm than they do in the corresponding hydrocarbon, and this fact has been cited in support of the homoaromatic character of the former.[40] Great care should be exercised, however, that suitable reference measurements are available; thus Trimitsis and Tuncay[41] have shown that anion **12a**, in which delocalization of the negative charge into the phenyl rings destroys homoaromatic delocalization (see preceding text), is still subject to an upfield chemical shift of H_6 and H_7 of 1 ppm! Clearly, the chemical shift data alone do not provide a totally convincing argument for homoaromaticity in this case.

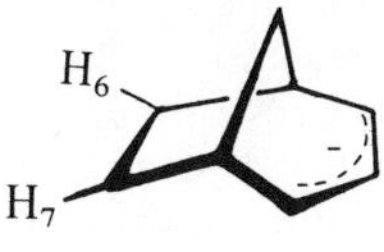

H_6 H_7 H_6 ϕ H_7 ϕ

12 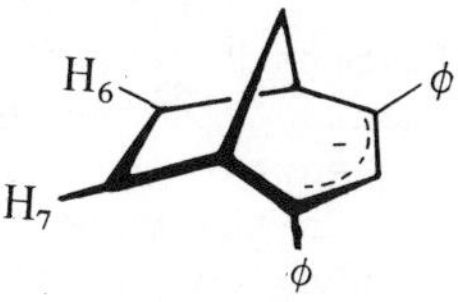 **12a**

One of the problems that workers in this field have not yet really overcome is that of pairing with cations, or, rather, variations in the degree of such pairing. There is no reason to assume that such variations do not occur. When they do, as we see in the text that follows, the NMR spectra may be considerably affected.

IV. ION PAIRING

Even in solvents of high dielectric constant, ions of opposite charge attract one another strongly enough so that the presence of ion pairs is easily detectable by means of spectroscopic and conductance measurements. In inorganic chemistry these observations spawned the Debye-Hückel theory; in organic chemistry they gave rise to the Winstein scheme of solvolysis.[42] Carbanions are particularly suited for studies of pairing, because the media in which the ion pairs can be generated are much more easily handled than those in which carbonium ions must be studied; furthermore, many pairs have visible or UV spectra (or ESR spectra in the case of radical anions) that are sensitive to small structural changes, so that, for example, the thermodynamic parameters defining the equilibria between tight and loose ion pairs can be precisely determined.[43]

Although our main concern here is the effect of pairing on the properties of anions, it is perhaps useful to point out that the electrostatic attraction responsible for the association of ions is neither as highly directional nor as capable of saturation as is the covalent bond, and hence both "isomerism" and further aggregation are expected and observed. Thus studies of the Raman spectrum of sodium cyanide dissolved in liquid ammonia recently revealed that no fewer than four bands are observable in the C≡N region; in some solutions, three of these are visible simultaneously.[44] The bands were attributed to free ions and contact, solvent-separated, and solvent-shared ion-pairs, but linkage-isomeric pairs ($CN^{\ominus}$- - -$Na^{\oplus}$ vs. $NC^{\ominus}$- - -$Na^{\oplus}$) or H-bonded versus non-H-bonded pairs cannot be ruled out. Variations in the sodium splitting of the ESR signals of sodium naphthalenide in organic media have, in fact, been interpreted as indicating multiple pairs that differ structurally.[45] Concerning further aggregation, the concentration dependence of both the UV and ESR spectra of metal fluorenone ketyls have revealed the presence of ion triplets and quadruplets and even higher clusters.[46] Zaugg has concluded from molecular-weight measurements that aggregations of about 100 ions must be occurring in benzene solutions (0.1*M*) of di-*n*-butyl sodiomalonate.[47] In the case of strontium difluorenide, tight-loose, doubly-tight, and doubly loose ion triplets have all been observed.[48] Ion triplet dilithium *Z*,*Z*-hexatriene **13** has been obtained in pure form, and its X-ray diffraction pattern has been reported.[49]

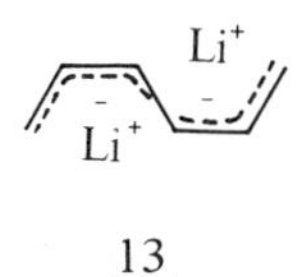

13

It also might be mentioned here that pairing is not restricted to ground-state anions; thus spectrofluorimetry in the case of alkali metal fluorenides reveals two emission maxima that can be attributed to the presence of tight and loose excited pairs. Interestingly, the equilibrium between them tends to favor the loose form more than does that of the corresponding ground-state pairs; for instance, the lithium salt, 70% tight in tetrahydropyran under certain conditions, is completely loose in the excited state.[50]

There are basically three means by which free carbanions can be produced: (a) dilution, (b) use of polar, aprotic solvents, and (c) use of cation complexing agents. In organic media, even the more polar ones, pairing is sufficiently extensive that high dilution; hence conductance techniques are required if the free anions are to be studied. A recent report[51] illustrates this point: the pairing constant for lithium *t*-butoxide in DMSO is 10^8. Pairing becomes less favored when heavier metals are considered; thus with cesium the constant has declined to 200. The opposite tendency has been noted in the tight loose equilibrium; in most instances, the heavier-metal pairs tend to be in the tight form.[43] This suggests that with the heavier cations, the loose pairs are not important energy minima at all.

In carbanion reactions carried out under preparative conditions the free anion, although present in relatively low concentration, is undoubtedly often the reactive species.[52] However, in other cases, the counterion influences the reaction; thus organic chemists have increasingly begun to use complexing agents to modify "anion" behavior. These agents, especially the crown ethers, have been extensively reviewed despite their brief history.[53]

An early example of the modification of anion behavior was described by Park and Simmons,[54] who found that halide anions could be tightly bound by the so-called *in,in*-configuration of the bicyclic cations 14; an X-ray-diffraction study has confirmed these structures.[55] A more extreme case was reported by

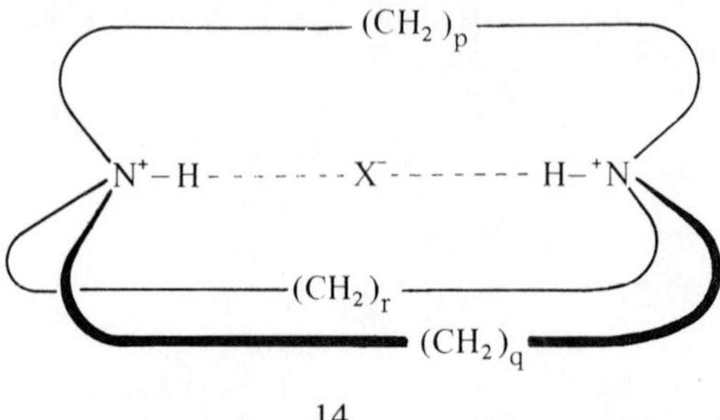

14

Graf and Lehn.[56] These authors studied the ^{13}C NMR spectrum of cryptating agent 15 as a function of the relative concentration of added acids. It was shown that the tetraprotonated compound

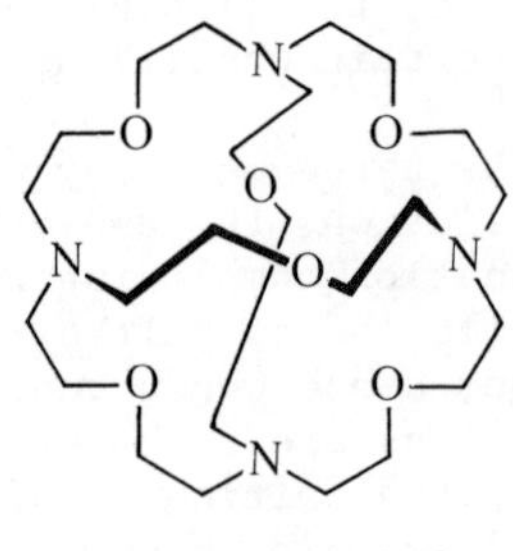

15

has two signals, the chemical shifts of which are independent of the nature of the anion if the latter is large (e.g., ClO_4^-, I^-, NO_3^-, CF_3COO^-); however, changes in chemical shift, indicative of complexation, are observed with the smaller halide anions. The stability constants for these complexes are enormous; hence that for the chloride in water is well in excess of 10^4. Even greater stabilities are observed in methanol. X-Ray diffraction has been used to confirm the (halide-inclusion) structures in these cases as well.[57]

In the vast majority of studies, the modification of anion behavior has been accomplished by the complexation of the counterion; in such cases, of course, the anion becomes more, rather than less, reactive. Some of these studies are now considered.

Sodium picrate, which in tetrahydrofuran is present in the form of a tight ion pair, is completely converted into a loose pair on the addition of one equivalent of dibenzo-18-crown-6.[58] The doubly tight barium difluorenide pair can be converted into either the tight-loose or doubly loose "isomer" by the addition of the appropriate crown ether,[59] and it is clear that such complexation is producing increased concentrations of free anions as well. This stabilization of the free or nearly free anion by use of counterion complexation can have important chemical consequences. Thus picryl halides are normally inert to fluoride in acetonitrile, but in the presence of 18-crown-6, potasium fluoride converts picryl fluoride into the Meisenheimer complex 16, which can then be observed at leisure; the ^{1}H and ^{19}F NMR spectra both exhibit 1:2:1 triplets with J_{HF} = 3.9 Hz.[60]

Even more remarkably, such agents can effect the dissolu-

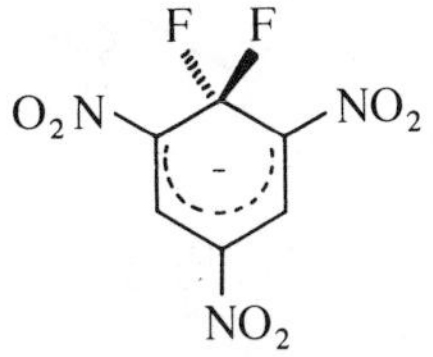

16

tion of alkali metals in aromatic hydrocarbons in which the metals are normally insoluble. Thus when dibenzo-18-crown-6 is added to benzene in contact with a potassium film, the metal is attacked and a blue solution is formed that exhibits the seven line ESR spectrum indicative of the benzene radical anion. If the cryptating agent 17 is added instead, only a single line is observed; evidently, rapid exchange of the electron between benzene molecules results when the cation is buried sufficiently deeply in the complexing agent to prevent strong interaction with the radical anion.[61]

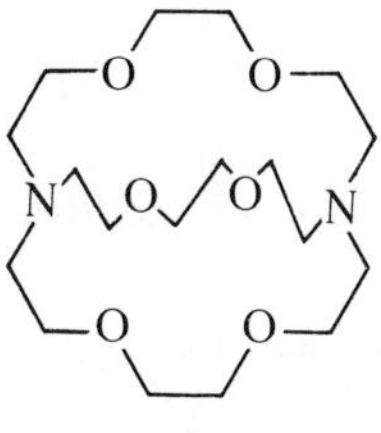

17

In one instance, coupling of the unpaired electron with some of the protons of the complexing agent has been observed[62] where the radical anion of mesitylene in the presence of 18-crown-6 (18) shows a basic 40-line spectrum (due to the nine methyl and three ring protons), which is, furthermore, weakly coupled (a = 0.2 g) with six other protons, presumably the six axial crown ether protons closest to the radical anion within the pair:

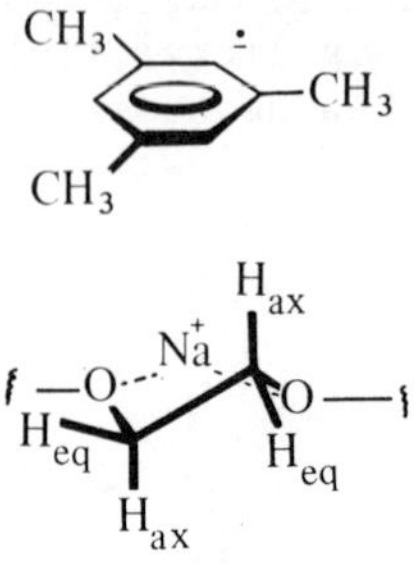

18

If no suitable molecule is present to capture the electron, the alkali anion is produced; in a recent instance, Tehan, Barnett, et al.[63] reported the isolation and X-ray-diffraction structure of compound 19, which contains a cryptated sodium cation and a sodium anion. It is

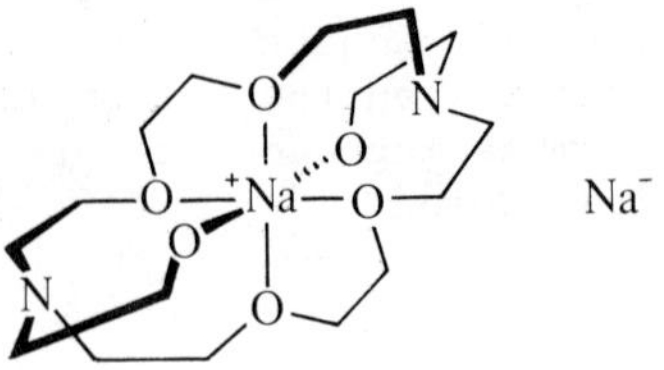

19

known from gas-phase thermodynamic data that the process is

$$2Na(g) \longrightarrow Na^+_{(g)} + Na^-_{(g)} \qquad (12)$$

endothermic by 4.5 eV. If crystal energies are added, then

$$2Na_{(s)} \longrightarrow (Na^+ \ Na^-)_{(s)} \qquad (13)$$

the overall process is still disfavored by 0.5 eV, but the presence of the cryptating agent is sufficient to stabilize the ion pair.

There is an alternative approach to the manipulation of ion pairs, in which the center of positive charge is more permanently interred in the center of an organic tomb; thus with quaternary salts such as **20** and **21**, the positively charged atoms are covalently linked to the organic layers that surround them. This approach also permits the introduction of relatively

$(\underline{n}\text{-Bu})_4N^+ Br^-$ $(\underline{n}\text{-Bu})_4P^+ Cl^-$

20   **21**

unencumbered anions into media in which the simple salts of hard cations would ordinarily not be soluble. One possible advantage of these quaternary ions over cryptating agents is that, in the former, the cations are locked in permanently; hence accidental ion extraction, handling, or disposal do not pose serious problems. In this connection, it might be noted that the crown ethers and related compounds are something of a health hazard.

As far as the synthetic organic chemist is concerned, the surface of this potentially enormous mountain of applications has only barely been scratched; however, some of the more intriguing beginnings may be mentioned here. In 1972 Sam and Simmons described the crown-ether promoted dissolution of potassium permanganate in benzene (< 0.06 *M*) and the efficient use of these solutions as oxidizing agents for olefins.[64] Benzene solutions of crown-ether-solubilized potassium hydroxide will hydrolyze mesitoate esters--a process not possible in water.[65] Herriott and Picker[66] described how **20** virtually completely extracts permanganate anion out of water and into benzene (the tetramethyl homolog is ineffective); more importantly, only a catalytic amount of the ammonium salt is necessary to carry out gentle, high-yield oxidations in benzene. Such oxidations normally would have to be conducted in harsh aqueous alkaline solutions. It has also been ascertained that certain anions, long considered poor nucleophiles, are unreactive only in hydroxylic solvents. Acetate is a good example, as it readily displaces bromide ion in high yield from alkyl bromides[67] and phenacyl bromides[68] in acetonitrile, if 18-crown-6 is present. Fluoride in the presence of this agent is able to displace other halide ions even from tertiary substrates.[69] Similarly, methoxide can displace chloride from chlorobenzenes. No benzynes are involved in the latter reaction; *o*- and *m*-dichlorobenzene react with potassium methoxide in the presence of crown ethers to give only *o*- and *m*-chloroanisole, respectively.[70]

One of the most outstanding observations concerning these "naked anions" is the change in the familiar nucleophilicity order $I^- > Br^- > Cl^- > F^-$. The Swain-Scott constants for these anions in methanol are 200, 16, 2, and 0.2, respectively; in acetonitrile containing 18-crown-6, they are 1.0, 1.3, 1.3, and 1.4, respectively.[71] Thus these nucleophilicities are purely a solvent-mediated property. This is not to say that acetonitrile does not affect the absolute rates of the reactions. On the contrary, the absolute rates of these reactions in the gas phase are some 10^{11} times greater,[72] but the relative nucleophilicities in the gas phase are similar to those reported by Liotta, Grisdale, et al.[71]

A closely related phenomenon, which occurred independently, is the development of the phase-transfer reaction. This is an ingenious solution to the perennial problem of how best to carry out reactions in which both neutral and charged reagents are involved, and for which solubility considerations make conflicting demands on the solvent. As a compromise, alcohols or aqueous alcohols are often used, but this creates another difficulty in that the products of the reaction are often not very stable in such media, so that poor yields result. Dipolar aprotic solvents are more expensive and pose recovery problems. To surmount such difficulties Makosza[73] and Brändström and Gustavi[74] began to use what is now commonly described[75] as the phase-transfer technique. Water is used, as well as an immiscible organic medium; hence the organic substrate and the anionic reagent are in different layers. A catalytic amount of a suitable quaternary salt is added; the cation escorts the desired anion into the organic phase or, rather, just across the phase boundary. The reaction then takes place and the displaced anion is transported back to the aqueous layer by the same route, while the organic product remains safely in the organic medium at all times. Many spectacular results of this approach have been reported;[76] one or two examples may suffice here. The well-known addition of dichlorocarbene to double bonds, conveniently produced from chloroform and base, occasionally leads not to dichlorocyclopropanes but to ring-opened products of further reaction. This problem can be nicely solved by the use of a large excess of chloroform and a quaternary salt as the phase-transfer reagent. Thus the CCl_2 adduct of cyclopentadiene, in aqueous medium, is converted into chlorobenzene, but under phase-transfer conditions the tricyclic product **22** is obtained in high yield;[77] cyclooctatetraene was even observed to give **23**.[78]

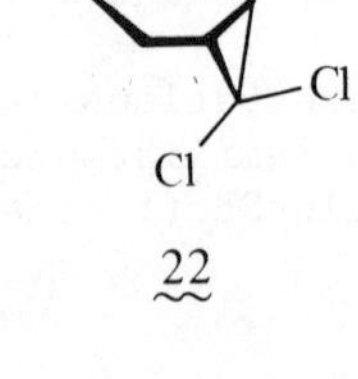

22 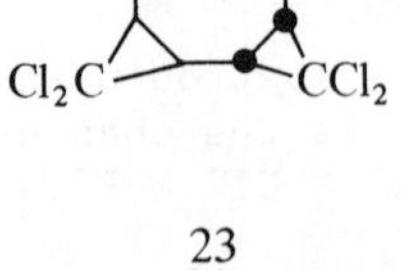 **23**

The mechanism of the phase-transfer reaction is difficult to study because of the presence of two phases. Starks and Owens have shown that the ion transport is not rate-controlling; modest stirring rates of a few hundred rpm suffice, and greater agitation does not help. They also found that the rate constant is proportional to the concentration of the quaternary ion in the organic phase.[79] Brändström was able to show that the reaction proceeds primarily through the ion pair and that free anions play only a minor role.[80]*

As indicated above, cryptating agents should also be usable as phase-transfer reagents, and experience accumulated thus far shows that they indeed are. In the reaction of 1-octyl bromide with iodide ion in a two phase system, the addition of dicyclohexyl-18-crown-6 increased the yield from 4% to 100%.[81] Sepp, Scherer, and Weber[82] noted that in the two-phase reaction of hydrazine with chloroform and alcoholic base (the only known reaction that produces diazomethane from economical precursors) the yield was doubled by the use of a catalytic quantity of 18-crown-6; tetra-*n*-butylammonium bromide also improved the yield, but not as much. A systematic comparison of phase-transfer reagents has been published by Herriott and Picker,[83] who concluded that combinations of rather bulky phase-transfer reagents and relatively polar organic media work best.

V. VARIOUS ANIONS

A. Ambident Anions

The reactions and reactivities of the anions of β-dicarbonyl compounds continue to provide useful insight into carbanion behavior. By 1972 it seemed clear[84] that virtually all available information supported the belief that *C*-alkylation is best promoted by encumbering the competing oxygen atom with hydrogen-bonding solvents or counterions.

These ideas have received further confirmation in many observations with alkylations carried out under phase-transfer conditions, or in the presence of cryptating agents. For example, the benzylation of β-naphthoxide ion normally occurs nearly exclusively (90%) at the α-carbon atom in either water, where the oxygen atom is strongly hydrogen bonded, or in nonpolar solvents where it is coordinated with the counter ion; however, the presence of one equivalent of tetra-*n*-butylammonium ion decreases the C/O ratio from 10 to 0.05. Similar observations were made with acetylacetone.[85] Kurts, Sakembaeva et al. have similarly observed that crown ethers strongly

*See Chapter III for further discussion of mechanism and applications.

promote *O*-alkylation, and by a judicious use of these agents, and of the concentration dependence, they were able to measure the limiting rate constants and O/C ratios for both the free anions and the ion pairs in the reaction of ethyl potassioacetoacetate with ethyl tosylate in dimethylformamide. The rate constant was about 100 times greater, and the O/C ratio six times greater, for the free anion than for the ion pair.[86] Tetraphenylarsonium ion as the counter ion has a similar effect.[87] In this study, information was also obtained concerning the *Z*/*E* ratio of the enol ether produced; increases in the size of the cation, from sodium to tetraphenylarsonium, led to corresponding decreases in the *Z*/*E* ratio from 0.6 to 0. This, of course, seems reasonable, because the smaller cations engage the anion more strongly and hence are more apt to induce the latter to assume the chelated configuration.

Similar observations have been reported elsewhere. Acylation of ethyl acetoacetate with propionyl chloride in benzene produces a small amount of enol ester which is purely the *Z*-isomer; in hexamethylphosphoramide (HMPA) the product was largely enol ester, and this in turn is largely the *E*-isomer. The addition of quaternary ammonium salt increases both the amount of enol ester and the percentage of *E*-product.[88] The same effect was reported for cryptating agent **17** when applied in ethylations.[89] On the other hand, Entenmann has observed that in acetylation with acetyl chloride, the 14% of enol acetate obtained is purely *Z*, whereas the 33% obtained with acetic anhydride, under otherwise identical conditions, was half *E* and half *Z*.[90] Thus it is clearly dangerous to simply equate the fraction of *Z,Z*-anion with *Z*-product.

It has become possible to test these apparent correlations more directly; it has been shown by means of proton NMR, for example, that sodium acetylacetonate does occur in both the *E,E* and *Z,Z*-conformations, **24** and **25**, and that equilibration is slow at temperatures below 0°. Thus in pyridine-d_5 containing 18-crown-6, two unequal signals are observed at 7.62 and 7.91 τ, and two others in the same ratio at 4.52 and 3.95 τ. Clearly both *E,E* and *Z,Z* salts are present, and *Z,E* salt **26** is absent (there are no equal-area signals in the methyl region). If the crown ether concentration is raised, the down-

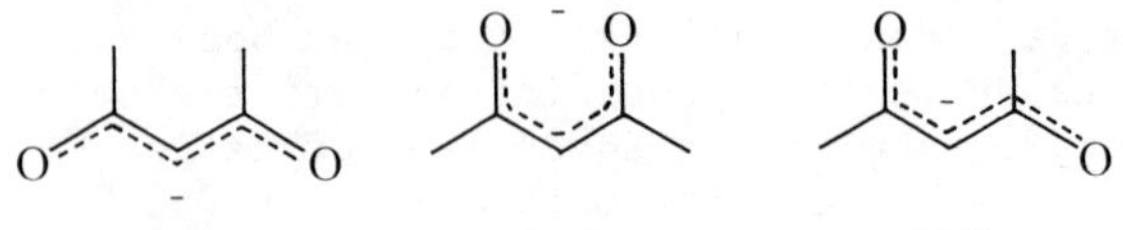

E,E **24**   *Z,Z* **25**   *Z,E* **26**

field signals increase in intensity; these evidently belong to the *E,E*-form.[91] In methanol the *Z,E* conformation has also been observed;[92] it was found that in that solvent the addition of sodium iodide led to more _Z,Z_-product. Potassium iodide has the same effect to a lesser, and lithium iodide to a greater, degree. These observations are in general agreement with the effects of these addends on both C/O and _E/Z_ (enol ether) ratios. On the other hand, if the equilibration between conformers is sufficiently facile to permit the operation of the Curtin-Hammett principle, one would not expect to observe any simple relation between the NMR spectrum of the anion and the _E/Z_ composition of the enol ether product. In a recent paper a Russian group of chemists undertook such a search, and their conclusion is that there is, indeed, no simple correlation.[93] This conclusion is, therefore, in agreement with that of Entenmann.[90]

B. Radical Anions

Recent studies of radical anions have provided perhaps the clearest examples yet available of how the counterion may affect the behavior of the anion. Specifically, the location of the cation is important. Merely because of its presence, for example, one half of an otherwise symmetrical anion may have properties different from those of the other. It is important to emphasize this fact, because it is often overlooked or neglected in discussions of solvolysis and carbonium ions. Thus many of the recent proposals in that area, such as carbon participation and the memory effect, are, in principle also explicable in terms of anion proximity effects.

Proximity effects are beautifully demonstrable by means of the ESR spectra of certain radical anions. As an example, consider [2.2]-paracyclophane. In HMPA, the radical anion (K^+ as cation) displays a sharp 81-line spectrum ascribable to two sets of eight equivalent protons each. In 1,2-dimethoxyethane (DME) the same spectrum is observable, but with somewhat broader lines, and in tetrahydrofuran (THF) the spectrum indicates interaction with four sets of four equivalent protons and one potassium nucleus.[94] Thus we have either a totally symmetrical ion **27** (or a very rapidly equilibrating ion pair) in HMPA, an anion, **28** unsymmetrical because of the nearby cation in THF, and averaging due to a loosely bound potassium cation in DME.

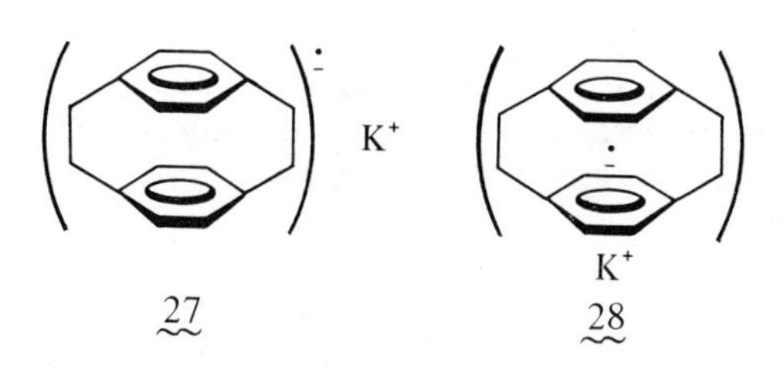

A similar situation has been encountered with spirobifluorene 29.[95] In THF the splitting pattern is that expected for a radical anion in which the unpaired electron is restricted to one half of the molecule; the addition of DME results in a new spectrum, signifying delocalization over the entire molecule. Again, the only reasonable explanation is that, in THF,

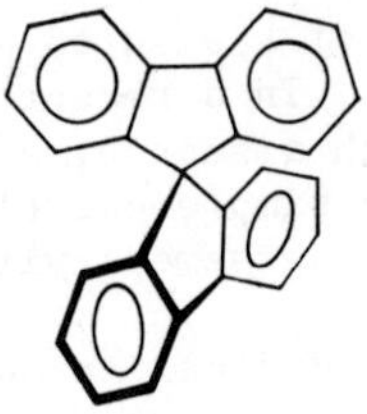

29

an ion pair forms in which the counterion differentiates one half of the anion from the other.

The aromatic radical anion (e.g., sodium naphthalenide) are very reactive and easily and cheaply available, so that it's little wonder that much effort has gone into studies of their possible chemical uses.[96] The radical anion appears to be capable of both electron transfer and reduction, as shown in Eq. (14).

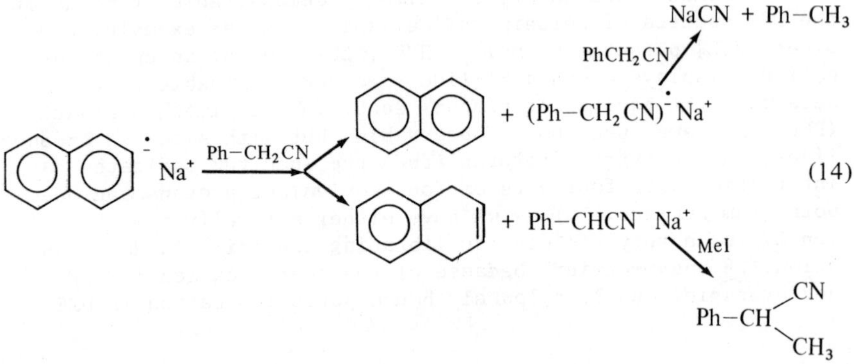

$$\text{C}_{10}\text{H}_8^{\dot{-}}\ \text{Na}^+ \xrightarrow{\text{Ph-CH}_2\text{CN}} \begin{cases} \text{C}_{10}\text{H}_8 + (\text{Ph-CH}_2\text{CN})^{\dot{-}}\ \text{Na}^+ \xrightarrow{\text{PhCH}_2\text{CN}} \text{NaCN} + \text{Ph-CH}_3 \\ \text{C}_{10}\text{H}_{10} + \text{Ph-CHCN}^-\ \text{Na}^+ \xrightarrow{\text{MeI}} \text{Ph-CH(CN)CH}_3 \end{cases} \quad (14)$$

Reduction is the primary reaction in THF, whereas electron transfer becomes important if glymes are added; the implication is strong that the nature of the ion pair dictates the course of the reaction.[97]

Radical anions have been postulated as transient intermediates in certain nucleophilic aromatic substitutions.[98] Thus the net photostimulated reaction, Eq. (15), is considered to

$$\text{1-chloronaphthalene} + CH_3-C(=O)=CH_2^- \longrightarrow Cl^- + \text{1-naphthylacetone} \quad (15)$$

proceed via the chain process shown in Eq. (16).

$$\text{ArCl} \xrightarrow{e^-} (\text{ArCl})^{\cdot -} \xrightarrow{-Cl^-} (\text{Ar})^{\cdot} \xrightarrow{CH_3-C(O)CH_2^-} (\text{ArCH}_2COCH_3)^{\cdot -} \xrightarrow{\text{ArCl}} \text{ArCH}_2COCH_3 \quad (16)$$

In support, it may be noted that, in the presence of potassium metal in solution, the same result can be produced in the dark.[99] Nucleophilic substitution at aliphatic sites also involves radical anion intermediates in certain instances, and studies of these reactions have produced many remarkable applications, such as substitution at tertiary sites and substitutions that circumvent poor leaving groups. An excellent and authoritative review of this field has recently appeared, and readers are referred to it.[100]

C. Cyclopropyl and Allylic Anions

Cyclopropyl anions readily open to produce allylic anions. This opening is predicted to occur in conrotatory fashion,[101] but the facts have been difficult to trace because the allylic anions are conformationally not very stable (however, see text that follows). Support for the prediction has nevertheless been found, based on kinetic data. Thus Ford has studied cyclopropane derivatives **30** through **32**.[102] It was found that solutions of all three compounds in THF, on treatment with a five- to ten-fold excess of lithium *t*-butylamide for 0.5 h at -78°, yield the cyclopropyl anions, as can be demonstrated by neutralization with D_2O and by the epimerization of 31. At -25° or above, all three anions open to allylic anions; the latter are colored and hence directly observable, and allylic hydrocarbons form on protonation. The opening of 32 requires a higher temperature and is estimated to be 10^4 times slower; hence it is subject to a barrier some 6 kcal/mol higher than those of the open-chain compounds. On this basis, conrotatory ring opening is strongly implicated. A careful study of the proton NMR changes that occur on the opening of

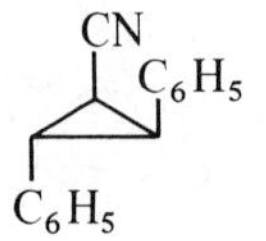

30 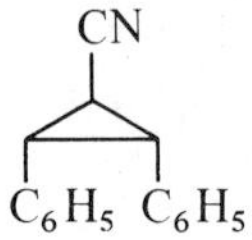 31 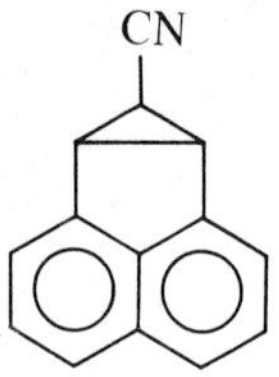 32

33 shows[103] that the opening is indeed conrotatory, because the 1- and 3-allylic protons of 34 have different chemical shifts [cf. Eq. (17)]:

(17)

33 34

The availability of 34 also afforded an opportunity to study its conformational stability.[103] Warming the solution to 62° reversibly led to coalescence, and hence this constitutes a demonstration of the process termed "topomerization" [see Eq. (18)].

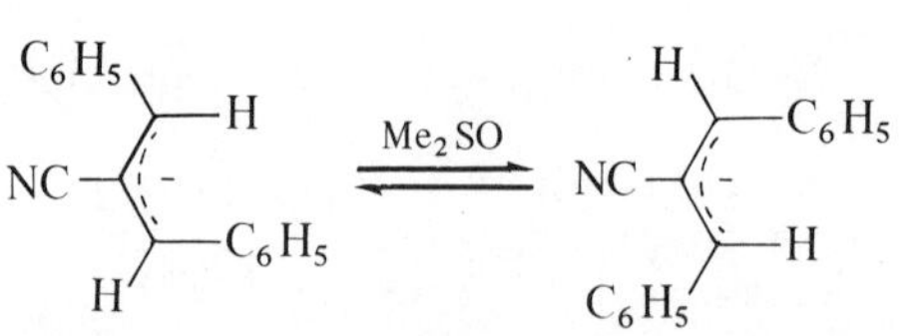

(18)

Several obvious mechanistic possibilities can be quickly ruled out: intermediate closure to a cyclopropane and reopening is impossible because the latter reaction is 100 times too slow, and intermediate formation of benzylcinnamonitrile is ruled out because deuterium incorporation, if given a chance (solvent

Me_2SO-d_6), does not occur. Metallated intermediates do not play a role because the reaction is unaffected by the choice (Li^+, Na^+, K^+) of counterion. Finally, a concerted reaction with synchronous double rotations is inconsistent with the low topomerization barrier ($\Delta G^{\ddagger}$ = 16.5 kcal/mol). This leaves only the stepwise isomerization by rotation occurring via 35.

35 36 37

It proved possible to capture this intermediate anion with acenaphthylene; a mixture of 36 and 37 was obtained.[103]

Further studies of these reactions have been based on the bicyclic trienes 38 (X = Cl or OMe), which, on reaction with potassium metal in THF at -30°, yield the aromatic cis,cis,cis,-

38

trans-cyclononatetraenide anion 39; the ^{1}H NMR spectrum (DMSO) of which shows a sharp triplet at +13.5 τ.[104] If 9-deuterio 38 is employed, the product's D-atom is found at the position neighboring the inside carbon (40). The intensity of the D-

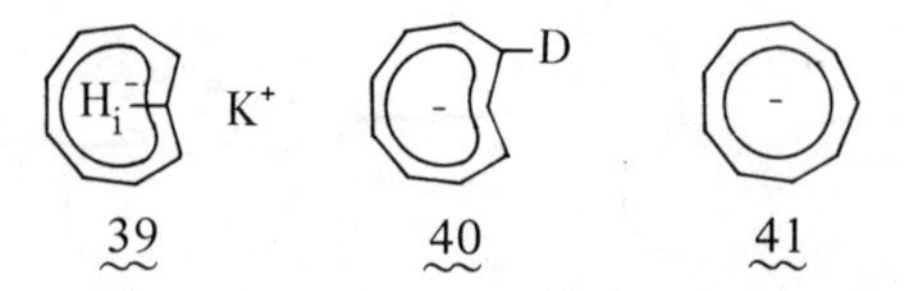

39 40 41

coupled signal slowly drops to one quarter of the original value, showing that exchange among the nine carbons is occurring; the topomerization in this instance is degenerate. At 135°, coalescence occurs; $\Delta H^{\ddagger}$ = 23.6 kcal/mol.[104] Eventually, the anion isomerizes to the all cis-anion 41. For this process $\Delta G^{\ddagger}$ is difficult to measure because proton catalysis is hard to avoid completely, but it appears to be well in excess of 30 kcal/mol. Thus the transition state 42 is much higher in

energy than 43; presumably, the former is more costly in resonance energy.[105] The benzo analogs of 39 and 41 are of

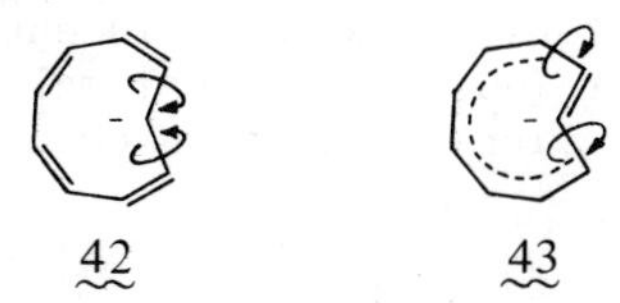

42 43

interest in that *trans*-isomer 44 is the more stable; this seems reasonable since *peri*-interactions are greater in the all *cis*-isomer.[106]

44

D. Zwitterions

Zwitterions are involved as intermediates in the cycloadditions of electron-rich olefins to less well-endowed unsaturated species. This conclusion follows inescapably from many observations made over the last few years.

Among these are such obvious indications as large solvent effects,[107] large negative entropies of activation,[108] lack of complete stereospecificity,[109] concurrent *E-Z* isomerizations of the olefins,[110] and the appropriate substituent effects.[111] Furthermore, rearrangements typical of carbonium ions are often observed in the electron-rich partner, as in the reactions of tetracyanoethylene (TCNE) with tricyclic hydrocarbons 45[112] and 46[113], [see Eqs. (19) and (20)].

TCNE

45

$(CN)_4$ $(CN)_4$

$(CN)_4$

(19)

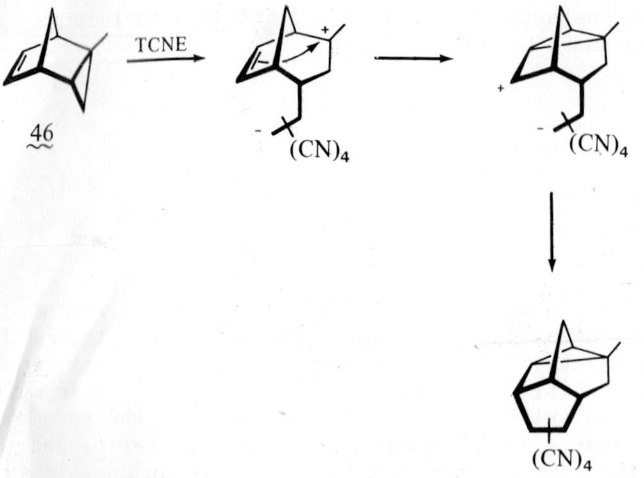

TCNE

46

$(CN)_4$

$(CN)_4$

(20)

$(CN)_4$

The ring-closure step is often reversible, so that the product may depend on whether kinetic or thermodynamic control applies. In one unique instance, the reaction of dicyclopropylfulvene with TCNE [Eq. (21)], no fewer than three different products can be obtained in succession. Depending on the vigor of the conditions, a [2 + 4]-, a [2 + 2]-, or a [2 + 6]-cycloadduct is obtained, and in each step the conversion is promoted by an increase in solvent polarity (the ultimate product also requires a 1,5-H shift).[114] These reversals of ring closure can go all the way, and olefin methatheses have been observed on that

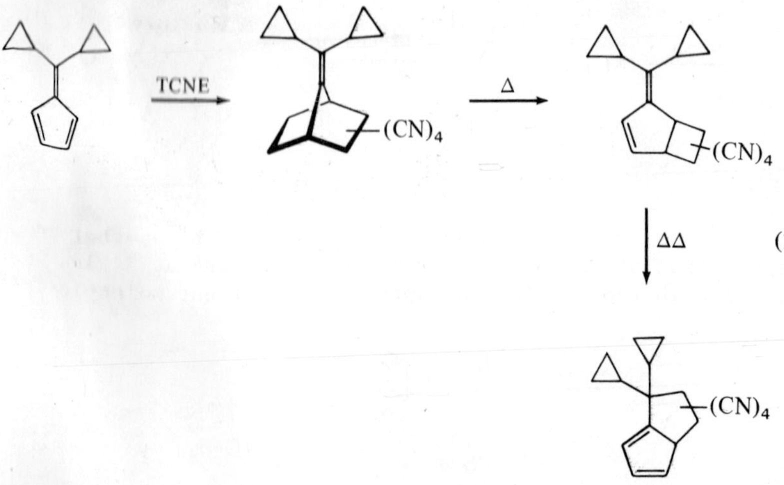

TCNE

$(CN)_4$

Δ

$(CN)_4$

ΔΔ

(21)

$(CN)_4$

basis; for instance, enamine 47 and tricyanoethylene produce olefins 48 and 49 [see Eq. (22)].[115]

(22)

47 48 49

Perhaps the most convincing evidence for an intermediate zwitterion comes from trapping experiments. Thus the reaction of TCNE with ethyl vinyl ether, which normally simply produces the corresponding cyclobutane, leads instead to 50 in methanol —a solvent in which the [2 + 2] adduct, under the same conditions, is stable.[116] By making use of enol ether 51 and (S)-2-butanol, Karl, et al., were able to isolate an acetal, shown

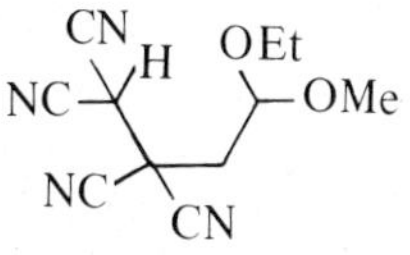

50

by X-ray diffraction to be the (S),(S),(S)-isomer 52, indicating that the intermediate zwitterion had the cis conformation and was captured from the outside [see Eq. (23)].[117]

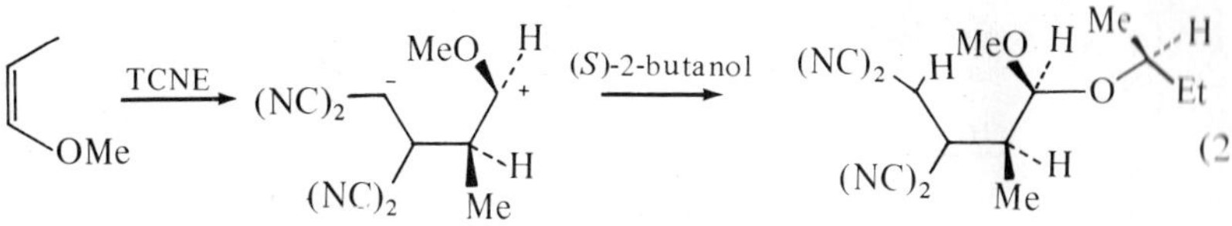

(23)

51 52

A remarkable example of trapping was reported by another group; the reaction of TCNE with hydrocarbon 53 in THF at 0° led to 54, in which the dipole has captured two solvent molecules:[118]

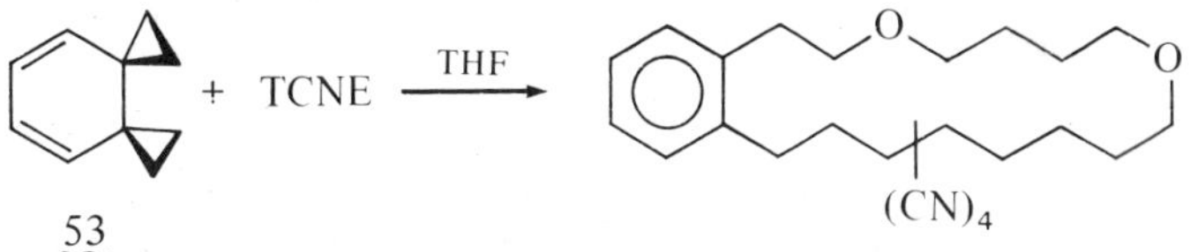

(24)

53 54

Pressure effects have been put to good use in this connection. Much work had been carried out to show that concerted cycloadditions are more accelerated by pressure than by stepwise analogs that involve radical intermediates, and hence that the transition states of the former reactions are smaller, due to more advanced bonding.[119] When stepwise, zwitterionic cycloadditions are compared with concerted cycloadditions, the result is the reverse; this is due to electrostriction by the ionic charges of the zwitterionic intermediate. Thus in the competing cycloadditions of tetrachlorobenzyne with norbornadiene, the "forbidden" product, 55, which is also favored by polar solvents, is promoted by the application of pressure [cf. Eq. (25)]:[120]

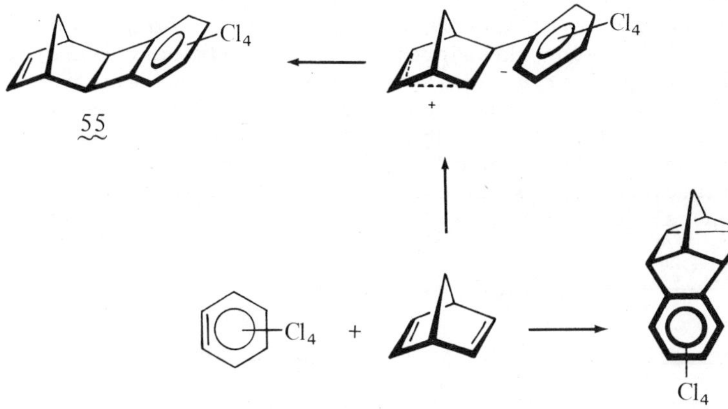

(25)

Likewise, the [2 + 2] cycloaddition of TCNE to enol ethers is accelerated much more by pressure than are concerted cycloadditions (such as the Diels-Alder reaction).[121] Perhaps more directly to the point, the reverse reaction (in this instance, the ring opening of 56 to give the zwitterion 57, which is then captured by methanol) is accelerated by pressure to an extent

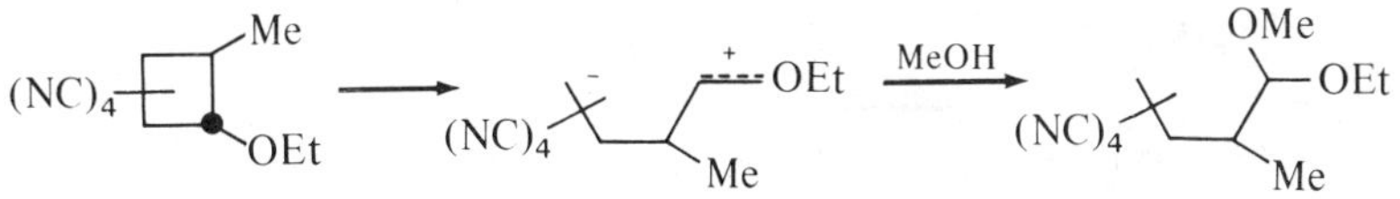

(26)

typical of solvolysis reactions. This is clear proof of the zwitterionic nature of the intermediate, because all other carbon-carbon bond cleavages studied under pressure have been retarded by pressure.[122]

With this array of evidence, it seemed inevitable that a zwitterion would eventually be directly observed, and this event has now come to pass: a group led by Osugi reports having observed it in the reaction of TCNE and styrene. Some UV observations could be traced to the zwitterion, and under 8-kbar pressure, its concentration was sufficient to permit the recording of the ^{1}H NMR spectrum as well.[123]

E. The S_N2 Transition State

There have recently been several new proposals regarding the detailed nature of the nucleophilic substitution reaction, [see Eq. (27)]. Among these are: (a) the preliminary transfer of an electron,[124] or at least the presence of an "electron-transfer component" in the transition state;[125] (b) the preliminary ionization of the substrate to give an intimate ion pair;[126] and (c) the intermediacy of a true pentacovalent intermediate, capable of pseudorotation in favorable instances.[127] Only proposal (c) is discussed here.

$$\mathrm{RX} \xrightarrow{(a)\ Y^-} (Y^\cdot, RX^{\overset{\cdot}{-}}) \longrightarrow YR + X^-$$

$$\mathrm{RX} \underset{}{\overset{(b)}{\rightleftharpoons}} (R^+, X^-) \xrightarrow{Y^-} (Y^-, R^+, X^-) \longrightarrow YR + X^- \qquad (27)$$

$$\mathrm{RX} \xrightarrow{(c)\ Y^-} Y{-}\overset{|}{C}{-}X \longrightarrow YR + X^-$$

It was noted in 1971 that S_N2 displacements at small-ring carbon atoms involve transition states of relatively high energy because of angle strain. If an intermediate species were formed, it could relieve some of this strain by pseudorotation [see Eq. (28)].[127]

$$Y^- + \text{(H, X-substituted cyclopropane with R)} \longrightarrow \left[Y \subset \overset{H}{\overset{|}{C}} \supset X \text{ (cyclopropane with R)} \right]^- \qquad (28)$$

$$\downarrow$$

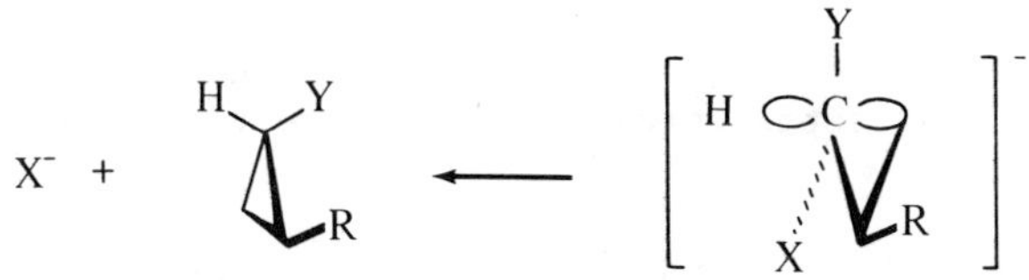

The result would be displacement with retention, an unprecented observation; however, in a subsequent publication, Ugi claimed to have found an example.[128] When an acetone solution of cis-3-ethoxycyclobutyl brosylate was refluxed with iodide ion, cis-3-ethoxycyclobutyl iodide was obtained after 5 days.

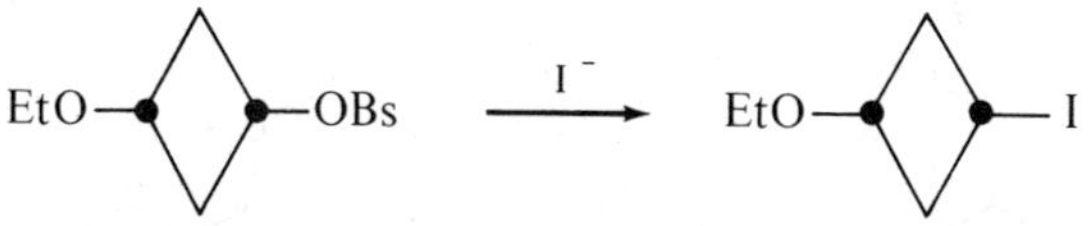

(29)

However, the substitution with retention interpretation was soon demolished. Various authors[129,130] reported that if a much shorter reaction time was employed, the trans-isomer was obtained, and only on prolonged heating did the cis-isomer take its place. Thermodynamically, the latter isomer is more stable. (The four-membered ring is nonplanar, and interactions between cis-1,3-substituents are minimized in this way.) Thus the initially formed trans-iodide undergoes a second iodide displacement, again with inversion, to yield the cis product actually observed by Ugi, et al.[128]

Even if displacements with retention are eventually found to take place at small-ring sites, it should be remembered that pseudorotation is not necessarily the only interpretation. Thus both the epimeric bromides 58 undergo bromide displacement by acetate ion with retention, as demonstrated by means of X-ray diffraction; the authors described how ring-opened intermediates 59 and 60 might be responsible:[131]

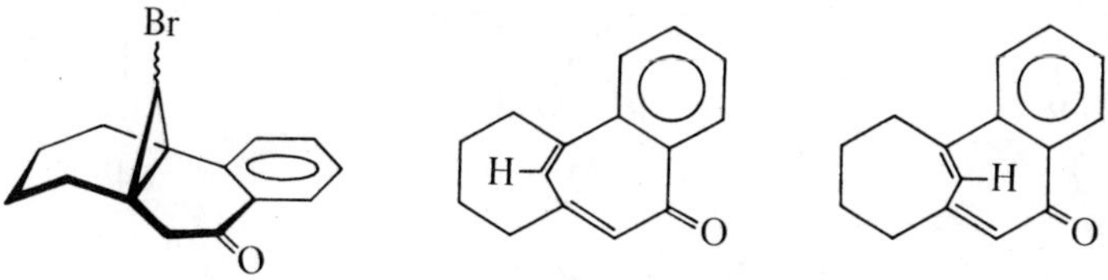

58 59 60

This may not seem likely, but such possibilities must not be dismissed out of hand in favor of some other unprecedented proposal. We may do well to remember that the anion CBr_5 has been known for some time; a recent study of this anion as the tetraphenylphosphonium salt has revealed the key carbon atom to have an uneventful tetrahedral configuration 61.[132]

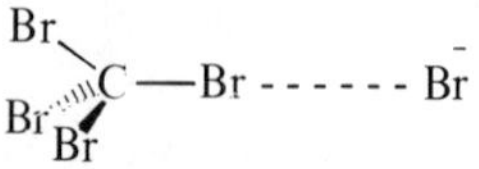

CBr_3—Br - - - - - - Br^-

61

This structural uncertainty is not present in the rapid degenerate rearrangement represented in Eq. (30).

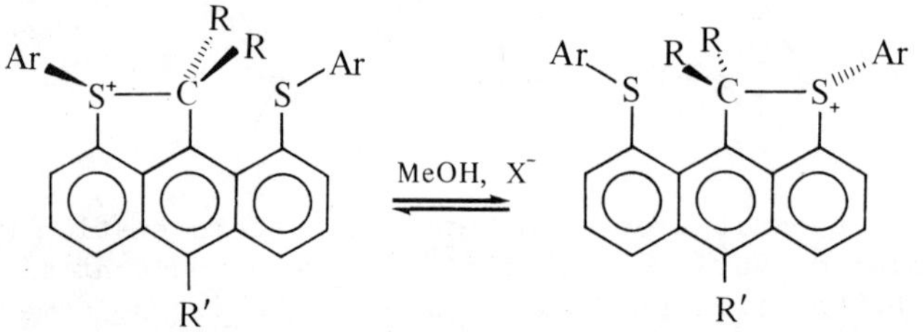

MeOH, X^- (30)

This reaction can be studied by means of the ^1H NMR coalescence of R and has been referred to as the bellclapper rearrangement.[133] It was observed that: (a) the chloride and tetrafluoroborate salts react with about equal rapidity, (b) there is virtually no solvent effect (e.g., CF_3COOH vs. $C_2H_2Cl_4$), (c) variation of R' has practically no effect, and (d) there is simultaneous NMR coalescence of protons of the aryl groups. These observations rule out such mechanistic alternatives as 62 through 65, or a simple sulfonium ion site inversion.

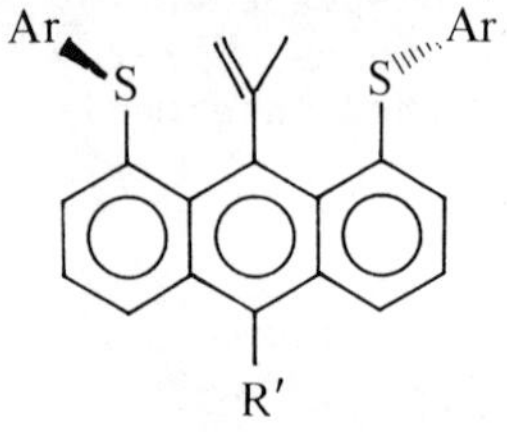

62 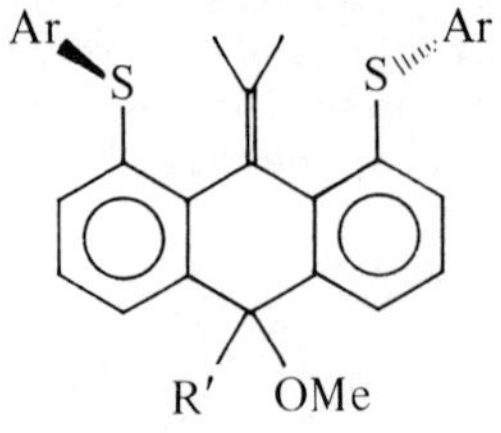 63 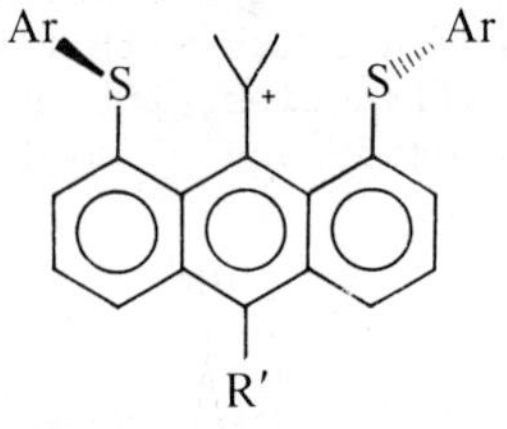 64

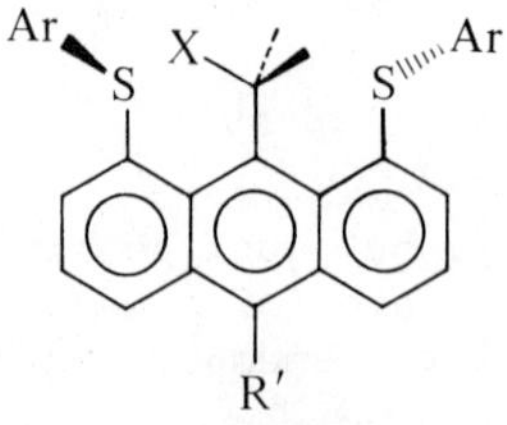

65

It was also observed that when $X^- = Cl^-$ and R = Me, product 66 formed very readily. This reaction does not occur with 67, or when R = H, and this curious difference in behavior has been tentatively ascribed to the trapping of hypervalent high-energy intermediate 68.[133]

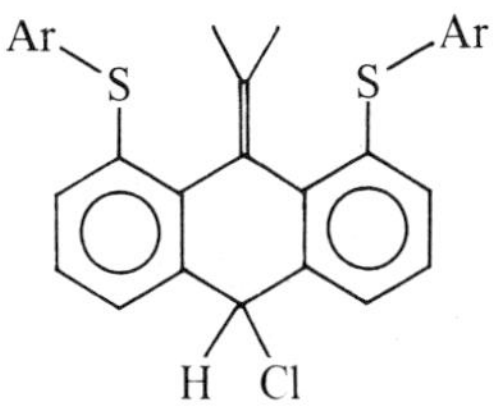

66 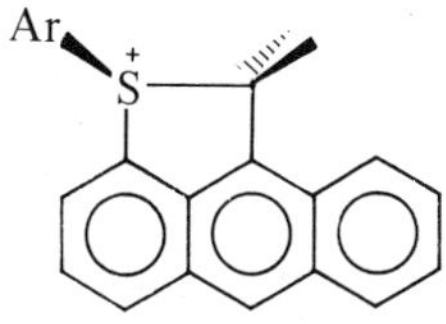 67 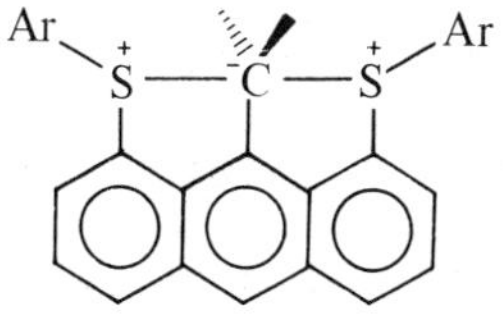 68

VI. REFERENCES

1.* H. C. Brown, D. H. McDaniel, and O. Häfliger, in Determination of Organic Structures by Physical Methods, E. A. Braude and F. C. Nachod, eds., Academic, New York, 1955, Chapter 14.

2. For further discussion and examples, see W. J. le Noble, Highlights of Organic Chemistry, Dekker, New York, 1974, Chapter 12.

3. Summarized in G. W. Wheland, Advanced Organic Chemistry, 3rd ed., Wiley, New York, 1960, Chapter 11.

4.* J. B. Conant and G. W. Wheland, J. Am. Chem. Soc., 54, 1212 (1932).

5. W. K. McEwen, J. Am. Chem. Soc., 58, 1124 (1936).

6. A. Streitwieser, Jr., and J. B. Hammons, Progr. Phys. Org. Chem., 3, 41 (1965).

7. D. E. Applequist and D. F. O'Brien, J. Am. Chem. Soc., 85, 743 (1963).

8. R. E. Dessy, W. Kitching, T. Psarras, L. Salinger, A. Chen, and T. Chivers, J. Am. Chem. Soc., 88, 460 (1966).

9. D. J. Cram, Fundamentals of Carbanion Chemistry, Academic Press, New York, 1965; see also O. H. Reutov, K. P. Butin, and I. P. Beletskaya, Russ. Chem. Rev. (Engl. transl.), 43, 17 (1974).

10.* A. Streitwieser, Jr., R. Caldwell, and M. Granger, J. Am. Chem. Soc., 86, 3578 (1964).

11. C. D. Ritchie, J. Am. Chem. Soc., 91, 6749 (1969).

12. R. Breslow and W. Chu, J. Am. Chem. Soc., 95, 411 (1973).

13. R. Breslow and R. Goodin, J. Am. Chem. Soc., 98, 6076 (1976).

14. C. A. Brown, Chem. Commun., 222 (1975); C. A. Brown and A. Yamashita, J. Am. Chem. Soc., 97, 891 (1975).

15. R. W. Alder, P. S. Bowman, W. R. S. Steele, and D. R. Winterman, Chem. Commun., 723 (1968).

16. R. A. Olofson and C. M. Dougherty, J. Am. Chem. Soc., 95, 581, 582 (1973).

17. G. B. Trimitsis and A. Tuncay, J. Am. Chem. Soc., 97, 7193 (1975).

18. M. V. Moncur and J. B. Grutzner, J. Am. Chem. Soc., 95, 6449 (1973).

19. M. J. Goldstein and R. Hoffmann, J. Am. Chem. Soc., 93, 6193 (1971).

20. M. J. Goldstein and S. Natowsky, J. Am. Chem. Soc., 95, 6451 (1973).

21.* C. D. Ritchie, "Interactions in Dipolar Aprotic Solvents," in Solute-Solvent Interactions, J. F. Coetzee and C. D. Ritchie, eds., Dekker, New York, 1969.

22. A. Streitwieser, Jr. and D. M. E. Reuben, J. Am. Chem. Soc., 93, 1794 (1971).

23. F. G. Bordwell and W. S. Matthews, J. Am. Chem. Soc., 96, 1214 (1974).

24.* J. I. Brauman and L. K. Blair, J. Am. Chem. Soc., 90, 6561 (1968); 92, 5986 (1970). For a review of the related negative-ion mass spectrometry, see V. I. Khvastenko and G. A. Tolstikov, Russ. Chem. Rev. (Engl. transl.), 45, 127 (1976).

25. S. Meyerson, Rec. Chem. Progr., 26, 257 (1965).

26. T. B. McMahon and P. Kebarle, J. Am. Chem. Soc., 98, 3399 (1976).

27.* P. Haberfield and A. K. Rakshit, J. Am. Chem. Soc., 98, 4393 (1976).

28. A. Streitwieser, Jr, and D. Holtz, J. Am. Chem. Soc., 89, 692 (1967).

29. A. Streitweiser, Jr. and F. Mares, J. Am. Chem. Soc., 90, 2444 (1968).

30. A. J. Schipperijn, Rec. Trav. Chim. Pays-Bas, 90, 1110 (1971).

31. H. M. Walborsky, F. J. Impastato, and A. E. Young, J. Am. Chem. Soc., 86, 2383 (1964).

32. H. M. Walborsky and F. M. Hornyak, J. Am. Chem. Soc., 77, 6026 (1955).

33. G. Köbrich, D. Merkel, and K. Imkampe, Chem. Ber., 106, 2017 (1973).

34.* H. M. Walborsky and M. P. Periasamy, J. Am. Chem. Soc., 96, 3711 (1974).

35.* G. J. Heiszwolf and H. Kloosterziel, Rec. Trav. Chim. Pays-Bas, 86, 807 (1967).

36. S. M. Esakov, A. A. Petrov, and B. A. Ershov, J. Org. Chem. U. S. S. R. (Engl. transl.), 11, 679 (1975).

37. R. B. Bates, S. Brenner, C. M. Cole, E. W. Davidson, G. D. Forsythe, D. A. McCombs, and A. S. Roth, J. Am. Chem. Soc., 95, 926 (1973).

38. D. A. Hutchison, K. R. Beck, R. A. Benkeser, and J. B. Grutzner, J. Am. Chem. Soc., 95, 7075 (1973).

39. R. J. Bushby and G. J. Ferber, Chem. Commun., 407 (1973).

40. J. M. Brown and J. L. Occolowitz, Chem. Commun., 375 (1965); S. Winstein, M. Ogliaruso, M. Sakai, and J. M. Nicholson, J. Am. Chem. Soc., 89, 3656 (1967).

41. G. B. Trimitsis and A. Tuncay, J. Am. Chem. Soc., 98, 1998 (1976).

42. See le Noble,[2] Chapter 21.

43.* J. Smid, Angew, Chem., Int. Ed. Engl., 11, 112 (1972); M. Szwarc, Science, 170, 23 (1970); G. C. Greenacre and R. N. Young, J. Chem. Soc., Perkin Trans. II, 1661 (1975); T. Takeshita and N. Hirota, J. Chem. Phys., 58, 3745 (1973).

44. P. Gans, J. B. Gill, and M. Griffin, J. Am. Chem. Soc., 98, 4661 (1976).

45. N. Hirota in Radical Ions, E. T. Kaiser and L. Kevan, eds., Wiley, New York, 1967.

46. S. W. Mao, K. Nakamura, and N. Hirota, J. Am. Chem. Soc., 96, 5341 (1974).

47. G. H. Barlow and H. E. Zaugg, J. Org. Chem., 37, 2246 (1972).

48. T. E. Hogen-Esch and J. Smid, J. Am. Chem. Soc., 94, 9240 (1972); J. Phys. Chem., 79, 233 (1975).

49. S. K. Arora, R. B. Bates, W. A. Beavers, and R. S. Cutler, J. Am. Chem. Soc., 97, 6272 (1975).

50. J. Plodinec and T. E. Hogen-Esch, J. Am. Chem. Soc., 96, 5262 (1974).

51. J. H. Exner and E. C. Steiner, J. Am. Chem. Soc., 96, 1782 (1974).

52. See, for example, S. G. Smith and D. V. Milligan, J. Am. Chem. Soc., 90, 2393 (1968); A. A. Solov'yanov, P. I. Dem'yanov, I. P. Beletskaya, and O. A. Reutov, J. Org. Chem. USSR, (Engl. transl.), 12, 714 (1976).

53. See, for example, G. W. Gokel and H. D. Durst, Synthesis, 169 (1976).

54.* C. H. Park and H. E. Simmons, J. Am. Chem. Soc., 90, 2431 (1968).

55. R. A. Bell, G. G. Christoph, F. R. Fronzeck, and R. E. Marsh, Science, 190, 151 (1975).

56.* E. Graf and J. M. Lehn, J. Am. Chem. Soc., 98, 6405 (1976).

57. B. Metz, J. M. Rosalky, and R. Weiss, Chem. Commun., 533 (1976).

58. K. H. Wong, M. Bourgoin, and J. Smid, Chem. Commun., 715 (1974).

59. U. Takaki and J. Smid, J. Am. Chem. Soc., 96, 2588 (1974).

60. F. Terrier, G. Ah-Kow, M. J. Pouet, and M. P. Simonnin, Tetrahedron Lett., 227 (1976).

61. B. Kaempf, S. Raynal, A. Collet, F. Schué, S. Boileau, and J. M. Lehn, Angew. Chem., Int. Ed. Engl., 13, 611 (1974).

62. A. G. V. Nelson and A. von Zelewsky, J. Am. Chem. Soc., 97, 6279 (1975).

63. F. J. Tehan, B. L. Barnett, and J. L. Dye, J. Am. Chem. Soc., 96, 7203 (1974).

64. D. J. Sam and H. E. Simmons, J. Am. Chem. Soc., 94, 4024 (1972).

65.* See, for example, C. J. Pedersen and H. K. Frensdorff, Angew. Chem., Int. Ed. Engl., 11, 16 (1972).

66. A. W. Herriott and D. Picker, Tetrahedron Lett., 1511 (1974).

67.* C. L. Liotta, H. P. Harris, M. McDermott, T. Gonzalez, and K. Smith, Tetrahedron Lett., 2417 (1974).

68. H. D. Durst, Tetrahedron Lett., 2421 (1974).

69. C. L. Liotta and H. P. Harris, J. Am. Chem. Soc., 96, 2250 (1974).

70. D. J. Sam and H. E. Simmons, J. Am. Chem. Soc., 96, 2252 (1974).

71. C. L. Liotta, E. E. Grisdale, and H. P. Hopkins, Tetrahedron Lett., 4205 (1975).

72. L. B. Young, E. Lee-Ruff, and D. K. Bohme, Chem. Commun., 35 (1973).

73.* M. Makosza, Tetrahedron Lett., 673, 677 (1969).

74.* A. Brändström and K. Gustavii, Acta Chem. Scand., 23, 1215 (1969).

75. C. M. Starks, J. Am. Chem. Soc., 93, 195 (1971).

76. For recent reviews, see: (a) E. V. Dehmlow, Angew. Chem., Int. Ed. Engl., 13, 170 (1974); (b) M. Makosza, Pure Appl. Chem., 439 (1975).

77. E. V. Dehmlow, Tetrahedron, 28, 175 (1972).

78. E. V. Dehmlow, H. Klabuhn, and E. C. Hass, J. Liebigs Ann. Chem., 1753 (1973).

79. C. M. Starks and R. M. Owens, J. Am. Chem. Soc., 95, 3613 (1973).

80. A. Brändström, Acta Chem. Scand., 30(B), 203 (1976).

81. D. Landini, F. Montanari, and F. M. Pirisi, Chem. Commun., 879 (1974).

82. D. T. Sepp, K. V. Scherer, and W. P. Weber, Tetrahedron Lett., 2983 (1974).

83. A. W. Herriott and D. Picker, J. Am. Chem. Soc., 97, 2345 (1975).

84. See le Noble,[2] Chapter 22. For a more recent discussion, and a proposal to extend the notion of "charge-control" to include substituent effects, see R. Gompper and H. U. Wagner, Angew. Chem., Int. Ed. Engl., 15, 321 (1976).

85. E. d'Incan and P. Viout, Tetrahedron, 31, 159 (1975).

86.* A. L. Kurts, S. M. Sakembaeva, I. P. Beletskaya, and O. A. Reutov, Doklady Phys. Chem. (Engl. transl.), 210, 380 (1973); J. Org. Chem. U. S. S. R. (Engl. transl.), 9, 1579 (1973). Even larger ratios are known to occur in anionic polymerization; see T. Shimomura, K. J. Tölle, J. Smid, and M. Szwarc, J. Am. Chem. Soc., 89, 796 (1967).

87. A. L. Kurts, S. M. Sakembaeva, I. P. Beletskaya, and O. A. Reutov, J. Org. Chem. U. S. S. R. (Engl. transl.), 10, 1588 (1974); A. L. Kurts, P. I. Dem'yanov, I. P. Beletskaya, and O. A. Reutov, ibid., 9, 1341 (1973).

88. R. Gelin, S. Gelin, and A. Galliaud, Bull. Soc. Chim. France, 3416 (1973).

89. C. Cambilleau, P. Sarthou, and G. Bram, Tetrahedron Lett., 281 (1976).

90.* G. Entenmann, Tetrahedron Lett., 4241 (1975).

91. E. A. Noe and M. Raban, J. Am. Chem. Soc., 96, 6185 (1974).

92. E. A. Noe and M. Raban, Chem. Commun., 165 (1976); see also A. A. Petrov, S. M. Esakov, and B. A. Ershov, J. Org. Chem. U. S. S. R. (Engl. transl.), 10, 1998 (1974).

93. A. A. Petrov, S. M. Esakov, and B. A. Ershov, J. Org. Chem. U. S. S. R. (Engl. transl.), 12, 774 (1976).

94.* F. Gerson, W. B. Martin, and S. C. Wydler, Helv. Chim. Acta, 59, 1365 (1976).

95. F. Gerson, B. Kowert, and B. M. Peake, J. Am. Chem. Soc., 96, 118 (1974).

96. S. Bank and W. D. Closson, Tetrahedron Lett., 1349 (1965); W. D. Closson, P. Wriede, and S. Bank, J. Am. Chem. Soc., 88, 1581 (1966); J. F. Garst, Acc. Chem. Res., 4, 400 (1971).

97. S. Bank and S. P. Thomas, Tetrahedron Lett., 305 (1973); for more recent work and references, see S. Bank and D. A. Juckett, J. Am. Chem. Soc., 97, 567 (1975).

98. J. K. Kim and J. F. Bunnett, J. Am. Chem. Soc., 92, 7463, 7464 (1970).

99. R. A. Rossi, R. H. de Rossi, and A. Lopez, J. Am. Chem. Soc., 98, 1252 (1976).

100.* N. Kornblum, Angew. Chem., Int. Ed. Engl., 14, 734 (1975).

101. R. B. Woodward and R. Hoffmann, Angew. Chem., Int. Ed. Engl., 8, 781 (1969).

102.* M. Newcomb and W. T. Ford, J. Am. Chem. Soc., 95, 7186 (1973); 96, 2968 (1974); W. T. Ford and M. Newcomb, ibid., 95, 6277 (1973).

103. G. Boche, D. Martens, and H. U. Wagner, J. Am. Chem. Soc., 98, 2668 (1976).

104. G. Boche, A. Bieberbach, and H. Weber, Angew. Chem., Int. Ed. Engl., 14, 562 (1975).

105. G. Boche and A. Bieberbach, Tetrahedron Lett., 1021 (1976).

106. A. G. Anastassiou and E. Reichmanis, Angew. Chem., Int. Ed. Engl., 13, 728 (1974).

107.* G. Steiner and R. Huisgen, J. Am. Chem. Soc., 95, 5056 (1973).

108. G. Steiner and R. Huisgen, Tetrahedron Lett., 3769 (1973).

109. P. D. Bartlett, Quart. Rev. Chem. Soc., 24, 473 (1970).

110. R. Huisgen and G. Steiner, J. Am. Chem. Soc., 95, 5055 (1973).

111. G. Steiner and R. Huisgen, Tetrahedron Lett., 3763 (1973).

112. M. A. Battiste, J. M. Coxon, R. G. Posey, R. W. King, M. Mathew, and G. J. Palenik, J. Am. Chem. Soc., 97, 945 (1975).

113. J. M. Coxon, M. de Bruijn, and C. K. Lan, Tetrahedron Lett., 337 (1975).

114. A. Cornelis and P. Laszlo, J. Am. Chem. Soc., 97, 244 (1975).

115. H. K. Hall, Jr. and P. Ykman, J. Am. Chem. Soc., 97, 800 (1975).

116. R. Huisgen, R. Schug, and G. Steiner, Angew. Chem., Int. Ed. Engl., 13, 80, 81 (1974); see also R. Schug and R. Huisgen, Chem. Commun., 60 (1975).

117. I. Karle, J. Flippen, R. Huisgen, and R. Schug, J. Am. Chem. Soc., 97, 5285 (1975).

118. D. Kaufmann, A. de Meijere, B. Hingerty, and W. Saenger, Angew. Chem., Int. Ed. Engl., 14, 816 (1975).

119. See le Noble,[2] Chapter 14.

120. W. J. le Noble and R. Mukhtar, J. Am. Chem. Soc., 96, 6191 (1974).

121. K. F. Fleischmann and H. Kelm, Tetrahedron Lett., 3769 (1973).

122. W. J. le Noble and R. Mukhtar, J. Am. Chem. Soc., 97, 5938 (1975).

123. M. Nakahara, Y. Tsuda, M. Sasaki, and J. Osugi, Chem. Lett., 731 (1976).

124. K. A. Bilevich and O. Y. Okhlobystin, Russ. Chem. Rev. (Engl. transl.), 37, 12 (1968).

125. S. Bank and D. A. Noyd, J. Am. Chem. Soc., 98, 8203 (1976).

126. R. A. Sneen, Acc. Chem. Res., 6, 46 (1973); D. J. McLennan, ibid., 9, 281 (1976).

127.* P. D. Gillespie and I. Ugi, Angew. Chem., Int. Ed. Engl., 10, 503 (1971).

128. T. ElGomati, D. Lenoir, and I. Ugi, Angew. Chem., Int. Ed. Engl., 14, 59 (1975).

129. C. A. Maryanoff, F. Ogura, and K. Mislow, Tetrahedron Lett., 4095 (1975).

130. T. Virgnani, M. Karpf, L. Hoesch, and A. S. Dreiding, Helv. Chim. Acta, 58, 2524 (1975).

131. F. Effenberger, W. D. Stohrer, and A. Steinbach, Angew. Chem., Int. Ed. Engl., 8, 280 (1969).

132. H. J. Lindner and B. K. von Gross, Chem. Ber., 109, 314 (1976).

133.* J. C. Martin and R. J. Balasay, J. Am. Chem. Soc., 95, 2572 (1973).

3

CARBENES

ROBERT A. MOSS

Rutgers University, New Brunswick, New Jersey 08903

MAITLAND JONES, JR.

Princeton University, Princeton, New Jersey 08540

I. PHYSICAL MEASUREMENTS

A. Singlet-Triplet Splitting in Methylene

One candidate for the most exciting paper in carbene chemistry for the last 2 years surely is the report by Lineberger and his co-workers of an experimentally determined value of 19.5 ± 0.7 kcal/mol for the singlet-triplet splitting in methylene.[1] To appreciate the impact of this work, it is necessary to review briefly the prior progress of theory and experiment in this area.[2,3]

Before 1972 most theoretical estimates of the singlet-triplet energy gap were in the range of 20-40 kcal/mol. By 1970 there were two experimental values[4,5] provided by different groups but each indicating that the gap was very small indeed (ca. 1-2 kcal/mol). Perhaps spurred on by this information, theorists provided ever-lower values as the years went on. Both *ab initio*[2] and semiempirical methods[6] began to converge on a value of roughly 10 kcal/mol. At the same time that theoretical values were declining, the experimentalists were providing numbers substantially higher than the early ones. Both Frey[7] and Simons[8] and their co-workers agreed on an energy gap of some 8-10 kcal/mol. A slightly smaller value, 6 kcal/mol, had already been advanced by Rowland, McKnight, and Lee.[9] So one had the comforting picture of experiment and both *ab initio* and semiempirical theory agreeing on the value of the singlet-triplet splitting for methylene.

Hence the impact of Lineberger's value of 19.5, which is roughly double the accepted experimental and theoretical values, is substantial, to say the least! Although we cannot resolve the discrepancy here, we can comment that Lineberger's value seems very solid, indeed. It derives from a measurement of the energies of the electrons produced by laser photodetachment from the methylene radical anion, CH_2^{-}. A beam of CH_2^{-} is crossed with an argon ion laser, and the electron-binding energies are measured simply by subtracting the observed kinetic energy from the energy of the laser photons. The whole process is calibrated by running a simultaneous photodetachment from O^{-}.

Theory is already compensating, and the calculated value has recently climbed to 16 kcal/mol. This number comes from a GVB/CI calculation in which more correlation effects were included.[10]

The earlier experimental estimates of 1-2 kcal/mol perhaps reflect an inaccurate estimate of the ΔH_f of triplet methylene. Furthermore, it has been pointed out that 3660-Å irradiation of ketene does not produce the 10% of singlet methylene once estimated.[2] However, the more recent experimental values of 8-10 kcal/mol are more difficult to dismiss and at this point we must simply wait for a reconciliation.[10a]

B. Kinetic Studies on Diarylcarbenes

Closs and Rabinow[10b] have used flash spectroscopy to derive the rates of several reactions of diarylcarbenes. Included are rate constants for the second-order dimerization of diphenylcarbene and substituted diphenylcarbenes and the absolute rate constants for the reaction of several diphenylcarbenes with 1,3-butadiene. Competition experiments led to a determination of the absolute rates of reaction of diphenylcarbene with styrene, diphenylethylene, and oxygen. Finally, an analysis of the competitive reactions of diphenylcarbene with methanol and oxygen led to an estimate of a value of ≤ 3 kcal/mol for the singlet-triplet splitting in diphenylcarbene.

II. GENERATION OF CARBENES

A. Copper-mediated Transfers of Carbenes

An example rich in implications is the Cu-catalyzed transfer of "CH_2" from diphenylsulfonium methylide, Eq.(1).[11] The

$$(C_6H_5)_2\overset{+}{S}-\overset{-}{C}H_2 + \text{(tetramethylethylene)} \xrightarrow[\text{THF, } 25^\circ]{Cu(AcAc)_2} \text{(tetramethylcyclopropane)} + (C_6H_5)_2S \qquad (1)$$

ylide is electronically similar to a principal canonical form of the diazoalkanes, compounds subject to Cu carbenoid transfer. The yields of the stereospecific addition [Eq.(1)] range from 48% (trans-2-octene) to 35% (cyclohexene) and are not competitive with those of the Simmons-Smith reaction or its analogs. However, Eq.(1) models the biosynthesis of certain cyclopropanes,[12] in which "CH_2" originates as the S-methyl group of S-adenosylmethionine, **1**. The latter could be enzymatically

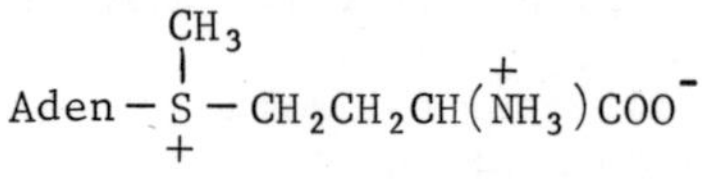

$$\text{Aden}-\overset{\overset{CH_3}{|}}{\underset{+}{S}}-CH_2CH_2CH(\overset{+}{N}H_3)COO^-$$

1

deprotonated and the resulting ylide made to yield CH_2 via a Cu-containing enzyme.[11]

The biosynthesis of presqualene pyrophosphate, **3**, from farnesyl pyrophosphate, **2**, was suggested to proceed as in Eq.(2), in which PP = pyrophosphate.[11] Note that presqualene alcohol has been synthesized by the ZnI_2-catalyzed addition of a farnesyl diazo compound (**2**, $CH_2OPP{=}CHN_2$) to farnesol (**2**, PP=H).[13]

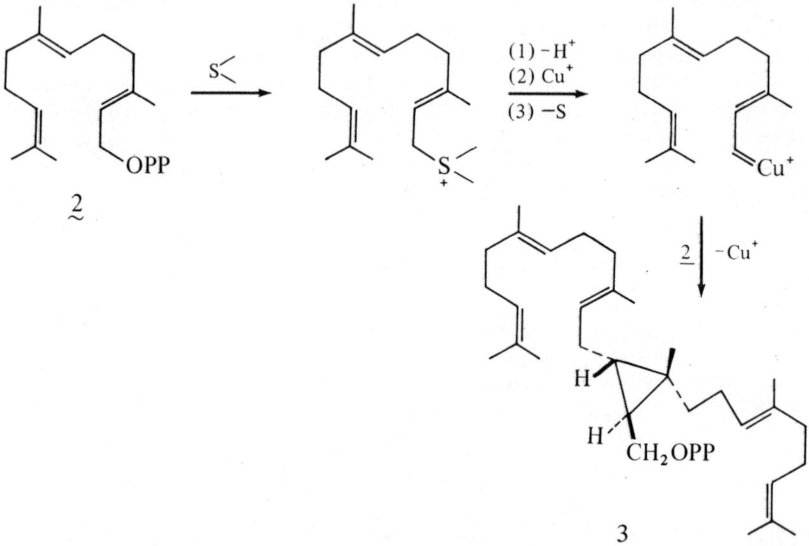

OPP

2

S<

(1) $-H^+$
(2) Cu^+
(3) $-S$

Cu^+

2 $-Cu^+$ (2)

H

H

CH_2OPP

3

However, dimethylsulfonium methylide does not transfer CH_2 under the conditions of Eq.(1),[11] so it is unlikely that this "model" reaction could be extended to 1. Moreover, there are other, noncarbenoid biosynthetic possibilities for the conversion of 2 to 3,[14] although the ylide-carbenoid route does subsume two major types of biological cyclopropanation under a single mechanistic concept.[11] Mechanistic, synthetic, and, possibly, biosynthetic elaborations of Eq.(1) are clearly desirable.[15]

In passing, we note the Cu-mediated carbenoid transfers of CH_2, CHCl, and $CHCOOCH_3$, mainly from precursor diiodides or dibromides, to cyclic and acyclic alkenes in moderate-to-good yields,[16] as well as the cyclopropylidene Cu-carbenoid studies of Hiyama and colleagues.[33,34] The Cu-catalyzed decomposition of diazoesters, as investigated by Wulfman, is discussed in Section V.

B. Methylene Carbenes

Methylene carbenes 4, have been extensively studied.[17] Of the numerous available generative methods, the alkoxide-induced decomposition of 5,5-disubstituted *N*-nitrosooxazolidones, 5,[18] and the alkoxide or alkyllithium-induced decomposition of vinyl halides[17] have been most frequently used. However, in terms of yield, simplicity, and flexibility, the treatment of vinyl trifluoromethanesulfonates (triflates), 6, with KO*t*-Bu is a superior source of 4[19] [cf. Eq.(3)]. The vinyl triflates, available from aldehydes via their trimethylsilyl enol ethers,

$$\mathbf{5}\ (\text{RR}'\text{C}(\text{CH}_2\text{–N–NO})\text{–O–C=O ring}) \xrightarrow{\text{ROLi}} \text{RR}'\text{C=C:}\ \mathbf{4} \xleftarrow{\text{KO}t\text{-Bu}} \text{RR}'\text{C=C(H)OTf}\ \mathbf{6} \qquad (3)$$

$$\text{RR}'\text{C=C(Hal)H} \xrightarrow[\text{or R}''\text{Li}]{\text{KO}t\text{-Bu}} \mathbf{4}$$

react readily at -20^0 to 0^0 with KO*t*-Bu. For 6, R=R'=alkyl, good yields of disubstituted methylenecyclopropanes result from addition to alkenes. These are always accompanied by minor quantities of 7, most likely a product of 4 and the *t*-butanol formed during deprotonation of 6. In the presence of excess *t*-BuOD, 6 ($R=R'=CH_3$) afforded partially deuterated 7 [$(k_H/k_D) \sim 9.9$ for competition between *t*-BuOD and *t*-BuOH)]. Neither 6 nor

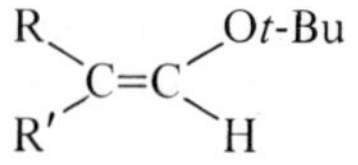

$$\underset{\mathbf{7}}{\mathrm{(R)(R')C{=}C(O}t\text{-Bu})(H)} \qquad \underset{\mathbf{8}}{\mathrm{(D)(}n\text{-}C_3H_7)C{=}C(OTf)(H)}$$

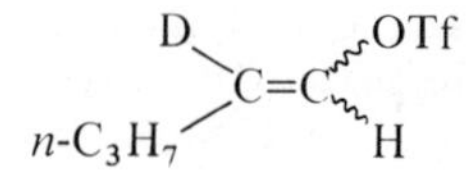

7 underwent H-D exchange, so that the deuterium incorporation experiment argues for a carbenic origin of 7 rather than addition of t-butoxide to 6 followed by elimination of triflate.[19a]

When, however, R or R' = H or phenyl, the reaction of 6 and KOt-Bu gave only the alkyne anticipated from 1,2-rearrangement of 4. No trapping was observed, even in neat olefinic solvents. Alkyne formation appears to be a carbenic reaction secondary to α-elimination and not the result of β-elimination. Thus the conversion of E- or Z-8 to pentyne-1 by KOt-Bu at -20° is kinetically independent of configuration, and no primary isotope effect can be detected during competitive decomposition of either Z- or E-8 and 8-H.[19b] Neither observation is consistent with β-elimination.

The carbene generated from 9 gave the products shown in [Eq.(4)]. The distributions of isomeric products appear to be independent of triflate configuration, suggesting that the cyclopropanes and ethers all arise from the same methylene carbene. The configurational assignments of the products (although not

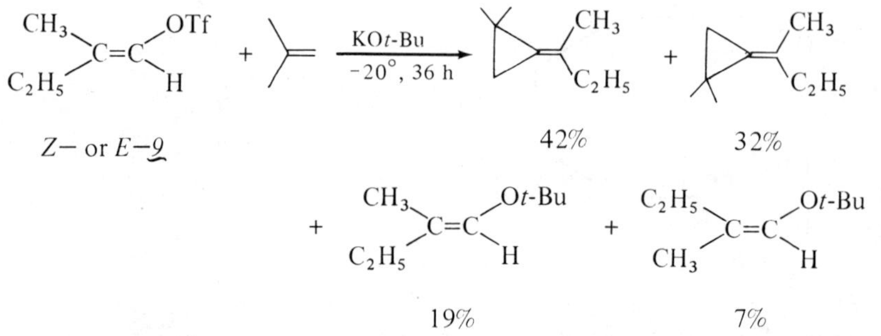

$$\underset{Z\text{- or }E\text{-}\mathbf{9}}{(CH_3)(C_2H_5)C{=}C(OTf)(H)} + (CH_3)_2C{=}CH_2 \xrightarrow[-20^\circ,\ 36\ h]{KOt\text{-Bu}} \underset{42\%}{\text{2,2-dimethyl-1-}[C(CH_3)(C_2H_5)]\text{cyclopropane}} + \underset{32\%}{\text{isomer}} + \underset{19\%}{(CH_3)(C_2H_5)C{=}C(Ot\text{-Bu})(H)} + \underset{7\%}{(C_2H_5)(CH_3)C{=}C(Ot\text{-Bu})(H)} \qquad (4)$$

essential to the argument) are not firm,[19a] so the simultaneous predominant formation of the apparently less sterically congested cyclopropane and more congested ether may not have mechanistic significance.

The additions of 4 (R=R'=CH_3) to cis- or trans-butene and to Z- or E-2-methoxy-2-butene were >99% and >98% stereospecific, respectively, and the addition to trans-butene remained so even in the presence of 95 mol % of perfluorocyclobutane. This led the authors to assign a singlet multiplicity to the reactant (also ground) state of 4.[19c] Although this conclusion is probably correct, cis-butene would have been the far more sensitive test olefin for the development of nonstereospecificity on dilution. Moreover, even for carbenes with spectroscopically authenticated triplet ground states, diluent concentrations of >98 mol % may be required to produce even a minimal display or nonstereospecific addition;[20] or, worse still, nonstereospecificity may not appear at all in response to dilution.[21]

Competitive additions of 4 (R=R'=CH_3), generated from 6 and KOt-Bu at -20° to substituted styrenes afforded relative reactivities that were successfully correlated by σ (not σ^+), giving $\rho = -0.75$. The kinetic data did not change when the carbene was generated in the presence of a 10% molar excess of either dicyclohexyl-18-crown-6 or 18-crown-6,[19e] suggesting, but not by itself proving (see text that follows), that free 4 is generated by the action of KOt-Bu on 6. Certainly, the reactive intermediate is not complexed with K^+, and because its reactivity is also independent of initial triflate configuration [cf. Eq.(4)],[19a] it is likely to be uncomplexed with an anion.

A problem arises in comparing the ρ-value (-0.75) of triflate-generated 4 (R=R'=CH_3) with literature values. From precursor 5, 4 yields $\rho = -3.4$ (σ^+),[18c] whereas from either [$(CH_3)_2C{=}CBr_2$ + (ethereal) CH_3Li, -40°] or [$(CH_3)_2C{=}CHBr$ + KOt-Bu·t-BuOH, -10° to 0°], $\rho = -4.3$ (σ^+).[22] The triflate data are not correlated by σ^+ constants, and neither the oxazolidone nor the vinyl halide data are correlated by σ constants. The latter studies[18c,22] used minimal 3-point Hammett plots (p-CH_3, p-Cl, and styrene itself), so that even though the distinction between σ^+ and σ correlations is apparent from comparative plotting, one would feel more comfortable had p-CH_3O been included (as in the triflate study).[19e]

Nevertheless, we are confronted with a dilemma. Except for possible complexation with t-butanol,[19e,23] 4 generated from 6 appears to be a free carbene. However, t-butanol cannot seriously perturb the reactivity of 4, because two quite different vinyl halide-based generations of 4, one employing t-butanol and the other not, give 4 of comparable reactivity,[22] albeit very different from, that of triflate-generated 4.

The modest electrophilic selectivity of the latter is more in keeping with that of such other carbenes as CCl_2, CF_2, and cyclopentadienylidene ($\rho = -0.62$, -0.57, and -0.76, respectively). Even $(CH_3)_2C{=}C{=}C{:}$ has $\rho = -0.95$.[22] In contrast, the very negative ρ-values of 4 generated from 5 or from vinyl halides are similar to those of ionic electrophilic additions to styrenes. Although the formation of methylenecyclopropanes by reaction of 5, lithium alkoxide, and alkenes[18] could conceivably involve cyclopropanation by an alcohol-activated diazo compound, 10,[25] or even by a diazonium ion, the similarity between 4 generated from 5 and from vinyl halides renders this unlikely. Moreover,

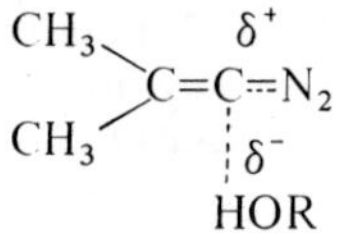

$$(CH_3)_2C{=}\overset{\delta^+}{C}{=}N_2 \quad \cdots \overset{\delta^-}{HOR}$$

10

the kinetic similarity of the highly electrophilic 4 generated from $(CH_3)_2C{=}CBr_2$ (LiBr leaving group) or from $(CH_3)_2C{=}CHBr$ (KBr leaving group) is most consistent with free dimethylmethylene carbene. Further work is most definitely required to harmonize the disparate results.

Finally, we note the use of triflate-derived 4 in the synthesis of cyclopropenium ions,[19d] and alternative generations of 4 from an organomercurial [$(CH_3)_2C{=}C(Br)HgBr$, 150°][27] and from $(CH_3)_2C{=}CClSi(CH_3)_3$ and $(CH_3)_4N^+F^-$.[28] The former method requires harsh thermal conditions. The latter, although perhaps not competitive with the triflate method in yield, avoids strong base and takes place at 25° in diglyme.

C. Phase-transfer Catalysis (PTC)

The use of phase-transfer catalysis (PTC) for the generation of (particularly) dihalocarbenes has attracted great attention.[29] Mechanistically, it now seems agreed that, in the classic case, hydroxide converts $CHCl_3$ to CCl_3^- at or near the aqueous-organic phase boundary; the quarternary alkylammonium ion (Q^+) then pairs with CCl_3^- and assists its penetration of the organic phase, where CCl_2 is subsequently generated.[29,30] It appears that $Q^+CCl_3^-$ does not penetrate deeply into the organic phase[30] and that the equilibrium, $Q^+CCl_3^- \rightleftarrows Q^+Cl^- + CCl_2$, lies to the left, alkene being needed to consume the carbene.[29,30]

Dehmlow and Lissel's systematic study of experimental conditions for the optimal PTC generation of CCl_2 is noteworthy,[30] as are the demonstrations that $R_3\overset{+}{N}-\overset{-}{C}Cl_2$ is not itself a source of CCl_2, but a precursor of $R_3N^+CHCl_2,Cl^-$ (the actual source of CCl_2) in the trialkylamine-catalyzed additions of CCl_2 to alkenes,[31] and that a polystyrene resin, functionalized with p-$CH_2N^+(CH_3)_2$-n-C_4H_9,Cl^- substituents, brings about the dichlorocyclopropanation of alkenes in a three-phase chloroform-alkene, aqueous hydroxide, resin system.[32] The mechanism of the latter reaction resembles that of PTC, although CCl_3^- must be anchored at the resin, so that CCl_2 generation and addition must also occur close to the solid-organic phase boundary.

The possibility of using functionalized phase-transfer catalysts to generate carbenes of modified reactivity[33] requires further discussion. Under comparable conditions (4-5 equiv of $CHCl_3$ per double bond, 50% aqueous NaOH, 0.025 equiv of Q^+, 55°, 3 h), it has been reported that CCl_2 additions are effectively catalyzed by 11 and 12. With 11, triene 13, limonene (14), and diene 15 afforded, respectively, tris-adduct (100%), bis-adduct ("only"), and a mixture of bis- (35%) and monoadducts (61%). However, with catalyst 12, the same alkenes reportedly gave monoadducts only, at the indicated sites (arrows), in 72%, 62%, and 93% yields.

$n\text{-}C_{16}H_{33}\overset{+}{N}(CH_3)_3,Br^-$ **11**

$C_6H_5CH_2\overset{+}{N}(CH_3)_2CH_2CH_2OH,\quad OH^-$ **12**

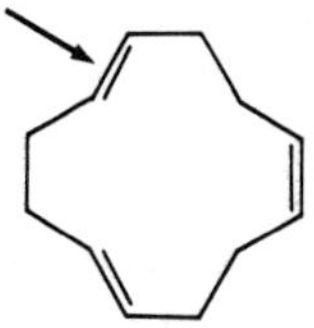

13 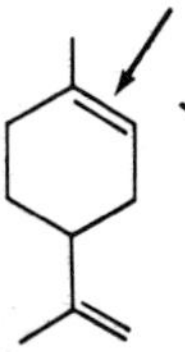 **14** 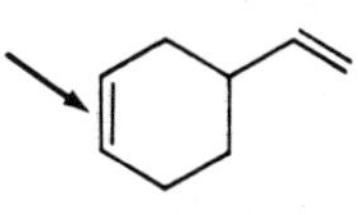 **15**

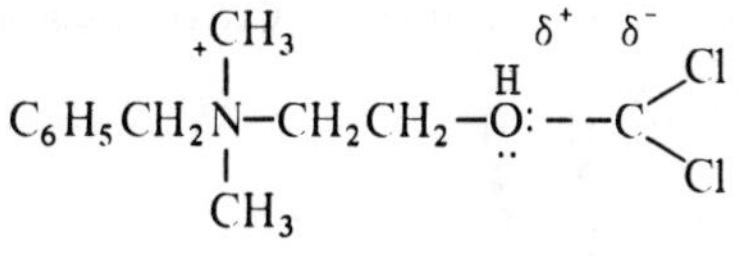

$C_6H_5CH_2\overset{+}{N}(CH_3)_2-CH_2CH_2-\overset{\delta^+}{\underset{H}{O}}:--\overset{\delta^-}{C}Cl_2$ **16**

The apparent regioselectivity exhibited by CCl_2 generated with 12 was attributed to complexation of the carbene by Q^+ (cf. 16).[33] Damping of its electrophilicity by complexation would exalt the ability of CCl_2 to discriminate between double bonds of varying nucleophilicity. One notes that the more highly alkylated positions of 14 and 15 appear to be preferred by the 12-CCl_2 reagent. In all three cases, once the inductively withdrawing dichlorocyclopropyl substituent has been introduced, the "moderated" CCl_2, 16, appears incapable of further attack.

It should be noted that monoaddition, with selectivity toward the more highly substituted double bond, is rather typical of the cyclopropanation of dienes and polyenes by classically generated CCl_2; the phase-transfer procedure using unfunctionalized catalysts is "atypical" in that multiple cyclopropanation occurs readily.[33] This is due to the ability of PTC to generate CCl_2 in an environment where the only possible acceptors are substrate double bonds. Moreover, the reversibility of the $Q^+CCl_3^- \rightleftarrows Q^+Cl^- + CCl_2$ step conserves CCl_2, ensuring that as many collisions of CCl_2 and (even deactivated) substrate sites as are necessary for product formation will occur with very little loss of CCl_2.[29] The opportunity for repeated collisions (effectively an enormous excess of CCl_2, even if the stoichiometric excess of $CHCl_3$ over substrate is only moderate) masks the innate selectivity of CCl_2, permitting, for example, the bis-dichloropropanation of a conjugated diene. Even though the second double bond is deactivated by the initially introduced dichlorocyclopropane group, the PTC-generated CCl_2 will eventually react with it because the carbene's isolation in the organic phase and low instantaneous concentration preclude reaction with base or dimerization, alternatives that would be preferred under more classical conditions of generation. In this context the use of functionalized catalysts (16) might restore to PTC the selectivity of the earlier generative methods while retaining the advantages of overall simplicity and high yield.

It was also reported that catalysis by chiral salt 17 of CCl_2 additions to styrene and trans-β-methylstyrene gave optically active dichlorocyclopropane derivatives.[33] In the latter case the optical yield was only ~0.9%, but with nonfunctional

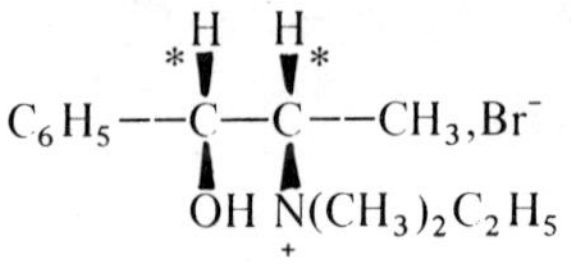

$C_6H_5-\overset{*}{C}H(OH)-\overset{*}{C}H(CH_3)-\overset{+}{N}(CH_3)_2C_2H_5,Br^-$

17

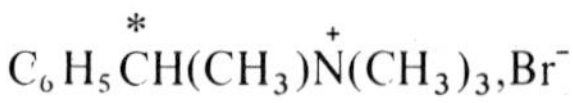

$C_6H_5\overset{*}{C}H(CH_3)\overset{+}{N}(CH_3)_3,Br^-$

18

H—C(C₆H₅)—O—C(H)—CH₃ (epoxide)

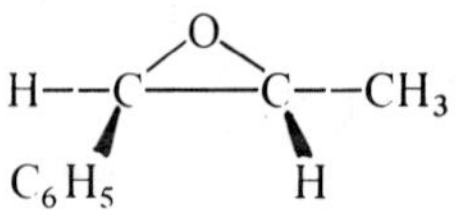

19

chiral 18 the optical yield in the addition to styrene was probably lower, and only ~4% of that obtained with catalyst 17, suggesting that the hydroxyl group is essential to significant asymmetric induction.

These stereochemical findings require cautious interpretation. Rotations of products obtained with 17 were small, and the lack of experimental detail makes it impossible to know if contamination by chiral oxirane 19, a decomposition product of

17, was avoided. Such contamination did attend the apparent asymmetric synthesis of 2-phenyloxirane from benzaldehyde, trimethylsulfonium iodide, and 17, under PTC conditions.[34] Indeed, a recent repetition of the styrene-CCl_2 asymmetric induction experiment suggests that contamination of the product by epoxide 19 was occurring. After the initially optically active, 1,1-dichloro-2-phenylcyclopropane product had been refluxed with ethereal $LiAlH_4$ (to destroy any 19 present) and then repurified, the final dichlorocyclopropane was optically inactive.[34a]

Similarly, caution must also be exercised in evaluating the reported asymmetric induction during PTC-CCl_2 addition to styrene (and several other alkenes) catalyzed by optically active N,N-dibutyl-2-methylpropylamine and N,N-dibutyl-1-phenylethylamine.[34b] The proferred mechanism is likely to be incorrect in detail (it proposes CCl_2 transfer from $R_2R'\overset{+}{N}\text{-}CCl_2^-$, whereas the ylide is more likely to be a precursor to $R_2R'\overset{+}{N}CHCl_2,CCl_3^-$, the actual source of CCl_2[31]); moreover, until the possibility of product contamination by traces of optically active trialkylamine catalyst is scrupulously excluded, the results too must be considered suspect.

Despite the difficulties encountered in the asymmetric induction experiments, the possibility of modifying carbenic selectivity by PTC with functional catalysts is exciting and worthy of thorough investigation.

D. Other Methods for Generating Dihalocarbenes

Brief citation is due the thermal gas-phase generation of CCl_2 from Cl_3CSiF_3. At 140° and $P_T \sim 1$ atm, 85 to 95% yields of dichlorocyclopropanes were realized with various alkenes, and additions to cis- or trans-butene occurred stereospecifically in >90% yields.[35a] The carbene-generative reaction has been shown to be first order in silane and independent of the nature of the acceptor alkene, providing good evidence for the intermediacy of free CCl_2.[35b] Carbon-hydrogen insertion reactions have also been studied under comparable conditions.[35c] We hope that these experiments can be coupled with spectroscopic methods so as to provide kinetic data for the CCl_2 reactions.

Finally, we note a new modification of decarboxylative CF_2 generation,[36] which is effective under mild, nonbasic conditions and avoids relatively expensive organometallic precursors[37] [cf. Eq.(5)]. With 4 equiv of alkene (80°), yields ranged from 30% (cyclohexene) to 100% (tetramethylethylene).

$$ClF_2CCOOCH_3 + LiCl(HMPA\ complex) \xrightarrow[\text{or triglyme, } 80^\circ]{\text{(tetramethylethylene), } CH_2Cl_2 \text{ reflux,}} \text{(1,1-difluoro-2,2,3,3-tetramethylcyclopropane)} + CO_2 + CH_3Cl \quad (5)$$

F F

E. Tetraphenylethylene as a Catalyst for Diazo Compound Decompositions

Ho, Conlin, and Gaspar have reported an extremely important new method of generating carbenes.[38] These authors made the peculiar observation that degassed solutions of diphenyldiazomethane and tetraphenylethylene evolved nitrogen. Solutions in air were stable and the end product of the reaction was benzophenone azine. The degassed solution exhibited an induction period before nitrogen evolution and developed a dark green color, persistent in the sealed tubes but discharged in air. The color was found to result from the benzophenone radical cation, a species also formed in the metal-catalyzed decomposition of diphenyldiazomethane. From these data the following mechanism was proposed:

1. Chain-initiating Steps:

$$Ph_2CN_2 + Ph_2C{=}CPh_2 \longrightarrow [Ph_2CN_2 \cdots Ph_2C{=}CPh_2]$$

$$[Ph_2CN_2 \cdots Ph_2C{=}CPh_2] + Ph_2CN_2 \longrightarrow Ph_2CN_2CPh_2^{\cdot +} + Ph_2C{=}CPh_2^{\cdot -} + N_2$$

2. Chain-carrying Steps:

$$Ph_2CN_2CPh_2^{\cdot +} + Ph_2CN_2 \longrightarrow Ph_2C{=}NN{=}CPh_2 + Ph_2CN_2^{\cdot +}$$

$$Ph_2CN_2^{\cdot +} + Ph_2CN_2 \longrightarrow Ph_2CN_2CPh_2^{\cdot +} + N_2$$

It was correctly deduced that tetraphenylethylene would act as a catalyst for the decomposition of other diazo compounds. Indeed, 10 mmol of diazoacetophenone decomposed in cyclohexene with 3 mmol of tetraphenylethylene gives 42% of 7-benzoylnorcarane. Similar reactions were observed with diazofluorene and diazomethane itself, Eq.(6)-(8).

$PhCOCHN_2$ —(cyclohexene, $Ph_2C{=}CPh_2$)→ 7-benzoylnorcarane (6)

CH_2N_2 —(cyclohexene, $Ph_2C{=}CPh_2$)→ norcarane (7)

diazofluorene —(cyclohexene, $Ph_2C{=}CPh_2$)→ spiro adduct (8)

However, the impact of that paper[38] is clearly not that it provides a new catalyst for the decomposition of diazo compounds, but rather in the questions it raises. If tetraphenylethylene can act as a catalyst, why not other formal carbene dimers? Could it be that those workers who carefully degassed their reaction mixtures were thereby building in a component of non-carbene reaction? On a more detached level, one wonders if adducts of other carbenes can be obtained, why not adducts of diphenylcarbene? This would be useful, as diphenylcarbene does not always react with olefins by addition, but more often prefers abstraction.[39]

Finally, one wishes for more examples to provide scope, and for an unraveling of the details of the mechanism of this new reaction pointed out in this remarkable and too-little-noted paper.[38]

F. Photolysis of Olefins

Over the last few years the groups of Kropp[40] and Hixson[41,42] have shown that one of the most common intramolecular reactions of carbenes, 1-2 carbon-hydrogen insertion, was reversible. The last 2 years have brought several significant developments in this still poorly understood reaction. Inoue, Takamuku, and Sakurai[43] have shown that this reaction, thought to be restricted to tetra- and trisubstituted olefins, also occurred in a cyclic disubstituted olefin, cycloheptene, Eq.(9). The intermediacy of the carbene was demonstrated by deuterium labeling by a method reminiscent of that earlier adopted by Kropp and Hixson.

(9)

The brief report that 3-phenylcycloheptene also rearranged through a carbene may be unrelated, as this photolysis was not direct but rather benzene-sensitized.[44]

Other recent results include the indirect evidence presented by Kropp, Fravel, and Fields[45] that it is the π,R(3 s) Rydberg state that is responsible for the formation of the carbene. This took the form of the demonstration that it is probably the π,σ^* state that is responsible for the intramolecular 1,3 shift of hydrogen that dominates the unsensitized photochemistry of alkenes. Thus it must be some other state that is involved in carbene formation, and a leading candidate is the Rydberg state.

To our knowledge the similarity of this reaction to the amazing thermal isomerization of phenylacetylene to phenylvinylidene seems not to have been remarked,[46] and one wonders if the thermal reversion of olefins to carbenes will one day emerge.

III. CARBENOIDS

A. The Effect of Crown Ethers on Base-catalyzed α-Eliminations

The widespread use of KO*t*-Bu and alkyl halides to generate carbenic intermediates has often led to ambiguity concerning the "freeness" of the reactive intermediate. Frequently, the formally identical carbenes, generated photochemically, exhibit markedly different abilities to discriminate between various alkenes. In the specific cases of phenylchloro- and phenylbromocarbenes, it has now been shown that these differences disappear when the base-catalyzed α-eliminations are done with

KO*t*-Bu-18-crown-6 rather than with KO*t*-Bu alone.[47] Generated as in Eq.(10), X = Cl or Br, identically discriminating phenylhalocarbenes may be produced from either precursor, despite the very different leaving groups.

$$C_6H_5CHX_2 \xrightarrow[-t\text{-BuOH, }-KX]{KOt\text{-Bu-18-crown-6}} C_6H_5\ddot{C}X \xleftarrow[-N_2]{h\nu} \begin{matrix} C_6H_5 \\ X \end{matrix}\!\!>\!C\!\!<\!\begin{matrix} N \\ \| \\ N \end{matrix} \qquad (10)$$

Conclusions and consequences include:

1. Photolyses of the diazirines generate free phenylhalocarbenes.

2. The same intermediates are produced by the benzal halide-KO*t*-Bu-18-crown-6 method.

3. The photogenerated species are neither vibrationally excited carbenes nor photoexcited diazirines or diazo compounds because their selectivities are identical to those of the thermally generated species.

4. In the absence of 18-crown-6, α-elimination from $C_6H_5CHX_2$ leads to phenylhalocarbenoids that are probably complexes of C_6H_5CX with either KX or KO*t*-Bu.

5. It should be possible to determine whether other base-induced α-eliminations afford carbenes or carbenoids by determining the olefinic selectivity of the carbenic species in the presence or absence of crown ether.

6. KOR-18-crown-6 α-eliminations should make free carbenes available when diazirine or diazoalkane precursors are hazardous or unavailable.

The crown-ether test for carbene "freeness" [item (5)] has been applied to the generation of **4** from triflate **6**[19e] and to the generation of CH_3SCCl from CH_3SCHCl_2.[48] In both cases KO*t*-Bu was used and no perturbation of the carbenes' selectivities toward alkenes was induced by the crown ether, implying that these carbenes could be generated as free species, even in the absence of crown. On the other hand, C_6H_5CF generated by the action of KO*t*-Bu on C_6H_5CHFBr showed a considerable selectivity change with the use of KO*t*-Bu-18-crown-6, indicating that a carbenoid was produced in the absence of crown.[48] Because diazo precursors to C_6H_5CF are unknown, the generation of free[49] C_6H_5CF from C_6H_5CHFBr with butoxide-crown is an example of item (6).

B. Cyclopropylidene Carbenoids

Much mechanistic and synthetic carbenoid chemistry has recently appeared. We concentrate on cyclopropylidene carbenoids, for here epimeric species have been isolated and aspects of configurationally dependent chemistry have emerged. Preparations of representative epimeric cyclopropylidene carbenoids appear in Eqs.(11)[50] and (12).[51] In the former, inductively withdrawing oxygen atoms preclude the normal cyclopropylidene → allene rearrangement; in the latter, the excessive strain associated with a 1,2-cycloheptadiene serves the same purpose.

Br, Br → (i-PrMgCl; −95°,−67°, 6 h THF) → Br, MgCl (21) → ($(CH_3)_3SnCl$; THF,−90°) → Br, $Sn(CH_3)_3$ (22-endo-Br)

22-endo-Br → (n-BuLi; −95°,THF) → Br, Li (23-endo-Br) (12a)

Br, Br → (0.95 equiv n-BuLi; −85°,THF-ether) → Li, Br (23-exo-Br) → (Cl_3CCCl_3) → Cl, Br

23-exo-Br → ($(CH_3)_3SnCl$; THF,−100°) → $(CH_3)_3Sn$, Br (22-exo-Br) (12b)

In Eq.(11a) favorable Li-oxygen interactions influence the kinetics of Li-Halogen exchange; 20-exo-Cl is the kinetic product. It is also the thermodynamic product, although here the favorable Li-O interactions[50a] are reinforced by the sterically favorable[51a] exo-halogen, endo-Li configuration. The less stable epimer, 20-endo-Cl, can be formed from an endo-Cl, exo-Br precursor [Eq.(11b)] because Li-Br exchange is apparently faster than Li-Cl exchange, even when the latter is assisted by internal oxygen solvation. The epimeric structures and stabilities of 20-exo-Cl and 20-endo-Cl were secured by high yield, highly stereoselective protolyses (CH_3OH), which afforded the epimeric monochlorides. Epimeric purities were estimated at ~99% for 20-exo-Cl and ~96% for 20-endo-Cl.

Carbenoid 21, the kinetic product of treating 7,7-dibromonorcarane with isopropylmagnesium chloride, gradually isomerizes to its more stable epimer.[51b] However, 21 is sterically able to react with $(CH_3)_3SnCl$, whereas its epimer apparently cannot, enabling carbenoid 22-endo-Br to be obtained epimerically pure.[51b] Transmetalation then affords 23-endo-Br, which remains configurationally stable for several hours at -95° (in the absence of dibromonorcarane), quenching with CO_2 gives only the endo-bromo, exo-acid.[51b] Direct Br-Li exchange with 7,7-dibromonorcarane initially gives 23-endo-Br. In the presence of unreacted starting material, however, an intermolecular exchange equilibrium, Eq.(13), permits 23-exo-Br to become the sole significant carbenoid present (at -85°);[52] quenching with $(CH_3)_3SiCl$ yielded only the endo-trimethylsilyl, exo-bromide in 76% yield.[51a]

Br, Li, 23-endo-Br + Br, Br $\xrightarrow{-85°}$ Br, Br + Li, Br, 23-exo-Br (13)

Under carefully controlled conditions the reactions of epimeric carbenoids 20 can be differentiated:[50b]

Br, Cl, O, O; CH_3Li-LiBr, ether, −78° 1 h, then 22°, 1.5 h → Li, Cl, O, O, 20-exo-Cl → H, CH_3, $CHOC_2H_5$, O, O, 25(3%) + O, O, 24(92%) (13a)

Cl, Br, O, O; CH_3Li, −78°, 2.5 h, ether-THF(10:1), then 10°, 1 h → O, O, Cl, Li, 20-endo-Cl → (O, O)$_2$ 26(61%) + 25(27%) + 24(1%) (13b)

Although the most stable conformation of 20-endo-Cl is probably the crown [Eq.(13b)] and that of its internally stabilized epimer, 20-exo-Cl, the chair [Eq.(13a)], it can be persuasively argued that suitable axial C-H bonds are available for intramolecular insertion by 20-endo-Cl.[50b] What is observed, however, is that the latter carbenoid displays almost entirely intermolecular chemistry, affording 25 by insertion into solvent and 26 by "dimerization." In contrast, epimer 20-exo-Cl reacts overwhelmingly by intramolecular insertion, yielding 24.

Noting that in each case the reactions occur rearside to the departing halogen, it is tempting to formulate them as nucleophilic attacks on intimate ion pairs [see Eqs.(14a) and (14b)]. The concept of C-Cl ionization in lithium carbenoids

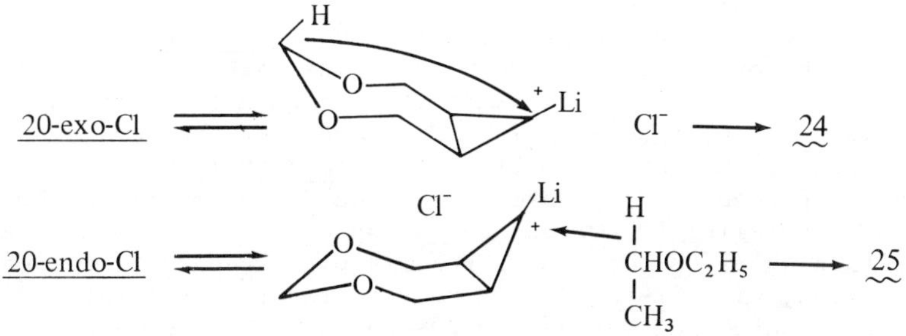

(14a)

(14b)

receives support from matrix-isolation studies of $LiCCl_3$,[53] and the S_N2-like formulation of lithium-carbenoid reactions is an old idea.[54] The nucleophiles in Eqs.(14a) and (14b) are internal and external α-oxygen-activated C-H σ bonds, respectively.[55]

At the very least the distinct reactivities of the 20 epimers exclude a free carbene intermediate and demand that, if Cl-ionized organolithium intermediates are involved, they are either formed irreversibly, or the chloride ion is so intimately bound to the carbenoid that endo-exo isomerization is precluded. What are lacking and clearly desirable are thermolyses of carbenoids 20 in alkenes. The prediction would be for facile intermolecular addition of 20-endo-Cl but continued (nondivertable) intramolecular insertion of 20-exo-Cl.

Problems with the perhaps-too-tempting interpretation offered above arise in connection with carbenoids 22-endo-Br [Eq.(12a)] and 22-exo-Br [Eq.(12b)].[51a] The configurations of these epimeric carbenoids were secured by stereochemically selective Sn-C cleavage reactions ($HCl/CHCl_3$).[51c] It was then found that 22-exo-Br decomposed considerably faster than 22-endo-Br in refluxing cyclooctene[51c] but that both carbenoids afforded the anticipated addition product.

Although the differences in product formation exhibited by epimeric carbenoids 20 are lost with epimers 22, decompositions of the latter remain kinetically distinct; 22-exo-Br requires 6 h of reflux in cyclooctene to afford 76% of adduct, whereas 22-endo-Br requires 23 h to afford 40%. In refluxing cyclohexene, 22-exo-Br requires ~750 h for complete decomposition and gives 33% of the addition product, but 22-endo-Br was completely stable under these conditions! Moreover, no interconversion of epimers was observed.[51c] Unfortunately, no attempt was made to detect intramolecular decomposition products from the 22 epimers. This should be done.

The situation with carbenoids 22 is thus less satisfactory than with carbenoids 20. The greater reactivity of 22-exo-Br may be due to strain induced by the endo-trimethyltin moiety, rendering the ensuing chemistry quite distinct from that of 20-exo-Cl, in which the endo-Li is actually stabilized by interaction with the oxygen atoms. Perhaps a free carbene is formed from 22-exo-Br, but, if so, then it either does not revert to starting material by reaction with $(CH_3)_3SnBr$, or the reversion is stereospecific because formation of 22-endo-Br is not observed. A carbenoid decomposition of the 22 epimers is doubtful because, parallel to 20, one would expect that intermolecular decomposition should proceed most readily with 22-endo-Br and not, as is observed, with 22-exo-Br. Moreover, reactions of the "7-norcaranylidene" generated from 22-exo-Br with triethylsilane (Si-H insertion) or methylenecyclohexane (C=C addition) gave ~ 1:1 mixtures of the anticipated isomeric products.[51c] The lack of stereoselectivity seems more consistent with reactions involving a free carbene than a sterically congested carbenoid transition state. Further study of the 22 epimers is clearly needed, and better still would be careful analysis of the intra- and intermolecular reactivity of the 23 epimers. The latter information, when compared with that derived from augmented studies of 20, could go a long way toward elucidating basic mechanisms of carbenoid chemistry.

Several themes illustrated above have recently found synthetic expression in the stereospecific endo-alkylation of 23-exo-Br.[56a] A related bis-alkylation involves Cu carbenoids [Eq.(15)];[56b] the implied mechanism is not established, but the ideas of initial exchange at the exo-halogen, followed by eventual formation of an endo-metallic intermediate, clearly echo the carbenoid reactions described here. Reactions analogous to Eq.(15) have been used to prepare precursors to d,ℓ-sesquicarene and d,ℓ-sirenin.[56b]

Finally, we note the carbenoid "cascade" experiments of Schlosser [Eq.(16)], in which various halomethanes were treated with ethereal CH_3Li, in 2-methyl-2-pentene at -78°.[57] Matched

$$\text{7,7-dibromonorcarane} \xrightarrow[\text{ether, }-48^\circ,\ 1\ \text{h}]{(1)\ (n\text{-Bu})_2\text{CuLi},} \text{7-bromo-7-}(\text{Cu-}n\text{-C}_4\text{H}_9)\text{norcarane} \longrightarrow \text{7-Cu-7-}n\text{-C}_4\text{H}_9\text{norcarane} \xrightarrow[-20^\circ]{(2)\ \text{CH}_3\text{I}} \text{7-CH}_3\text{-7-}n\text{-C}_4\text{H}_9\text{norcarane} \qquad (15)$$

82%, stereoselectivity >99:1

pairs of halomethanes, capable of affording "identical" carbenoids by either Li-halogen or Li-proton exchanges, gave species that, in fact, showed minor but reproducibly divergent reactivity. In cases where "CCl_2" or "CFCl" were generated initially, reasonable yields of cyclopropanes were obtained by olefinic trapping of these species, little product was obtained from the carbenic progeny (i.e., "CH_3CX"), and no product of the more remote descendant, "CH_3CCH_3," was observed. In contrast, the four halomethanes that appear at the top right of Eq.(16) preferentially gave "CF_2" as the initial species. Here little CF_2-addition was observed; F-CH_3 exchange predominated and the products were mainly derived from "CH_3CF" and "CH_3CCH_3." A reasonable interpretation is that CCl_2 or CFCl are reactive enough to be trapped by the alkene, whereas CF_2 is sufficiently stabilized (see Section IV) to be more selective and to react preferentially with the more nucleophilic CH_3Li. Although carbenoids are implicated in these reactions, they may not be the exclusive product-determining intermediates; free dihalocarbenes may also arise from lithium trihalomethide precursors.[58]

IV. REACTIVITY OF CARBENES

A. Carbene-to-Carbene Rearrangements

A reasonable mechanistic description of the reaction by which substituted phenylcarbenes interconvert has appeared,[59] along with a fine review of the whole process.[60] Addition reactions of carbenes are generally regarded as electrophilic processes, and it was suggested early that the gas-phase ring expansion of the naphthylcarbenes proceeded through addition to the bond of greater bond order and not necessarily toward the

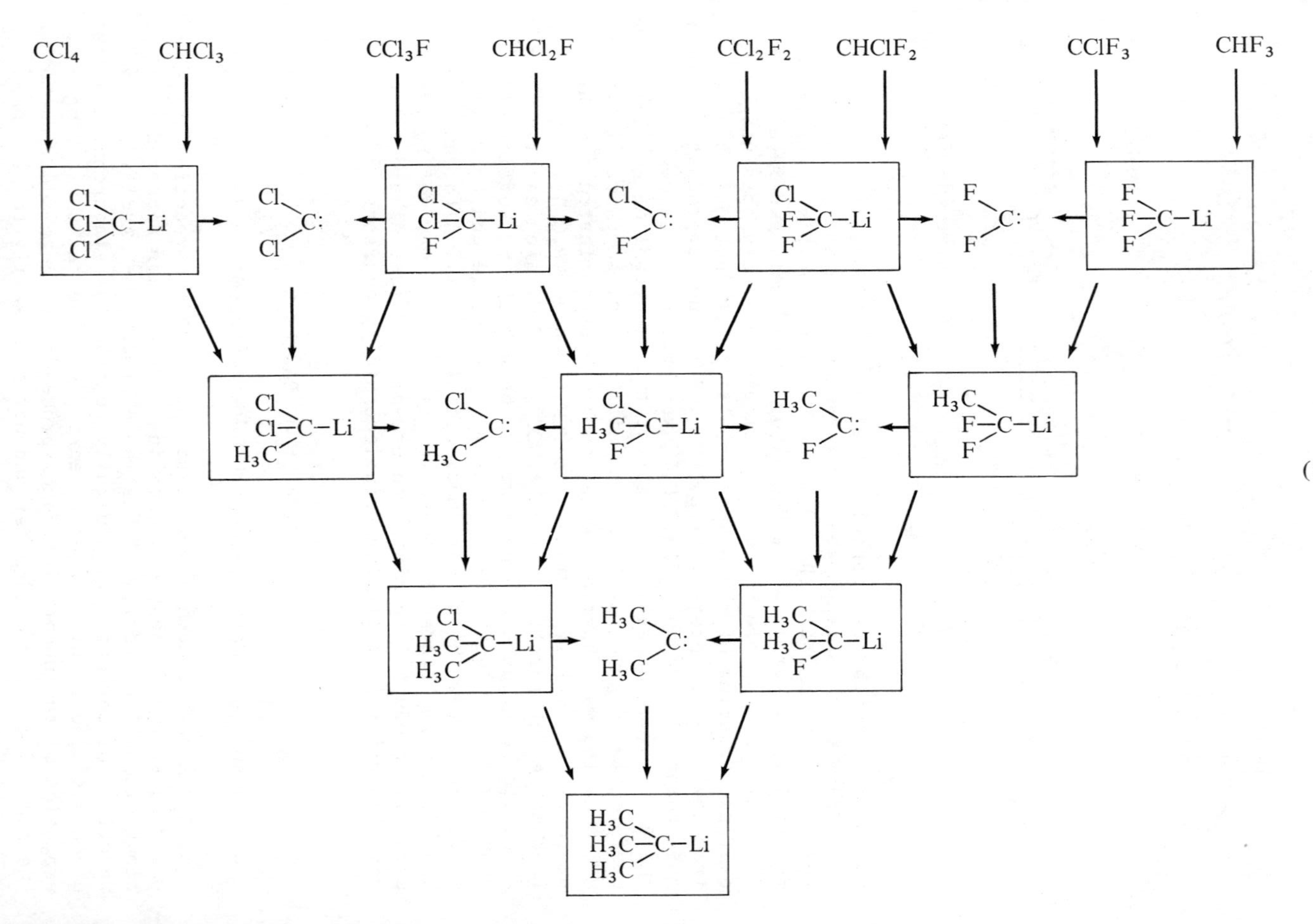

(16)

weaker bond.[61] But this appears not to be the whole story, and the following picture has been advanced by Wentrup.[59,60] Any carbene addition must be best described as a simultaneous transfer or delocalization of the electrons in the carbene sp^2 orbital into the LUMO of the π-system and of the electrons in the HOMO of the π-system into the vacant p-orbital of the carbene. A low-lying LUMO with a large coefficient would favor such a process in which the carbene and the π-system act simultaneously as nucleophiles and electrophiles.

For instance, the seemingly capricious rearrangements of the pyridylphenylcarbenes **27**, **28**, and **29** can be rationalized in this way. 4-Pyridylphenylcarbene **27** gives overwhelmingly the

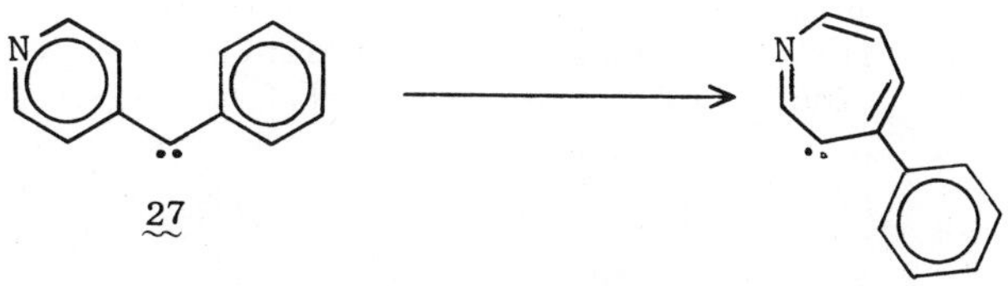

(17)

product of initial attack on the pyridine ring [Eq.(17)]. In contrast, 3-pyridylphenylcarbene **28** gives at least 93% attack on

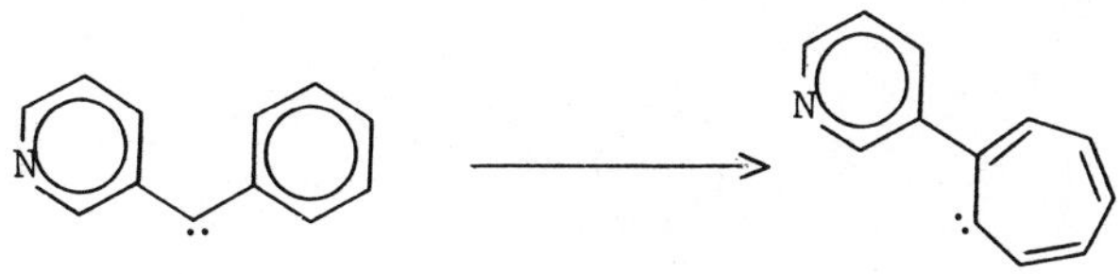

(18)

the benzene ring [Eq.(18)]. 2-Pyridylphenylcarbene **29** attacks

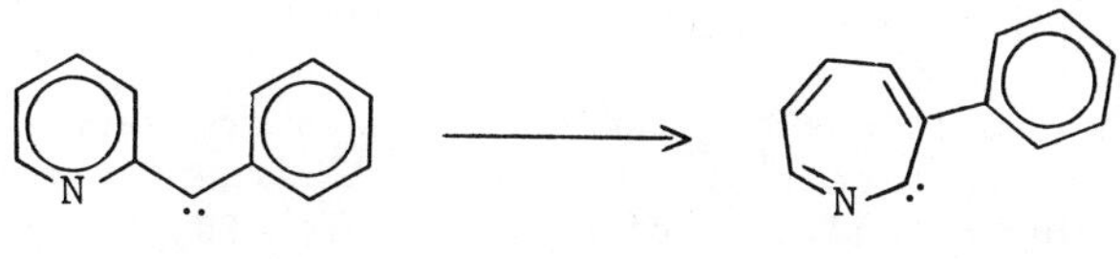

(19)

the pyridine ring exclusively [see Eq.(19)].

Wentrup describes these reactions as involving simultaneous electrophilic attack by the empty p-orbital of the carbene on the <u>ortho</u> position of a ring as the carbene's σ-electrons interact with the LUMO of the more electrophilic ring. Thus ring expansion is favored by a low-energy LUMO with a high coefficient at the position of carbene attachment and a high-lying HOMO with electron density at the *ortho* position. In 3-pyridylphenylcarbene 28 the carbene is at a position of low electrophilicity and the HOMO electron densities at the positions ortho to the carbene (2 and 4 in the pyridine) are low. Thus both nucleophilic and electrophilic interactions with the pyridine ring are poor. Reaction occurs with the phenyl ring overwhelmingly.

In contrast, in 2- and 4-pyridylphenylcarbenes 27 and 29 the carbene is attached to relatively electrophilic positions and attack now occurs in the pyridine ring. If one transforms 29 into 30, thus lowering the LUMO energy of the "benzene" part and creating a favorable position for attack on the carbene p-orbital (the α position of naphthalene), addition to naphthalene is favored by a factor of 2 [see Eq.(20)].

→ 64% Products (20)

30

→ 30% Products

Other examples of this intellectually satisfying description are given by Mayor and Wentrup[59] and Wentrup.[60]

B. 1,4-Additions of Carbenes

Concerted 1,4-additions of singlet carbenes are rare, as are their stepwise counterparts. Qualitatively, the key interaction in the conjugate addition of, for example, CH_2 to *s-cis*-butadiene, 31, is between the carbenic p-orbital (LUMO) and the π_2(HOMO) orbital of the diene. The π_1 orbital of the diene is symmetry-forbidden from interacting with p and, in fact, because the carbene has to come quite close to the diene to make the p-π_2 overlap significant, repulsion between its filled σ-orbital and π_1 offsets much of the favorable p-π_2 interaction. No such difficulty attends the 1,2-addition of CH_2 to butadiene; both π_1 and π_2 can effectively donate to the vacant

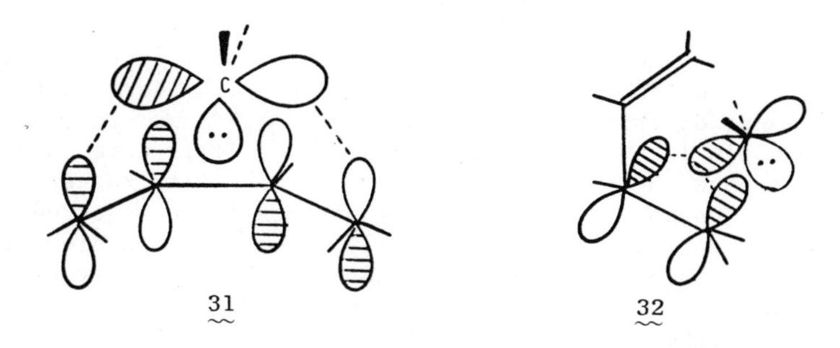

carbenic p-orbital (cf. 32). These qualitative considerations are reinforced by molecular orbital (MO) calculations.[62,63]

Nevertheless, numerous examples of 1,4-dihalocarbene additions to norbornadiene and its derivatives have now been discovered.[64,65] The addition of CF_2 to norbornadiene [Eq.(21)] is representative. It is worth noting that in the original experiments[64b,c] CF_2 had to be generated from $C_6H_5HgCF_3$ and NaI

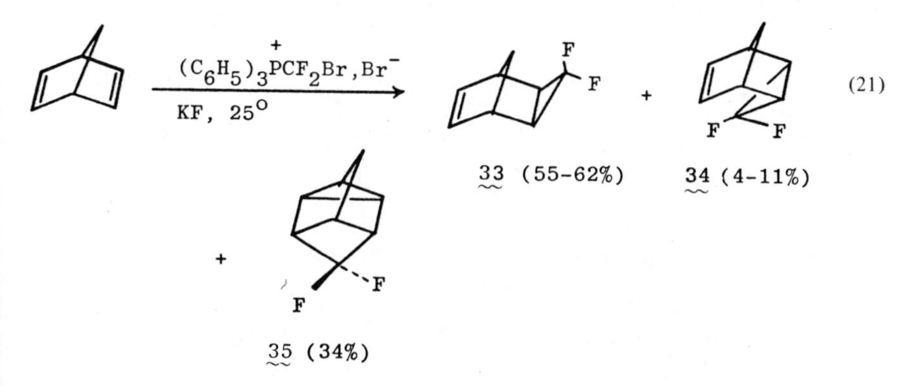

at 80°, and endo-adduct 34 was not observed. It was feared that 34 might have formed but then thermally rearranged to 35. The introduction of the gentle phosphonium route to CF_2,[66] however, made possible both the detection of 34 and the demonstration that it rearranges at 60°-80°, not to 35, but to 33,[65e] thus establishing the primary nature of the 1,4-product, 35, in Eq.(21). The outcome of the competition between 1,2- and 1,4-addition depends on the particular derivatives of norbornadiene used as the substrate and also on the carbene. Steric factors appear to be important in determining the balance, but the role of electronic factors is as yet unknown.

1,4-Additions to various norbornadiene substrates have now been established for CF_2,[64] CFCl,[64] CCl_2,[65a,b,d] and CBr_2,[65b] so that the reaction mode seems general for dihalocarbenes. It is not clear whether other carbenes will exhibit 1,4-addition to norbornadiene; dicarbomethoxycarbene does not.[67] We see no theoretical problem in the way of such additions; it is necessary, however, to have the proper carbene. A carbene with rather small substituents and moderate selectivity appears to be required. The dihalocarbenes are the most readily accessible examples of these species, but may not uniquely possess the necessary attributes. Such carbenes as CH_3SCCl, CH_3OCH, $(CH_3)_2C{=}C{:}$, or $(CH_3)_2C{=}C{=}C{:}$ might profitably be studied.

exo-1,2-Carbene addition to norbornadiene is sterically less demanding than 1,4-conjugate addition and becomes increasingly favored as the carbenic substituents are made large;[65a,b] specifically, the 1,4/1,2-addition ratios decrease in the order CF_2 (0.51, 25°)[64a] $>$ CFCl (0.18, 25°)[64b] $>$ CCl_2 (0.12[65b] - 0.15,[65a] 40-50°) $\gg$ CBr_2 (0.0, only 1,2-addition is observed).[65a,b] However, 1,4-addition may be enhanced by default if *exo*-1,2-addition is disfavored by the judicious introduction of blocking groups in the substrate. Several interesting observations emerge from a comparison of substrates **36**-**40**:

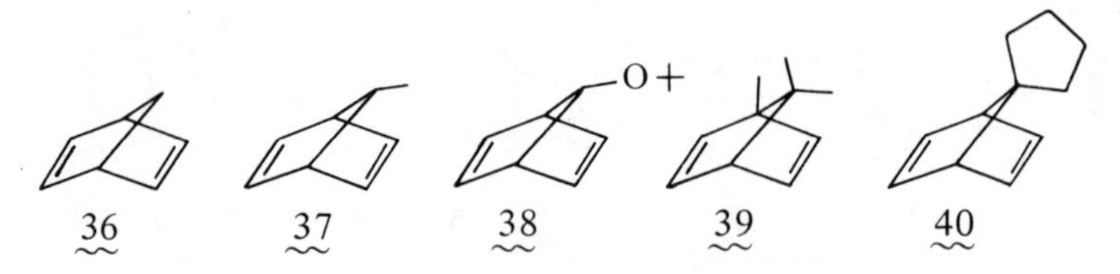

1. As the steric encumbrance of the substrate's *exo* face increases, the 1,4/1,2-addition ratio for CF_2 increases: **36**, 0.51 (25°)[64a] or 0.85 (80°);[64b] **37**, 1.6 (25°)[64a] **39**, 24 (only *endo*-1,2-addition, 25°);[65d] **40**, "∞" (only 1,4-addition is detected, 80°).[65d]

2. The behavior of CCl_2 is similar: **36**, 0.12-0.15 (40°-50°);[65a,b] **38**, 0.43 (temperature not given);[65b] **39**, 0.82 (minimum value, 50°).[65d]

3. With common substrates the 1,4/1,2-addition ratio is smaller for the bulkier, more sterically demanding CCl_2 than for CF_2.

4. With the still more hindered CBr_2, 1,4-addition is observed only with a hindered norbornadiene; that is, 1,4/1,2 is 0.14 for addition to **38**.[65b]

It has been noted that these [2 + 2 + 2] chelotropic cycloadditions of σ^2 carbenes to norbornadienes are allowed under orbital symmetry considerations.[64a] Moreover, the observation of such processes for CCl_2 and CBr_2[65b] demonstrates that fluorine is not a necessary substituent and that 1,4-addition is not restricted to carbenes of quite weak electrophilicity and (presumed) high nucleophilicity, such as CF_2.

It remains to be demonstrated that dienes other than those of the norbornadiene family can undergo the conjugate carbene addition. Bicyclo[2.2.2]octadiene, for instance, gives only 1,2-additions with CF_2.[65c] Norbornadiene is a special substrate in that there is not only extensive through-space interaction of the formally isolated π-bonds,[68] but the molecular architecture favors conjugate addition by bringing the termini of the π-systems close together;[69] the strain incurred in reaching the transition state is thus largely built in with **36**, whereas even with bicyclo[2.2.2]octadiene this is not so.[65c] Additionally, "removing" an electron from π_2 (SA) of norbornadiene should promote bond formation between C_3 and C_5 (i.e., favor 1,4-addition across C_2 and C_6).[68a] The electrophilic attack of a singlet carbene on **36** accomplishes just such a change, electron density being transferred from π_2 to p.[62] (Indeed, this suggests that 1,4-addition to **36** will be observed only with moderate to substantially selective carbenes, for which $\pi_2 \rightarrow p$ transfer will be significant.)

Finally, we note the recent suggestions of possible 1,4-additions of the nucleophilic carbenes, cycloheptatrienylidene[70,71] and 1,2-diphenylcyclopropenylidene[71] to 2,3-dicyanobicyclo[2.2.2]octatriene[70] and to tetraarylcyclopentadienones.[71] Further experiments are needed to determine whether these represent the initial examples of singlet 1,4-additions to substrates other than norbornadienes.

C. Quantitative Characterization of Carbenic Selectivity

Addition of a carbene to an alkene is usually formulated in terms of (electrophilic) attack of its vacant p-orbital (LUMO) on the filled π-orbital (HOMO) of the alkene.[72] One must also consider donation of the carbene's lone pair, contained in its sp^2 (HOMO) orbital, to the alkene's π^* (LUMO) orbital. This nucleophilic action of the carbene is normally minor and overshadowed by the dominant electrophilic behavior, but as noted earlier, it is discernible during certain aromatic carbene-carbene rearrangements[59,60] and clearly visible in the addition reactions of cycloheptatrienylidene and 1,2-diphenylcyclopropenylidene.[73]

At present the most widely applicable quantitative characterization of carbenic selectivity[74] relies on linear free-energy relationships between the olefinic selectivities[75] of various carbenes, CXY, and of CCl_2. In this approach a standard set of alkenes is selected: $Me_2C{=}CMe_2$, $Me_2C{=}CHMe$, $Me_2C{=}CH_2$ (standard alkene), c-MeCH=CHMe, and t-MeCH=CHMe. Relative reactivities ($k_i/k_{isobutene}$) are determined for CXY and for CCl_2 at 25°, and the "carbene selectivity index," m_{CXY}, is defined as the least-squares slope of $\log(k_i/k_{iso})_{CXY}$ plotted against $\log(k_i/k_{iso})_{CCl_2}$. Considering only singlet carbenes (i.e., those added stereospecifically to cis- and trans-butene), 10 m_{CXY} values are available:[74] CF_2, 1.48; CFCl, 1.28; CCl_2, 1.00; CH_3SCCl, 0.91; C_6H_5CF, 0.89; C_6H_5CCl, 0.83; C_6H_5CBr, 0.70; CBr_2, 0.65; CH_3CCl, 0.50; and $BrCCOOC_2H_5$, 0.29. Excepting CH_3SCCl, for which we lack a well-behaved value of $(\sigma_R^+)_{SCH_3}$, the nine other m values are well correlated by Eq.(22).[74b]

$$m_{CXY} = -1.10\,\Sigma_{X,Y}\,\sigma_R^+ + 0.53\,\Sigma_{X,Y}\,\sigma_I - 0.31 \qquad (22)$$

In terms of the transition state for carbene-alkene addition, 41, note that increasing π-electron donation and increasing inductive withdrawal (by X and Y) both augment the electrophilic selectivity of CXY; that is, the coefficients of σ_R^+ and σ_I are negative and positive, respectively. Electrophilic selectivity

R_1, R_2, R_3, R_4; δ+; δ−; X; Y; C

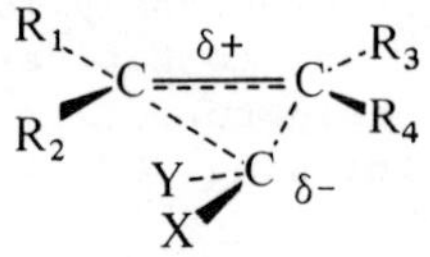

41

is greatest (m_{CXY} is largest) when strong resonance interactions of X and Y with the carbenic center necessitate correspondingly strong π-electron donation by the olefin; electron-releasing alkyl substituents moderate the resulting accumulation of positive charge on the olefinic centers, whereas inductively withdrawing carbenic substituents mitigate the accumulation of negative charge on the carbenic center.

In a distantly related correlation of carbenic reactivities, the ρ-values for the Si-H insertions of $HCCOOC_2H_5$ and of CCl_2 into aryldimethylsilanes were found to be proportional to the sums of the σ^*-values for the substituents of each carbene.[76] Further generalization will require standardization of carbene generative conditions and an alternative choice of substituent constants.

Of those carbenes correlated by Eq.(22), the observed reactivity of CBr_2 ($\underline{m}^{obsd} = 0.65$) deviated most seriously from the calculated value ($\underline{m}^{calc} = 0.82$). This deviation is most reasonably attributed to the greater steric hindrance opposing CBr_2, as compared to CCl_2 additions to alkenes.[77] Nevertheless, the correlative ability of Eq.(22) is not markedly improved by the addition of a steric term.[74b] That the equation works so well without such a term must reflect the high reactivity of even the more selective carbenes, which with simple olefinic substrates must lead to early, relatively "loose" transition states in which differential steric effects are of minor importance.

Carbenes more selective than CF_2 ($\underline{m}^{obsd} = 1.48$ [74]) may be readily identified using Eq.(22); one selects carbenic substituents for which $\sigma_R^+ \ll 0$, leading to $\underline{m}^{calc}_{CXY} = 2.22$, 1.85, and 1.74 for $C(OCH_3)_2$, CH_3OCF, and C_6H_5OCF, respectively. Additions of the latter two carbenes have not yet been observed with the standard alkenes.[74b] Dimethoxycarbene is discussed in the text that follows.

D. Dimethoxycarbene

Dimethoxycarbene, **42**, is a paradigmatic example of a carbene that derives so much resonance stabilization by p-p overlap of the carbenic center with its lone-pair substituents that, relative to the π-orbital of a potential alkene substrate, the carbenic p-orbital is no longer "vacant" in a chemically significant way. The pyrolysis of **43** has long been known to afford **42**,[78] but the carbene could not be trapped by cyclohexene, ketene diethyl acetal, or ethyl acrylate.[79] Now it has been

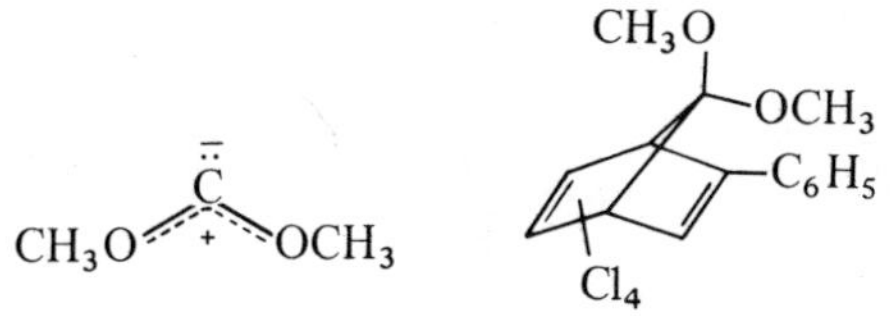

CH_3O–$\ddot{C}$–OCH_3 (+) CH_3O, OCH_3, C_6H_5, Cl_4

42   **43**

found that **42** can cyclopropanate diethyl fumarate, diethyl maleate, styrene, 1,1-diphenylethylene, and ethyl cinnamate.[80,81]

In its disdain for electron-rich alkene π-bonds and its facile reaction with electron-poor olefins, **42** resembles such nucleophilic carbenes as cycloheptatrienylidene,[73] 1,2-diphenylcyclopropenylidene,[73] and 2-oxa-3,3-diphenylsilacycloheptanylidene.[84] However, in its acyclic structure and nonaromatic p-p overlap, 42 is perhaps the clearest example of extreme attenuation of carbenic electrophilicity by the simple manipulation of substituents.

At 125° pyrolysis of 43 in 50% excess diethyl fumarate, followed by careful fractional distillation, afforded 70% of adduct 44, as well as 27% of 45 ($E = COOC_2H_5$).[80] The yields are based on that portion of 43 that, on pyrolysis, eliminates 42 and simultaneously gives 2,3,4,5-tetrachlorobiphenyl; in a competitive and dominant reaction, 43 eliminates $CHCl_3$ with the

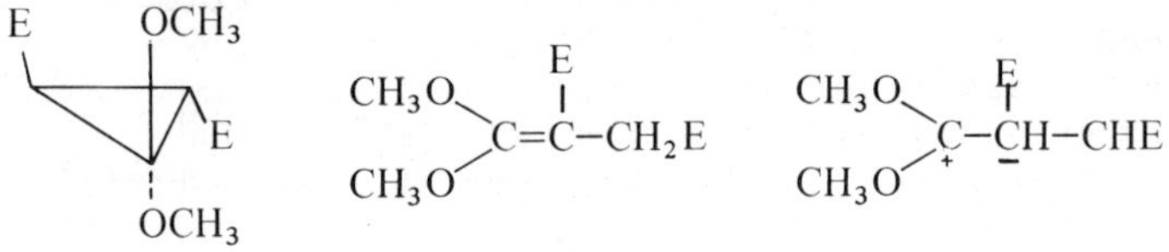

E OCH₃ / E / OCH₃ — 44

$(CH_3O)_2C{=}C(E){-}CH_2E$ — 45

$(CH_3O)_2\overset{+}{C}{-}CH(E){-}\overset{-}{C}HE$ — 46

formation of isomeric trichlorocarbomethoxybiphenyls. The genesis of 45 is reasonably traced to stabilized 1,3-dipole 46, which can undergo an (acid-mediated) proton shift. However, it is unclear whether 46 forms directly by nucleophilic addition of 42 to diethyl fumarate, or whether 42 first cyclopropanates the substrate, yielding 44 from which dipole 46 arises thermally. Various possibilities appear in Eq.(23).

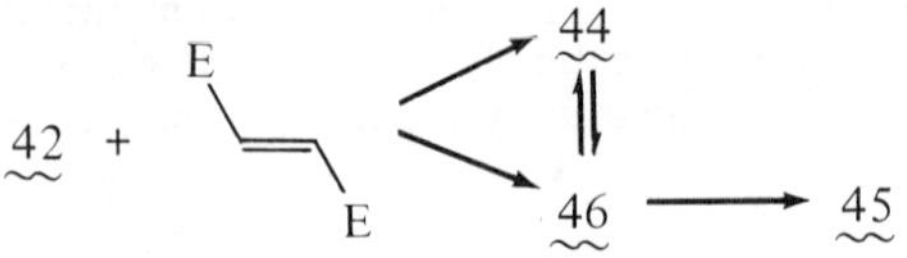

$$42 + E{-}CH{=}CH{-}E \rightarrow 44 \rightleftharpoons 46 \rightarrow 45 \qquad (23)$$

Pyrolysis of 43 in 50% excess diethyl maleate at 145° afforded only 44 (not its cis-isomer) in 81% yield, as well as 45 (11%).[80] It was shown that the maleate did not isomerize to fumarate under these conditions. Although 46 again appears to be a necessary intermediate to explain the overall conversion of cis-E substrate to trans-E product, it is unclear whether 46 is formed directly from 42 and substrate or from an unstable and nonisolable cis-isomer of 44. There is no a priori bar to the stereospecific addition of a nucleophilic carbene to an acceptor olefin. Indeed, cyclopropanations by 42 of the isomeric β-deuteriostyrenes occur stereospecifically.[85] In this case a 1,3-dipole analogous to 46 would be considerably less resonance-stabilized (C_6H_5 instead of E on the carbanionic center of 46) and does not intervene, at least not to interconvert stereoisomeric cyclopropane products.

What is needed is a photolytic precursor to 42, which could be examined at temperatures low enough to preclude thermal isomerizations of sensitive products. This would very likely help resolve the ambiguities associated with Eq.(23). Additionally, similar precursors are required for CH_3OCF and C_6H_5OCF, so that Eq.(22) can be more thoroughly tested. Finally, it is possible that these carbenes, because of their particular substituents, may be so well balanced between electrophilic and nucleophilic propensities that either might dominate the carbene's selectivity over a narrow range or substrate type. Generally, carbenes are predominantly electrophilic (e.g., CF_2 or nucleophilic $[(CH_3O)_2C]$ toward a substrate class such as alkenes. It remains to be seen whether a given carbene can be demonstrably both an electrophile and a nucleophile within one substrate class.

E. α-Silylcarbenes

Over the last 2 years there have been increasing accounts of the chemistry of α-silylcarbenes. Much of this activity derives from the obvious possibilities of such species as sources of carbon-silicon double bonds (silenes) and silacyclopropanes. We review this area here with emphasis on two papers that seem the most definitive of the lot.[86,87]

First of all, more than one claim of an exotic organosilane was discredited. It was shown, for instance, through a ^{13}C-labeling experiment, that phenyltrimethylsilylcarbene does not give its major product 47 through a silacyclopropane, but rather via a carbene-to-carbene rearrangement as shown in Eq.(24).[88]

Si Ph ⟶ CH Si ⟶ Si (24)

47

Similarly, doubt was cast on claims[89] that carbomethoxytrimethylsilylcarbenes gave silenes. It was shown[90] that the ketene 48 was the real source of compound 49, which might reasonably have been attributed to silene 50, Eq.(25). It is very important to note that this work[90] does not remove the possibility that silenes are involved in the reaction, as silene 50 could be the precursor of ketene 48. What this work does do is to raise the strong possibility that they are not.

(25)

No such ambiguities exist in either the article by Kreeger and Shechter[86] or Ando, Sekiguchi, et al.[87] Here it seems is the best evidence to date that α-silylcarbenes do rearrange to silenes. Kreeger and Shechter isolated, along with other products, the disilacyclobutanes 51 and 52 (38%), and it is very difficult to see any source of these molecules other than the carbon-silicon double bonds of a silene intermediate, Eq.(26).

$$(CH_3)_3Si\text{-}\ddot{C}H \longrightarrow (CH_3)_2Si{=}CHCH_3 \longrightarrow \mathbf{51} + \mathbf{52} \qquad (26)$$

Both groups[86,87] were able to trap the silene with added alcohols. Deuterated alcohols provided the expected pattern of substitution in the isolated alkoxysilanes, and this alcohol trapping adds further weight to the already formidable evidence that silenes are involved.

In a related study the Ando group was able to trap the silene produced from phenyltrimethylsilylcarbene through a Diels-Alder reaction with 2,3-dimethylbutadiene to give 53, Eq.(27).[91]

$$(CH_3)_3Si\ddot{C}Ph \longrightarrow (CH_3)_2Si{=}C(CH_3)Ph \longrightarrow \mathbf{53}$$

Also isolated were the cyclopropanes formed by conventional addition of the α-silylcarbene to the diene and the cyclopentene produced by thermal rearrangement of the cyclopropanes. It has now been shown that neither of these conventional products rearranges to **53** under the reaction conditions.[92] Thus this claim of silene formation seems very solid as well.

One must mention, if only in passing, that two descriptions of the IR spectrum of trimethylsilene have appeared back-to-back in the *Journal of the American Chemical Society*.[93,94] Warm-up of the matrix containing the silene does produce the known dimers, so one must assume that the IR spectra are of matrices containing the silene. However, it is our opinion that little of substance can be gained from an inspection of the published spectra. Moreover, we are somewhat discomforted by the differences in the two published spectra and interpretations. It is extremely important to note, however, that the original claim[95] of a 1407 cm^{-1} band for the carbon-silicon double bond was withdrawn in one of these papers.[93]

F. Singlet and Triplet Carbenes

Many of the significant developments in the chemistry of singlet and triplet carbenes involve applications of CIDNP. We do not mention this work in detail, except to point out that the contributions of H. D. Roth have been summarized in a very recent and well-written review.[96]

Forrester and Sadd have adapted the technique of "spin-trapping" in order to detect triplet carbenes as long-lived iminoxyl radicals (**54**), Eq.(28).[97] Triplet carbenes are formed

$$^{3}Ph-\ddot{C}H \quad + \quad NO \quad \longrightarrow \quad \underset{\mathbf{54}}{Ph-CH=NO\cdot} \qquad (28)$$

in a solution saturated with nitric oxide and are converted to iminoxyl radicals detectable by ESR spectroscopy. The carbenes detected in this way are dimethylcarbene, phenylcarbene, phenylethylcarbene, phenylanisylcarbene, fluorenylidene, tetraphenylcyclopentadienylidene, cyclohexanylidene, benzoylcarbene, dicarboethoxycarbene, and thioxanthenylidene S,S-dioxide. Of particular interest is the detection of triplet dimethylcarbene and cyclohexanylidene, as these species are known to react in solution as singlets. Thus indirect evidence for the equilibration of singlet and triplet alkyl carbenes can be gleaned from this paper.

Cyclohexanylidene can be trapped by additions to olefins in solution, albeit in very low yield.[98] It is tempting to speculate that it is the triplet cyclohexanylidene that is trapped by external addition, whereas the singlet is converted entirely to cyclohexene. The stereochemistry of addition should be examined.

V. THE OXIDATION STATE OF COPPER CATALYSTS

Among various copper complexes employed as catalysts, those in which the copper ion is relatively uncomplexed are the most effective in promoting cyclopropanation. For example, copper(I) triflate[99] is much more active than chloro(trialkylphosphite)-copper(I)[100] in catalyzing the reaction between ethyl diazoacetate and olefins. This difference underscores the care that must be exercised in interpreting the effect of adventitious peroxides on the catalytic rate. For example, Wolfman, Thinh, et al.[101] have used the observation of a peroxide effect to conclude that copper(II) species are the actual catalysts in cyclopropanation promoted by chloro(trialkylphosphite)copper(I), since it is known[102] that peroxides oxidize copper(I) to copper(II), Eq.(29).

$$(RO)_3PCu^{I}X + \text{peroxide} \longrightarrow (RO)_3PO + Cu^{II}X^{+}... \qquad (29)$$

However, it is also known that trialkyl phosphites are readily oxidized by peroxides to trialkyl phosphates,[103] which are very poor ligands. Thus the subsequent rereduction of copper(II) by diazo compounds[104] affords an altered copper(I) species in which the phosphite ligand has been stripped, Eq.(30). The net effect

$$Cu^{II}X^{+} + R_2CN_2 \longrightarrow Cu^{I} + N_2 + \tfrac{1}{2}(R_2CX)_2 \qquad (30)$$

of peroxides is to generate a more active copper(I) catalyst than the one added. Such an explanation is also in accord with the absence of an observable peroxide effect[101] when copper acetylacetonate (which contains no readily oxidizable ligand) is used. In the absence of definitive evidence to the contrary, it still appears that copper(I) species are the catalysts in the cyclopropanation of olefins with diazo compounds.

VI. TRANSITION-METAL CARBENE COMPLEXES

Since Fischer and Maasböl's[105] discovery in 1964 of transition-metal carbene complexes, many such compounds have been synthesized and characterized. Most of these, which may be represented as $L_nM[C(X)Y]$, have had X,Y groups with heteroatoms capable of π-bonding with the electron-deficient carbene carbon atom, such as X,Y = OR, NR_2, SR, Cl, or Ph.[106] More recently, however, simple alkylidene complexes such as $(Me_3CCH_2)_3TaCHCMe_3$[107] and the carbene complex $Cp_2Ta(CH_2)CH_3$[108] have been isolated. There is also a structural characterization of a vinylidene complex $CpMn(CO)_2(C{=}CHPh)$.[109] In all of these carbene compounds lacking X,Y groups capable of π-donation, the metal must be the sole source of stabilization by metal *d*-to-carbon *p* π-bonding. Structural support for metal-to-carbon π-bonding is seen in the M-C bond distances in the compounds $Cp_2Ta({=}CH_2)(CH_3)$[110] and $(Me_3CCH_2)_3Ta{\equiv}CCMe_3^-$ (Li^+, N,*N*'-dimethylpiperazine):[111] $Ta\text{-}CH_3$ = 2.246(2) Å, $Ta{=}CH_2$ = 2.026(10) Å, and $Ta{\equiv}CCMe_3$ = 1.76(2) Å. The significance of metal-to-carbon π-bonding is also indicated by the variable temperature ^1H NMR spectra of the compound $(\eta^5\text{-}C_5H_5)(\eta^5\text{-}C_5H_4CH_3)Ta(CH_2)(CH_3)$. The carbene protons are inequivalent in the ground state (**55**) and appear inequivalent on

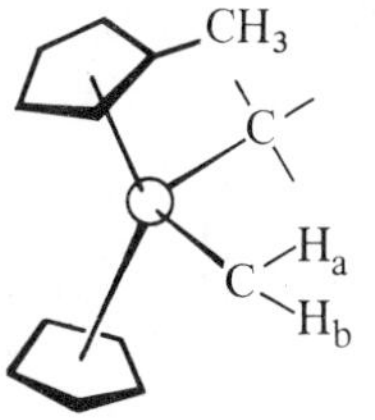

55

the NMR time scale even at 100° C.[110] Based on these observations, $\Delta G^{\ddagger}$ for rotation about the $Ta\text{-}CH_2$ bond can be estimated as $\geq$ 21.4 kcal mol^{-1}, which sets a lower limit on the strength of the Ta-to-C π-bond.

No general reactivity pattern emerges for carbene ligands in compounds of the general formula $L_nM[C(X)Y]$. The reactivity depends on the nature of the metal and the ligands bonded to the metal, namely, the L_nM moiety, the X,Y groups of the carbene ligand, and the substrate molecule with which reaction takes place.

In the chemistry of the later transition metals the metal-carbene carbon bond appears kinetically and thermodynamically rather stable. For example, cationic alkoxycarbene complexes of platinum(II) enter into a large number of reactions in which the

Pt-C carbene is retained. Examples of these reactions are shown in Eqs.(31)-(33).[112,113]

$$Pt^{+}{-}C(OMe)(R) + HNR_2 \longrightarrow Pt^{+}{-}C(NR_2)(R) + MeOH \quad (31)$$

$$Pt^{+}{-}C(OMe)(R) + Nu{:} \longrightarrow Pt{-}C({=}O)(R) + \overset{+}{N}uMe \quad (32)$$

$$Pt^{+}{-}C(OMe)(CH_2R) + B{:} \rightleftharpoons Pt{-}C(OMe){=}CHR + B\overset{+}{H} \quad (33)$$

A platinum-stabilized vinyl carbonium ion or a cationic vinylidene intermediate has been proposed[113] as a common reactive intermediate in the formation of the platinum alkoxycarbene moiety [see Eq.(34)].

$$Pt{-}C{\equiv}CR + HX \rightleftharpoons Pt{-}^{+}C{=}C(H)(R)\;X^{-} \rightleftharpoons Pt^{+}{\leftarrow}(RC{\equiv}CH)\;X^{-} \quad (34)$$

$$Pt{-}^{+}C{=}C(H)(R)\;X^{-} \rightleftharpoons Pt{-}C(X){=}C(H)(R)$$

$$Pt{-}^{+}C{=}C(H)(R)\;X^{-} \overset{+\,MeOH}{\rightleftharpoons} Pt{-}C(OMe){=}CHR + HX \rightleftharpoons Pt^{-}{-}C^{+}(OMe)(CH_2R)\;X^{-}$$

A single-crystal structural determination[114] of trans-$Pt(CCl{=}CH_2)_2(PMe_2Ph)_2$, which was found to be exceedingly labile toward HCl elimination by an E_1 mechanism,[115] revealed a large Pt-C-C angle, 133^0 to 135^0 and a long C_{vinyl}-Cl bond distance of

1.81 Å, supportive of the incipient carbonium ion reactivity of the $Pt\text{-}CCl{=}CH_2$ moiety:

$$Pt\text{-}CCl{=}CH_2 \rightleftarrows Pt\text{-}\overset{+}{C}{=}CH_2,\ \ Cl^- \rightleftarrows Pt\text{-}C{\equiv}CH + HCl$$

The early transition-metal carbene complexes $(\eta^5\text{-}C_5H_5)_2\text{-}Ta(CH_2)(CH_3)$ and $(Me_3CCH_2)_3TaCHCMe_2$ have been shown to enter into a number of reactions that parallel phosphorus ylide reactions; in other words, the methylene carbon atom appears nucleophilic. For example,[116] $(Me_3CCH_2)_3TaCHCMe_3$ reacts with ketones $R_2C{=}O$ to give $(Me_3CCH_2)_3Ta{=}O$ and $R_2C{=}CHCMe_3$; similar reactions occur with esters and amides.

A number of synthetically useful reactions involving metal-carbene bond cleavage are now known[117] and are represented in Eq.(35). All of these reactions, even the thermal decomposition

$$CH_2{=}C(OR_2)R_1 \xleftarrow{CH_2N_2} M{=}C(OR_2)R_1 \xrightarrow{\Delta} (OR_2)(R_1)C{=}C(R_1)(OR_2)$$
$$M{=}C(OR_2)R_1 \xrightarrow{O_2} O{=}C(OR_2)R_1 \qquad M{=}C(OR_2)R_1 \xrightarrow{CH_2{=}CHCOOCH_3} \underbrace{CH_2{-}CHCOOCH_3}_{C(OR_2)R_1} \quad (35)$$

reactions and the alkene addition reactions (for which the intervention of free carbenes seems plausible), have been shown not to proceed via free carbenes.

The olefin metathesis reaction, which may be represented by Eq.(36), has recently received considerable attention.[118]

$$\begin{matrix}(a)\ C = C\ (b)\\ + \\ (c)\ C = C\ (d)\end{matrix} \rightleftharpoons \begin{matrix}(a)\\ C\\ \|\\ C\\ (c)\end{matrix} + \begin{matrix}(b)\\ C\\ \|\\ C\\ (d)\end{matrix} \quad (36)$$

Although a number of mechanisms could conceivably account for this remarkable reaction, most recent work[119] has been consistent with the carbene chain mechanism shown in Eq.(37).

$$\mathrm{M{=}C(a)} \leftarrow \mathrm{C(b){=}C(c)} \rightleftharpoons \mathrm{M}\!\left\langle\begin{matrix}\mathrm{C(a)}\\ \mathrm{C(c)}\end{matrix}\right\rangle\!\mathrm{C(b)} \rightleftharpoons \mathrm{C(a){=}C(b)} \rightarrow \mathrm{M{=}C(c)} \qquad (37)$$

Further support for Eq.(37) has been found in the reactions of thermally stable metallocyclobutanes of molybdenum[120] and platinum.[121] For example, the observed isomerization reaction, Eq.(38), may be achieved by the mechanism of Eq.(37).

$$\mathrm{(Py)_2Cl_2Pt}\!\left\langle\begin{matrix}\mathrm{C^1H(Ph)}\\ \mathrm{C^3H_2}\end{matrix}\right\rangle\!\mathrm{C^2H_2} \rightleftharpoons \mathrm{(Py)_2Cl_2Pt}\!\left\langle\begin{matrix}\mathrm{C^2H_2}\\ \mathrm{C^3H_2}\end{matrix}\right\rangle\!\mathrm{C^1HPh} \qquad (38)$$

VII. REFERENCES

1.* P. F. Zittel, G. B. Ellison, S. V. ONeil, E. Herbst, W. C. Lineberger, and W. P. Reinhardt, *J. Am. Chem. Soc.*, **98**, 3731 (1976).

2. J. F. Harrison, *Acc. Chem. Res.*, **7**, 378 (1974).

3. P. P. Gaspar and G. S. Hammond, "Spin States in Carbene Chemistry," in *Carbenes*, Vol. 2, R. A. Moss and M. Jones, Jr. (Eds.), Wiley-Interscience, New York, 1975.

4. M. L. Halberstadt and J. R. McNesby, *J. Am. Chem. Soc.*, **89**, 3417 (1967).

5. R. W. Carr, Jr., T. W. Eder, and M. G. Topor, *J. Chem. Phys.*, **53**, 4716 (1970).

6. M. J. S. Dewar, R. C. Haddon, W.-K. Li, W. Thiel, and P. K. Weiner, *J. Am. Chem. Soc.*, **97**, 4540 (1975) and references cited therein.

7. H. M. Frey, *J. Chem. Soc., Chem. Commun.*, 1024 (1972).

8. W. L. Hase, R. J. Phillips, and J. W. Simons, Chem. Phys. Lett., 12, 161 (1971).

9. F. S. Rowland, C. McKnight, and E. K. C. Lee, Ber. Bunsengess, 72, 236 (1968).

10. See footnote 10 in J. H. Davis, W. A. Goddard III, and R. G. Bergman, J. Am. Chem. Soc., 98, 4015 (1976); see also A. H. Pakiari and N. C. Hardy, Theor. Chim. Acta, 40, 17 (1975), where a value of 15.4 kcal/mol is reported. (a) Perhaps such has appeared; see L. B. Harding and W. A. Goddard III, J. Chem. Phys., 67, 1777 (1977); (b) G. L. Closs and B. E. Rabinow, J. Am. Chem. Soc., 98, 8190 (1976).

11.* T. Cohen, G. Herman, T. M. Chapman, and D. Kuhn, J. Am. Chem. Soc., 96, 5627 (1974).

12. For review, see J. H. Law, Acc. Chem. Res., 4, 199 (1971).

13. L. J. Altman, R. C. Kowerski, and H. C. Rilling, J. Am. Chem. Soc., 93, 1782 (1971).

14. E. E. van Tamelen and M. A. Schwartz, J. Am. Chem. Soc., 93, 1780 (1971); H. C. Rilling, C. D. Poulter, W. W. Epstein, and B. Larsen, J. Am. Chem. Soc., 93, 1783 (1971).

15. Cohen et al. (private communication from Professor Cohen, December 23, 1976) have now shown that yields in Eq.(1) can be improved 15% to 18% by using an ammonia complex of $Cu(AcAc)_2$; that the ability of the Cu carbenoid intermediate to discriminate among various alkenes parallels that of the Simmons-Smith reagent; that the intermediate has probably shed the Ar_2S group of its progenitor because its selectivity toward alkenes is experimentally identical to that of the carbenoid generated from $(p\text{-}ClC_6H_4)_2S$; and, most interestingly, that $(CH_3)_2C{=}CH(CH_2)_2\bar{C}(CH_3){=}CHCH_2\text{-}S(CH_3)_2$ derived from nerol, self-cyclopropanates, giving 2-carene under the conditions of Eq.(1). The latter reaction occurs in low yield due to competitive side reactions, but does represent a significant advance in scope.

16. N. Kawabata, M. Naka, and S. Yamashita, J. Am. Chem. Soc., 98, 2676 (1976).

17. For review, see H. D. Hartzler, in Carbenes, Vol. II, R. A. Moss and M. Jones, Jr. (Eds), Wiley-Interscience, New York, 1975, pp. 44-57.

18. (a) M. S. Newman and M. C. Van der Zwan, J. Org. Chem., 39, 761 (1974); (b) M. S. Newman and Z. ud Din, J. Org. Chem., 38, 547 (1973); (c) M. S. Newman and T. B. Patrick, J. Am. Chem. Soc., 91, 6461 (1969); (d) M. S. Newman and A. O. M. Okorodudu, J. Am. Chem. Soc., 90, 4189 (1968); (e) M. S. Newman and A. O. M. Okorodudu, J. Org. Chem., 34, 1220 (1969).

19.* (a) P. J. Stang, M. G. Mangum, D. P. Fox, and P. Haak, J. Am. Chem. Soc., 96, 4562 (1974); (b) P. J. Stang, J. Davis, and D. P. Fox, J. Chem. Soc., Chem. Commun., 17 (1975); (c) P. J. Stang and M. G. Mangum, J. Am. Chem. Soc., 97, 1459 (1975); (d) P. J. Stang and M. G. Mangum, J. Am. Chem. Soc., 97, 3854 (1975); (e) P. J. Stang and M. G. Mangum, J. Am. Chem. Soc., 97, 6478 (1975).

20. R. A. Moss and J. R. Przybyla, J. Org. Chem., 33, 3816 (1968).

21. M. Jones, Jr., A. M. Harrison, and K. R. Rettig, J. Am. Chem. Soc., 91, 7462 (1969).

22. T. B. Patrick, E. C. Haynie, and W. J. Probst, J. Org. Chem., 37, 1553 (1972).

23. A priori, it is conceivable that such complexation could make ρ more positive; the related carbene $(CH_3)_2C{=}C{=}C{:}$ shows $\rho = +0.52$ for insertion into the α-C-H of p-substituted benzyl alcohols, a striking phenomenon that requires the presence of OH groups in the substrate (benzylmethyl ether does not react).[24]

24. T. B. Patrick and D. L. Schutzenhofer, Tetrahedron Lett., 3259 (1975).

25. $RR'C{=}CHN_2^+$ and $RR'C{=}C{=}N_2$ are presumed to form during the decomposition of 5,[7e] and the acid-catalyzed reactions of aryldiazomethanes with alkenes stereospecifically afford cyclopropanes as major products.[26]

26. G. L. Closs and S. H. Goh, J. Org. Chem., 39, 1717 (1974).

27. D. Seyferth and D. Dagani, J. Organomet. Chem., 104, 145 (1976).

28. R. F. Cunico and Y.-K. Han, J. Organomet. Chem., 105, C29 (1976).

29. For reviews, see: (a) M. Makosza, Pure Appl. Chem., 43, 439 (1975); (b) E. V. Dehmlow, Angew. Chem., Int. Ed. Engl., 13, 170 (1974).

30. E. V. Dehmlow and M. Lissel, Tetrahedron Lett., 1783 (1976).

31. M. Makosza, A. Kacprowicz, and M. Fedorynski, Tetrahedron Lett., 2119 (1975).

32. S. L. Regan, J. Am. Chem. Soc., 97, 5956 (1975).

33.* T. Hiyama, H. Sawada, M. Tsukanaka, and H. Nozaki, Tetrahedron Lett., 3013 (1975).

34. T. Hiyama, T. Mishima, H. Sawada, and H. Nozaki, J. Am. Chem. Soc., 97, 1626 (1975); J. Am. Chem. Soc., 98, 641 (1976); (a) E. V. Dehmlow, M. Lissel, and J. Heider, Tetrahedron, 33, 363 (1977); (b) Y. Kimura, Y. Ogaki, K. Isagawa, and Y. Otsuji, Chem. Lett., 1149 (1976).

35. (a) J. M. Birchall, G. N. Gilmore, and R. N. Haszeldine, J. Chem. Soc., Perkin Trans. 1, 2530 (1974); (b) F. Anderson, J. M. Birchall, R. N. Haszeldine, and B. J. Tyler, J. Chem. Soc., Perkin Trans. 2, 1051 (1975); (c) J. M. Birchall, R. N. Haszeldine, and P. Tissington, J. Chem. Soc., Perkin Trans. 1, 1638 (1975).

36. G. A. Wheaton and D. J. Burton, J. Fluorine Chem., 8, 97 (1976); G. A. Wheaton and D. J. Burton, J. Fluorine Chem., 9, 25 (1977) for details and limitations.

37. For review, see D. Seyferth, in Carbenes, Vol. II, R. A. Moss and M. Jones, Jr. (Eds.), Wiley-Interscience, New York, 1975, pp. 101-122.

38.* C.-T. Ho, R. T. Conlin, and P. P. Gaspar, J. Am. Chem. Soc., 96, 8109 (1974).

39. W. J. Baron, M. R. DeCamp, M. E. Hendrick, M. Jones, Jr., R. H. Levin, and M. B. Sohn, in Carbenes, Vol. I, M. Jones, Jr., and R. A. Moss (Eds.), Wiley-Interscience, New York, 1973, Chapter 1.

40. T. R. Fields and P. J. Kropp, J. Am. Chem. Soc., 96, 7559 (1974).

41. S. S. Hixson, J. Am. Chem. Soc., 97, 1981 (1975).

42. S. S. Hixson, J. C. Tausta, and J. Borovsky, J. Am. Chem. Soc., 97, 3230 (1975).

43.* Y. Inoue, S. Takamuku, and H. Sakurai, J. Chem. Soc., Chem. Commun., 577 (1975).

44. S. J. Cristol and C. S. Ilenda, abstracts of the 167th ACS Meeting, April, 1974, No. ORGN 113.

45. P. J. Kropp, H. G. Fravel, Jr., and T. R. Fields, J. Am. Chem. Soc., 98, 840 (1976).

46. R. F. C. Brown, K. J. Harrington, and G. L. McMullen, J. Chem. Soc., Chem. Commun., 123 (1974).

47.* R. A. Moss and F. G. Pilkiewicz, J. Am. Chem. Soc., 96, 5632 (1974).

48. R. A. Moss, M. A. Joyce, and F. G. Pilkiewicz, Tetrahedron Lett., 2425 (1975).

49. The use of more than 1 equiv of crown per equiv of KO*t*-Bu induced no further change in the selectivity of the intermediate.

50.* (a) K. G. Taylor and J. Chaney, J. Am. Chem. Soc., 98, 4158 (1976); (b) K. G. Taylor, J. Chaney, and J. C. Deck, J. Am. Chem. Soc., 98, 4163 (1976).

51.* (a) D. Seyferth, R. L. Lambert, Jr., and M. Massal, J. Organomet. Chem., 88, 255 (1975); (b) D. Seyferth and R. L. Lambert, Jr., J. Organomet. Chem., 88, 287 (1975); (c) D. Seyferth and R. L. Lambert, Jr., J. Organomet. Chem., 91, 31 (1975).

52. Operation of the equilibrium was demonstrated by converting 3-*t*-butyl-7,7-dibromonorcarene to its lithium carbenoids (-113^0). The latter were allowed to react (-93^0) with 7,7-dibromonorcarane, whereupon protolysis furnished 48% of *anti*-7-bromonorcarane and 34% of 3-*t*-butyl-*anti*-7-bromonorcarane, as well as 34% of 7,7-dibromonorcarane and 59% of its *t*-butyl analog. The two monobromides are products of the corresponding *endo*-Li, *exo*-bromo carbenoids.[51a]

53. D. A. Hatzenbühler, L. Andrews, and F. A. Carey, J. Am. Chem. Soc., 97, 187 (1975).

54. G. L. Closs and R. A. Moss, J. Am. Chem. Soc., 86, 4042 (1964) and references cited therein.

55. It would be extraordinarily interesting to have parallel thermolytic results for the carbocyclic epimeric carbenoids, 23. These results, although crucial, are sadly lacking.

56. (a) K. Kitatani, T. Hiyama, and H. Nozaki, J. Am. Chem. Soc., 97, 949 (1975); (b) K. Kitatani, T. Hiyama, and H. Nozaki, J. Am. Chem. Soc., 98, 2362 (1976).

57. M. Schlosser, B. Spahić, and Le Van Chau, Helv. Chim. Acta, 58, 2586 (1975). Equation (12) is reprinted from this publication with the kind permission of the Swiss Chemical Society. See also M. Schlosser, Le Van Chau, and B. Spahić, Helv. Chim. Acta, 58, 2575 (1975).

58. See, for example, G. Köbrich, H. Büttner, and E. Wagner, Angew. Chem., Int. Ed. Engl., 9, 169 (1970); P. S. Skell and M. S. Cholod, J. Am. Chem. Soc., 91, 6035, 7131 (1969).

59.* C. Mayor and C. Wentrup, J. Am. Chem. Soc., 97, 7467 (1975).

60.* C. Wentrup, Topics in Current Chemistry, 62, 173 (1976).

61. W. M. Jones, R. C. Joines, J. A. Myers, T. Mitsuhashi, K. E. Krajca, E. E. Waali, T. L. Davis, and A. B. Turner, J. Am. Chem. Soc., 95, 826 (1973).

62. H. Fujimoto and R. Hoffmann, J. Phys. Chem., 78, 1167 (1974).

63. References 39 and 67 cite possible cases of 1,4 carbene additions. The early examples are best formulated as 1,2-addition followed by rearrangement.

64.* (a) C. W. Jefford, J. Mareda, J.-C. E. Gehret, nT. Kabengele, W. D. Graham, and U. Burger, J. Am. Chem. Soc., 98, 2585 (1976); (b) C. W. Jefford, nT. Kabengele, J. Kovacs, and U. Burger, Helv. Chim. Acta, 57, 104 (1974); (c) C. W. Jefford, nT. Kabengele, J. Kovacs, and U. Burger, Tetrahedron Lett., 257 (1974).

65. (a) C. W. Jefford, V. de los Heros, and U. Burger, Tetrahedron Lett., 703 (1976); (b) P. M. Kwantes and G. W. Klumpp, Tetrahedron Lett., 707 (1976); (c) C. W. Jefford, A. Delay, T. W. Wallace, and U. Burger, Helv. Chim. Acta, 59, 2355 (1976); (d) C. W. Jefford, W. D. Graham, and U. Burger, Tetrahedron Lett., 4717 (1975); (e) C. W. Jefford, J.-C. E. Gehret, J. Mareda, nT. Kabengele, W. D. Graham, and U. Burger, Tetrahedron Lett., 823 (1975).

66. D. J. Burton and D. G. Naae, J. Am. Chem. Soc., 95, 8467 (1973).

67. M. Jones, Jr., W. Ando, M. E. Hendrick, A. Kulczycki, Jr., P. M. Howley, K. F. Hummel, and D. S. Malament, J. Am. Chem. Soc., 94, 7469 (1972). An interesting general discussion of abortive 1,4-additions is included in this article.

68. (a) R. Hoffmann, Acc. Chem. Res., 4, 1 (1971); (b) P. Bischof, J. A. Hashmall, E. Heilbronner, and V. Hornung, Helv. Chim. Acta, 52, 1745 (1969).

69. The C_2-C_6 separation in 36 is ~2.4 Å, whereas it is ~2.8 Å in s-cis-butadiene. A singlet carbene approaching the endo face of 32 in the proper orientation for 1,4-addition[41a] may, therefore (at a given molecular separation), experience stronger overlap of its vacant p (LUMO) orbital with norbornadiene's HOMO (SA) orbital, relative to the closed-shell σ-π_{SS} repulsion, than is possible in the (putative) comparable addition to butadiene, 31.

70. K. Saito, Y. Yamashita, and T. Mukai, J. Chem. Soc., Chem. Commun., 58 (1974).

71. T. Mitsuhashi and W. M. Jones, J. Chem. Soc., Chem. Commun., 103 (1974).

72. R. Hoffmann, D. M. Hayes, and P. S. Skell, J. Phys. Chem., 76, 664 (1972); R. Hoffmann, J. Am. Chem. Soc., 90, 1475 (1968); W. R. Moore, W. R. Moser, and J. E. LaPrade, J. Org. Chem., 28, 2200 (1963).

73. W. M. Jones, M. E. Stowe, E. E. Wells, Jr., and E. W. Lester, J. Am. Chem. Soc., 90, 1849 (1968); L. W. Christensen, E. E. Waali, and W. M. Jones, J. Am. Chem. Soc., 94, 2118 (1972) and references cited therein.

74.* (a) R. A. Moss and C. B. Mallon, J. Am. Chem. Soc., 97, 344 (1975); (b) R. A. Moss, C. B. Mallon, and C.-T. Ho, J. Am. Chem. Soc., 99, 4105 (1977).

75. W. von E. Doering and W. A. Henderson, Jr., J. Am. Chem. Soc., 80, 5274 (1958); P. S. Skell and A. Y. Garner, J. Am. Chem. Soc., 78, 5430 (1956).

76. H. Watanabe, N. Ohsawa, M. Sawai, Y. Fukasawa, H. Matsumoto, and Y. Nagai, J. Organomet. Chem., 93, 173 (1975).

77. R. A. Moss, M. A. Joyce, and J. K. Huselton, Tetrahedron Lett., 4621 (1975).

78. R. W. Hoffmann and H. Häuser, Tetrahedron Lett., 197 (1964); D. M. Lemal, E. P. Gosselink, and A. Ault, Tetrahedron Lett., 579 (1964); R. W. Hoffmann and H. Häuser, Tetrahedron, 21, 891 (1965).

79. D. M. Lemal, E. P. Gosselink, and S. D. McGregor, J. Am. Chem. Soc., 88, 582 (1966).

80.* R. W. Hoffmann, W. Lilienblum, and B. Dittrich, Chem. Ber., 107, 3395 (1974).

81. Structure 42 can also be intercepted by dimethyl acetylenedicarboxylate,[80] phenylacetylene,[80] diphenylketene,[80] benzoyl chloride,[80] aryl isocyanates,[82] and aryl isothiocyanates.[82] Each of these reactions involves initial nucleophilic attack of 42. For reaction at 140^0 with aryl isocyanates (relative to phenyl isothiocyanate), 42 shows $\rho = +2.0 \pm 0.5$.[83]

82. R. W. Hoffmann, K. Steinbach, and B. Dittrich, Chem. Ber., 106, 2174 (1973).

83. R. W. Hoffmann and M. Reiffen, Chem. Ber., 109, 2565 (1976).

84. A. G. Brook, H. W. Kucera, and R. Pearce, Can. J. Chem., 49, 1618 (1971).

85. R. A. Moss and J. K. Huselton, J. Chem. Soc., Chem. Commun., 950 (1976).

86.* R. L. Kreeger and H. Shechter, Tetrahedron Lett., 2061 (1975).

87.* W. Ando, A. Sekiguchi, and T. Migita, Chem. Lett., 779 (1976).

88. T. J. Barton, J. A. Kilgour, R. R. Gallucci, A. J. Rothschild, J. Slutsky, A. D. Wolf, and M. Jones, Jr., J. Am. Chem. Soc., 97, 657 (1975).

89. W. Ando, T. Hagiwara, and T. Migita, J. Am. Chem. Soc., 95, 7518 (1973).

90. W. Ando, A. Sekiguchi, T. Migita, S. Kammula, M. Green, and M. Jones, Jr., J. Am. Chem. Soc., 97, 3818 (1975).

91. W. Ando, A. Sekiguchi, J. Ogiwari, and T. Migita, J. Chem. Soc., Chem. Commun., 145 (1975).

92. W. Ando, A. Sekiguchi, A. J. Rothschild, R. R. Gallucci, M. Jones, Jr., T. J. Barton, and J. A. Kilgour, J. Am. Chem. Soc., 99, 6995 (1977).

93. O. L. Chapman, C.-C. Chang, J. Kolc, M. E. Jung, J. A. Lowe, T. J. Barton, and M. L. Tumey, J. Am. Chem. Soc., 98, 7844 (1976).

94. M. R. Chedekel, M. Skoglund, R. L. Kreeger, and H. Schechter, J. Am. Chem. Soc., 98, 7846 (1976).

95. T. J. Barton and C. L. McIntosh, J. Chem. Soc., Chem. Commun., 861 (1972).

96. H.-D. Roth, Acc. Chem. Res., 10, 85 (1977).

97.* A. R. Forrester and J. S. Sadd, J. Chem. Soc., Chem. Commun., 631 (1976).

98. A. J. Rothschild, A.B. thesis, Princeton University, 1975.

99. R. G. Salomon and J. K. Kochi, J. Am. Chem. Soc., 95, 3300 (1973).

100. W. R. Moser, J. Am. Chem. Soc., 91, 1135, 1141 (1969).

101. D. S. Wolfman, N. v. Thinh, R. S. McDaniel, B. W. Peace, C. W. Heitsch, and M. T. Jones, Jr., J. Chem. Soc., Dalton Trans., 522 (1975).

102. J. K. Kochi, Free Radicals, Vol. I, Wiley, New York, 1973, Chapter 11.

103. R. Hiatt, Organic Peroxides, Vol. 2, D. Swern (Ed.), Wiley, New York, 1973, p. 875; D. C. Nonhebel and J. C. Walton, Free-Radical Chemistry, Cambridge University Press, Cambridge, 1974, p. 551.

104. T. Shirafuji, Y. Yamamoto, and H. Nozaki, Tetrahedron, **27**, 2353 (1971) and reference 99.

105. E. O. Fischer and A. Maasböl, Angew. Chem., Int. Ed. Engl., **3**, 580 (1964).

106. For reviews of transition-metal carbene complexes, see: E. O. Fischer, Pure Appl. Chem., **30**, 353 (1972); F. A. Cotton and C. M. Lukehart, Progr. Inorg. Chem., **16**, 487 (1973); D. J. Cardin, B. Cetinkaya, and M. F. Lappert, Chem. Reviews, **72**, 545 (1972).

107. R. R. Schrock, J. Am. Chem. Soc., **96**, 6796 (1974).

108. R. R. Schrock, J. Am. Chem. Soc., **97**, 6577 (1975).

109. A. N. Nesmayanov, G. G. Aleksandrov, A. B. Antonova, K. N. Anisimov, N. E. Kolobova, and Y. T. Struchkov, J. Organomet. Chem., **110**, C35 (1976).

110. R. R. Schrock and L. J. Guggenberger, J. Am. Chem. Soc., **97**, 6578 (1975).

111. R. R. Schrock and L. J. Guggenberger, J. Am. Chem. Soc., **97**, 2935 (1975).

112. M. H. Chisholm, H. C. Clark, W. S. Johns, J. E. H. Ward, and Y. Yasafuku, Inorg. Chem., **14**, 900 (1975).

113. R. A. Bell, M. H. Chisholm, D. A. Couch, and L. A. Rankel, Inorg. Chem., **16**, 677 (1977).

114. R. A. Bell, M. H. Chisholm, and G. G. Christorph, J. Am. Chem. Soc., **98**, 6046 (1976).

115. R. A. Bell and M. H. Chisholm, Inorg. Chem., **16**, 698 (1977).

116. R. Schrock, J. Am. Chem. Soc., **98**, 5399 (1977).

117. C. P. Casey, in Transition Metal Organometallics in Organic Synthesis, Vol. 1, Academic Press, New York, 1976, Chapter 3, p. 214.

118. For recent reviews dealing with olefin metathesis, see: W. B. Hughes, Organomet. Chem. Synth., 1, 341 (1972); R. J. Haines and G. J. Leigh, Chem. Soc. Rev., 4, 155 (1975); N. Calderon, Acc. Chem. Res., 5, 127 (1972).

119. T. J. Katz and J. McGinnis, J. Am. Chem. Soc., 97, 1592 (1975); T. J. Katz and R. Rothchild, J. Am. Chem. Soc., 98, 2519 (1976) and references cited therein.

120. M. Ephritikhine and M. L. H. Green, J. Chem. Soc., Chem. Commun., 926 (1976).

121. R. J. Puddephatt, M. A. Quyser, and C. F. H. Tipper, J. Chem. Soc., Chem. Commun., 626 (1976).

4

CARBOCATIONS

D. BETHELL

Robert Robinson Laboratories, University of Liverpool
Liverpool, England

I. INTRODUCTION

The spate of publications in the area of carbocation* chemistry has seen no decline in the decade that has passed since the appearance of the first monograph on the subject.[1] Indeed, the new terminology implies a broadening of the subject.

Trends in methodology apparent in 1967 have continued; for example, the use of superacidic media, once restricted to NMR studies of carbocations, has been extended not only by instrumental advance but also to other quantitative techniques. The newer methods of studying carbocations in the gas phase (e.g., ion cyclotron resonance spectroscopy, high-pressure mass spectrometry) have attained a sophisticated state of development, and the results are finding two important applications over and above their intrinsic interest. First, they provide the link between experiment and quantum-mechanical calculation, itself a field of carbocation chemistry that has grown dramatically in recent years. Second, the gas-phase results provide a firm baseline against which carbocation behavior in solution can be gauged. Thus understanding of the interaction of carbocations with the solvent seems now to be approaching the point of breakthrough; Section III is devoted to an attempt to weave together some of the threads of endeavor in this area.

Yet, despite the technical advances, many of the problems that were being forcefully debated 10 years ago still lack universally accepted solutions. The mechanisms of solvolysis and the structures of the 2-norbornyl and cyclopropylmethyl cations

---

* Throughout this chapter we adopt the convention of referring to all cationic species in which the positive charge is located principally on carbon as carbocations and differentiating carbenium ions, R_3C^+, from true carbonium ions, R_5C^+. Individual carbenium ions will be named as alkyl cations and individual carbonium ions, where their structure is beyond doubt, as cationated (often protonated) alkanes.

are notable examples.

This chapter is devoted to three themes in carbocation chemistry, each concerned with structure and its relation to reactivity. The selection of the themes and of the work with which to illustrate them is based on a purely personal assessment of the important trends in the subject. It is hoped that the chapter will make a small contribution to the understanding and development of carbocation chemistry, and not, by its omissions or emphasis, merely add to the rancor that has bedeviled the subject over the years.

II. NUCLEAR MAGNETIC RESONANCE SPECTROSCOPY OF CARBENIUM IONS

Nuclear magnetic resonance spectroscopy, arguably the organic chemist's most powerful and versatile structural tool, has found widespread use in the study of carbenium ions, usually generated at low temperature in superacidic media. Chemical shifts, particularly of carbon, and to a smaller extent nuclear-nuclear coupling constants, are now being studied to obtain information about charge distribution in the ions. Furthermore, by extrapolation to other situations, the findings are being used to assist the understanding of questions of stability and reactivity of carbenium ions. But doubts have been voiced about the relevance of NMR studies in superacidic media to the transient carbenium ions generated as intermediates in, for example, solvolysis.[2]

A. Chemical Shift and Charge Distribution

The subject has been critically reviewed recently,[3] but a number of important studies have appeared subsequently and justify inclusion of the present topic.

The chemical shift of a nucleus A in an organic species can be characterized by a screening constant, σ_A, which can be conceptually subdivided as in Eq. (1) into contributions from diamagnetic shielding by electrons localized on A $\left(\sigma_D^{AA}\right)$, from a paramagnetic term $\left(\sigma_p^{AA}\right)$ that accounts for any deviations from spherical symmetry of the circulation of these localized electrons, from electrons localized on other nuclei B $\left(\sigma^{AB}\right)$, and from delocalized electrons $\left(\sigma^{A,deloc}\right)$. In proton NMR, the last two terms can affect chemical shifts by up to 2 ppm, and whereas σ_D^{AA} causes changes of about 10 ppm per unit charge, σ_p^{AA} has very little effect.

$$\sigma_A = \sigma_D^{AA} + \sigma_p^{AA} + \sum_{A \neq B} \sigma^{AB} + \sigma^{A,deloc} \tag{1}$$

For heavier nuclei (^{13}C, ^{15}N, ^{19}F) σ_p^{AA} becomes the dominant term because its magnitude shows an inverse dependence on the energy separation of ground and excited electronic states of the atom, a quantity that is large for hydrogen but decreases for heavier elements. Thus for ^{13}C, σ_D^{AA} is thought to have an effect of 13 to 18 ppm per unit charge and σ_p^{AA}, 140 to 150 ppm per unit charge. The effect of the last two terms, which are determined by other atoms, is still no more than 2 ppm and is thus negligible. For this reason, it might appear that ^{13}C chemical shifts $\left(\delta_C\right)$ in carbenium ions would be simply related to the charge on the carbon atom, but this does not appear to be so[3] (see text that follows).

One important problem is the choice of the quantum-mechanical method by which to calculate the charge distribution.[4] In addition, Fliszár[5] has argued that the charge distribution is usually based on a partitioning of overlap population terms that is invalid for nonhomonuclear interactions. The alternative semiempirical procedure that he offers may not command wide acceptance, but his fundamental criticism deserves careful examination. In part, at least, the poor correlation of

δ_C and charge distribution probably arises from slightly different dependences on charge of the diamagnetic and paramagnetic terms of Eq. (1). Significantly, in ^{19}F NMR σ_D^{AA} is also small enough compared with σ_p^{AA} to be neglected, and ^{19}F-shifts have been claimed to provide a better guide than δ_C to charge distribution in the somewhat limited area of fluorine-containing carbenium ions.[6]

Recent work on the correlation of δ_C with charge distribution has laid greater emphasis on empirical approaches that might define areas within which ^{13}C shifts can be used directly as charge and stability probes. Following Hammett, the series of benzyl cations, $Ph\overset{+}{C}RR'$, is the system of choice for initial detailed study since the geometrical and conformational possibilities that can alter the stereoelectronic situation are strictly limited. Olah, Westerman, et al.[7] have reported the complete ^{13}C spectra of 22 benzylic cations generated by cleavage (12 ions, e.g., $Ph\overset{+}{C}Cl_2$ from $PhCCl_3$) or by protonation (10 ions, e.g., $Ph\overset{+}{C}HOH$ from PhCHO) in SO_2ClF solution in the temperature range -60° to -80° C, using a variety of superacids (SbF_5, SbF_5/HF, SbF_5/FSO_3H). The chemical shifts of carbon atoms at various positions in the ions were compared with charge densities evaluated by the CNDO/2 procedure. As a guide to estimating the degree of correlation for the carbenium ions, literature values of δ_C for 16 uncharged monosubstituted benzenes were also included in the study.

The correlation of the downfield shift from TMS of _para_-carbon atoms with total charge density on that atom was excellent, as shown in Figure 4.1, both for the uncharged systems and for the carbenium ions formed by heterolysis. The carbocations generated by protonation of heteroatoms showed, in general, a slightly smaller shift than predicted from the best straight line through the other two groups of points. The origin of these deviations is not clear, but they may arise from inadequacies in the geometries assumed for the CNDO/2 calculations, or as a result of hydrogen bonding (see Section III). An important conclusion, however, is that interaction of the carbenium ions with their counter ions has a negligible effect on the correlation.* The correlation of $\delta_{C-\underline{p}}$ with π-charge density is as good as that in Figure 4.1, and the slope corresponds to a shift of 166.6 ± 5.8 ppm per unit charge, very close to the value of 160 which has been a rule of thumb among organic chemists for some years.

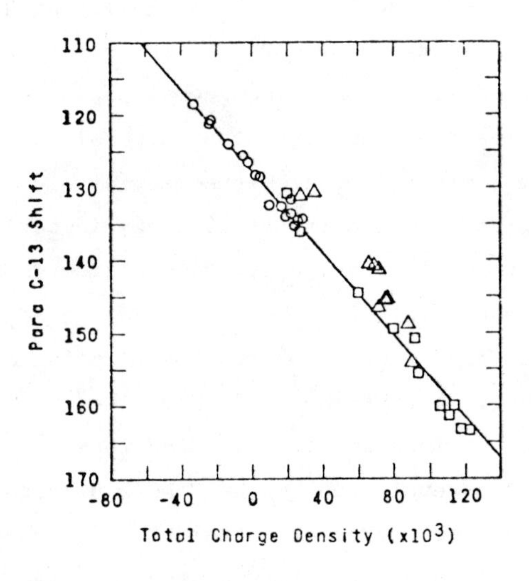

FIGURE 4.1. Correlation of $\delta_{C-\underline{p}}$ with total charge density

Rotation about the Ph—$\overset{+}{C}$ bond is slow on the NMR time scale and, with unsymmetrical substitution at the carbenium center, two signals are observed for the ortho-carbon atoms. As expected, the values of $\delta_{C-\underline{o}}$ cover roughly the same range as

---

* Hydrogen bonding between a protonated heteroatom and the counter ion could be responsible for some of the deviations from the straight line in Figure 4.1.

the $\delta_{C-\underline{p}}$ values, but they correlate less well with total charge density (Figure 4.2a). The range of shifts for the <u>meta</u>-carbon atoms is only about 4 ppm, and the correlation with charge density is poor. However, the downfield shift for the carbenium ions is always greater than for uncharged aromatics, in line with the appreciable positive charge predicted by CNDO/2 for <u>meta</u>-positions.

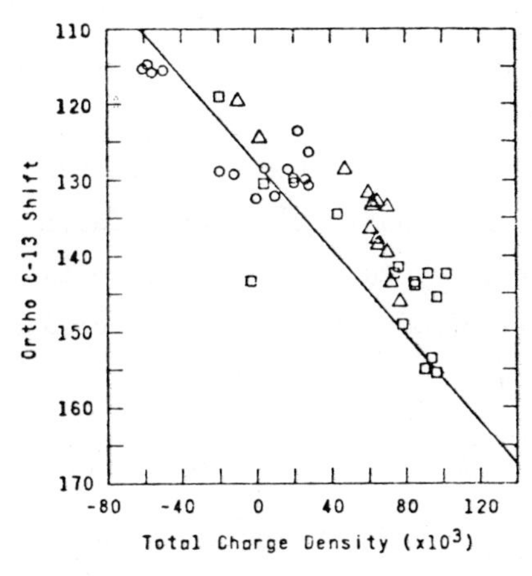

FIGURE 4.2a. Correlation of $\delta_{C-\underline{o}}$ with total charge density

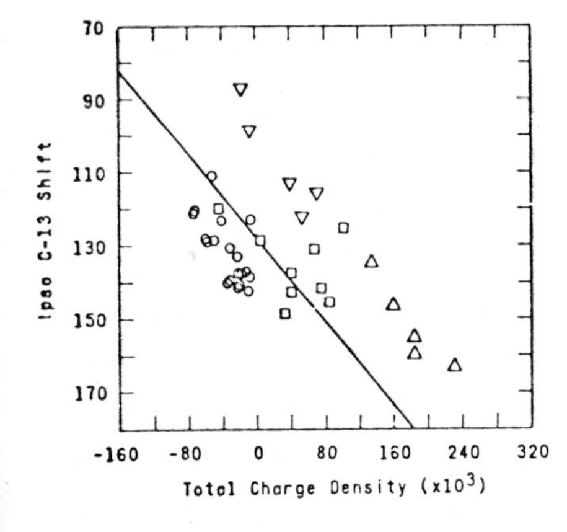

FIGURE 4.2b. Correlation of $\delta_{C-\underline{ip}}$ with total charge density

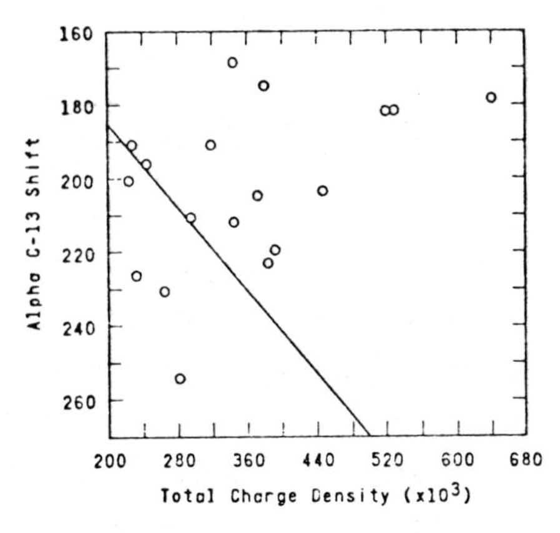

FIGURE 4.2c. Correlation of $\delta_{C-\alpha}$ with total charge density

Perhaps the most notable feature of the spectra is that δ_C for carbon atoms at or near the formal carbenium center, the α-carbon atom, show a large variation with carbenium ion structure but correlate poorly with the calculated charge density (Figures 4.2b and c). Clearly, δ_C cannot be used directly as a guide to the extent of charge delocalization from the α-carbon atom, nor to the chemical behavior of the ion. Indeed, it was suggested that, because calculated π-charge densities on the α-carbon atom and on the *p*-carbon of the ring are related in a roughly linear fashion, $\delta_{C\text{-}p}$ is a better indicator of charge density at the α-carbon atom than is $\delta_{C\text{-}\alpha}$. This suggestion has been applied[8] to carbocations of types **1**, **2**, **3**, and **4** (e.g., $X = OCH_3$), and leads to the conclusion that charge delocalization into the ring decreases in the sequence **1** > **2** > **3** > **4**.

(1) 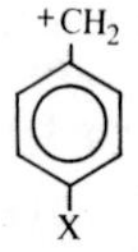 (2) 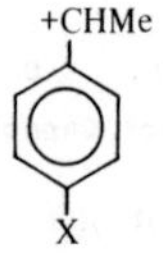 (3)  (4)

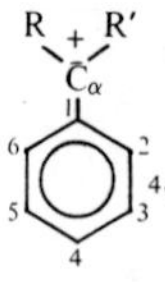

(5) 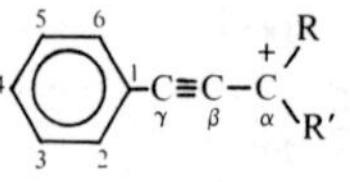 (6) 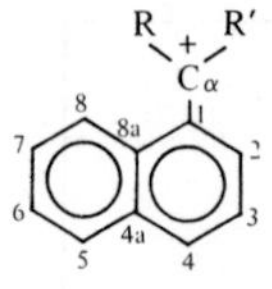 (7) 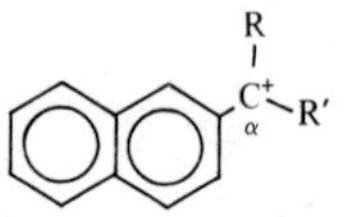 (8)

A rapid and convenient means of examining the response of δ_C to carbenium ion structure has been developed using a type of linear free-energy relation.[9] Values of δ_C for a given carbon atom in carbenium ions of types **5** - **8** were plotted against $\delta_{C\text{-}4}$ in the corresponding benzyl cation (**5**), that is, the series of shift values that correlate best with the calculated π-charge densities. Linearity in such plots is taken to indicate that there is a systematic dependence of chemical shift on the electronic effect of the substituents R and R', and this is plausibly related to π-charge density. The slope of such a linear plot would thus be a measure of the sensitivity of the chemical shift to such electronic influences, and some justification for this view is provided by the linear dependence of the slopes on the difference in π-charge density [best calculated by self-consistent-field molecular orbital (SCF MO) methods] at that carbon atom between the arene and the related arylmethyl cation (i.e., ArH and $ArCH_2^+$). The chemical shift response patterns for systems 5 through 8 are given in Table 1, from which it can be seen that correlations are in general very good for carbon atoms remote from the site of structural change, the carbenium center. However, for carbon atoms close to the formal carbenium center, and especially for the *ipso*-carbon atom of the ring system, the correlation is poor, indicating the operation of "special" influences that might include one or more of the following:[10] σ inductive effects, substituent magnetic anisotropies, steric interactions, or diamagnetic shielding by neighboring groups.

Although the authors do not claim any great sophistication in their approach, it serves to display in a particularly clear way the limitations of ^{13}C NMR as a guide to electronic structure. Only shifts of carbon atoms of the same hybridization and substitution pattern, at positions remote from the site of structural change, are likely reliably to reflect charge density. If these conditions cannot be fulfilled in the relatively simple arylmethyl cations, even poorer correlations between δ_C and charge distribution are to be anticipated in wholly aliphatic systems. Thus it is not surprising that there has been controversy regarding the interpretation of ^{13}C NMR

TABLE 1. Least-squares Analysis of Chemical Shift Response Patterns in Arylcarbenium Ions[a]

Position	Slope	Intercept	SD[b]	cc[c]	n[d]
		(a) Phenyl System			
1	0.124	113.6	8.47	0.174	25
2,6	0.669	39.7	4.10	0.879	36
3,5	0.197	102.4	0.52	0.977	28
4	1.000	0.0	0.00	1.000	25
		(b) Phenylethynyl System			
β	1.287	-91.0	9.39	0.768	10
γ	4.159	-464.8	9.13	0.970	10
1	-0.053	125.2	1.79	-0.251	10
2,6	0.350	86.8	0.35	0.993	10
3,5	0.086	117.8	0.32	0.922	10
4	0.558	57.5	0.62	0.992	10
		(c) 1-Naphthyl System			
1	0.214	98.9	8.74	0.277	7
2	0.681	44.4	5.61	0.814	8
3	0.112	110.0	0.51	0.929	8
4	1.052	-5.8	0.94	0.997	8
5	0.152	108.9	0.29	0.986	8
6	0.081	116.8	0.47	0.893	8
7	0.301	89.3	0.75	0.977	8
8	-0.089	138.0	1.35	-0.607	8
4a	-0.054	141.6	0.25	-0.927	8
8a	0.135	112.3	1.39	0.744	8
		(d) 2-Naphthyl System			
1	1.172	-24.3	3.72	0.965	7
2	0.244	95.3	8.84	0.308	7
3	0.153	102.3	1.93	0.682	7
4	0.144	110.0	0.28	0.987	7
5	0.075	118.2	0.26	0.958	7
6	0.605	48.2	0.57	0.997	7
7	0.150	107.2	0.20	0.994	7
8	0.339	84.1	1.10	0.964	7
4a	0.399	82.8	0.52	0.994	7
8a	0.029	128.5	0.85	0.366	7

TABLE 1. (footnotes)

[a] Reprinted from Forsythe, Spear, et al.[9] with permission of the American Chemical Society.

[b] Standard deviation

[c] Correlation coefficient

[d] Number of chemical shifts included in the analysis.

data for aliphatic systems for which σ-bridging (nonclassical carbenium ion structures) has been suggested[11] (however, see Section II.B).

Nevertheless, Farnum and Botto[12] have attempted to use δ_C as a guide to charge distribution in 2-aryl-2-norborn-5-enyl cations (**9**), and hence to interpret kinetic data for solvolyses of the related *exo*- and *endo*-*p*-nitrobenzoates (**10**; R = Me). The chemical shift of C-5 in a series of *meta*- and *para*-substituted cations was plotted against the substituent constant σ^+. For the unsubstituted compound (X = H) and for electron-repelling substituents (three points only), a linear relation exists, but powerful electron-attracting substituents (CF_3) give rise to much larger downfield shifts than would be expected on the basis of σ^+. The deviation from linearity for the two trifluoromethyl-substutited cations was taken to indicate that for these compounds rehybridization occurs at the carbenium center due to homoallylic interaction with the double bond. Consistently (but not too convincingly in the light of the foregoing discussion), the downfield trend in δ_{C-2} with increasing σ^+ is reversed for the trifluoromethyl-substituted cations. Powerful electron withdrawal also results in a marked change in the *exo*/*endo* rate ratio for the solvolysis of **10** (R = Me) in 80% acetone-water at 25° (although not for the series of esters (**10**; R = H) corresponding to **9**.[13] Since this change was also interpreted as evidence for the onset of π-interaction with the developing carbenium center in solvolysis of the *exo*-compound, these reports encourage the belief that NMR studies of carbenium ions in superacidic media are directly relevant to solvolyses.

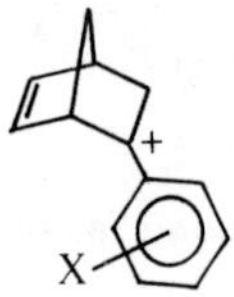

(**9**) 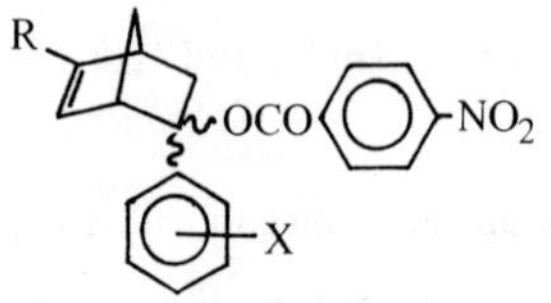 (**10**) 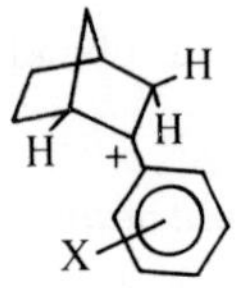 (**11**)

Such a conclusion should be approached with caution. Even in the case reported by Farnum and Botto, an additional methyl group is necessary at C-5 to make the homoallylic interaction detectably large using kinetic methods. Moreover, for 2-aryl-2-norbornyl cations (**11**) where strongly electron-attracting aromatic substituents lead to breaks in plots of δ_{H-1} against δ_{H-3} (average of *exo* and *endo*), ascribed to delocalization of the 2,6-σ-bond electrons,[14] no corresponding change in solvolytic *exo*/*endo* rate ratios is observable.[13a,15] Kinetic and spectroscopic methods undoubtedly have quite different sensitivities to structural changes, the environment is totally different in the two types of experiment, and, as recent studies have shown (see Section III.B), gross solvolysis rates can sometimes be misleading guides to carbenium ion stability and structure. It is to be hoped that much more effort will be devoted to examining the reality of the correspondence between δ_C in carbenium ions and the solvolytic generation of the same cations and, if the correspondence is demonstrable, to define the structural limits within which such correspondence is to be expected. Certainly, system **9** is a particularly appropriate one for study because it fulfills all the requirements for a satisfactory correlation of δ_{C-5} with charge density.

B. Nuclear-Nuclear Coupling Constants

The coupling constant between a ^{13}C nucleus and a directly attached proton $\left(^1J_{CH}\right)$ is known to increase markedly with increasing s character of the carbon atom; thus $^1J_{CH}$ is greater for sp^2-hybridized carbon than for sp^3 carbon. This phenomenon has been used for some years now to provide information on charge distribution in carbenium ions. However, $^1J_{CH}$ is also sensitive to other factors such as the electronegativity of attached groups and also strain.[16] Kelly, Underwood, et al.[17] have examined structural effects on $^1J_{CH}$ for carbon atoms adjacent to tertiary carbenium centers in a variety of situations. To allow for the effects of strain, $^1J_{CH}$ for the ketone structurally related to the carbenium ion by replacing $\rangle C^+ - CH_3$ by $C=O$ was subtracted from $^1J_{CH}$ for the carbenium ion. It was found that ΔJ is dependent on the dihedral angle (θ) between the C—H bond and the vacant p-orbital at the carbenium center. The relationship is expressed in Eq. (2), in which the constant term represents the enhancement of $^1J_{CH}$ due to induction differences, and the angular dependence term represents the hyperconjugative diminution. From Eq. (2) it can be seen that the range of ΔJ values is predicted to be roughly +23 to -11 Hz for static systems. Carbenium ions with $\theta = 0$ are not static, but generally undergo rapid 1,2-hydride shifts; the observed value of J_{CH} is then roughly half of $^1J_{CH}$ because it is the average of $^1J_{CH}$ and $^2J_{CH}$. For other equilibrating ions, where the C—H bond is adjacent to only one of the two cationic centers, Eq. (2) again holds.

$$^1J_{CH}\left(\rangle \overset{+}{C} - CH_3\right) \; - \; ^1J_{CH}\left(\rangle C=O\right) \; = \; 22.5 - 33.1\cos^2\theta \qquad (2)$$

Equation (2) is potentially capable of making a valuable contribution to conformational studies of carbenium ions. It is desirable, however, that the method be shown to reproduce known conformations of carbenium ions. The authors show that Eq. (2) predicts conformation **12** as the most stable for the *t*-pentyl cation, rather than **13**, predicted by *ab initio* methods.[18] Nor has the procedure proved capable of throwing new light on the vexing question of the structure of the cyclopropylmethyl cation,[17,19] which is known to be an equilibrating species. On the basis of $^{1}J_{CH}$, the ion would be twisted in a way consistent neither with the normal bisected conformation (**14**), expected for a classical structure, nor with the idealized eclipsed arrangement (**15**), anticipated if σ-bridging is important. Obviously, coupling constants are unable to provide a clear structural answer in the way that ^{13}C chemical shifts appear to do for the unsubstituted[20] and 1-methylcyclopropylmethyl cations.[21] In the latter case, rapid equilibration, reaction (3), between 1-methylcyclobutyl and three 1-methylcyclopropylmethyl cations occurs. Part of the evidence

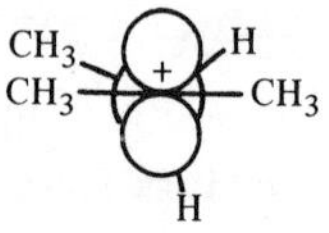

(**12**) 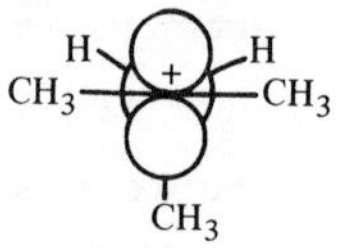 (**13**)

for this is that the observed shift for C-1 (163.3 ppm) lies between that found for C-1 in the static α,α-dimethylcyclopropylmethyl cation and that predicted for C-1 in a static 1-methylcyclobutyl cation. Interestingly, this prediction was based on a remarkable linear correlation found between $\delta_{C\text{-}\alpha}$ and calculated (STO-3G) π-charge density for *t*-butyl, α,α-dimethylcyclopropylmethyl, 1-phenylcyclobutyl, *t*-cumyl, and

2-methoxyisopropyl cations.[21] Although the authors did not find this linear correlation surprising, the observation underlines the need for further investigation of the circumstances under which meaningful δ_C-charge-density correlations can be expected.

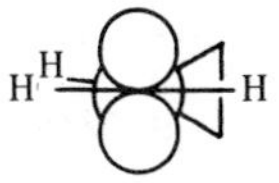

(14)

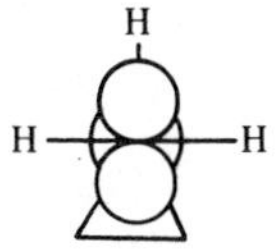

(15)

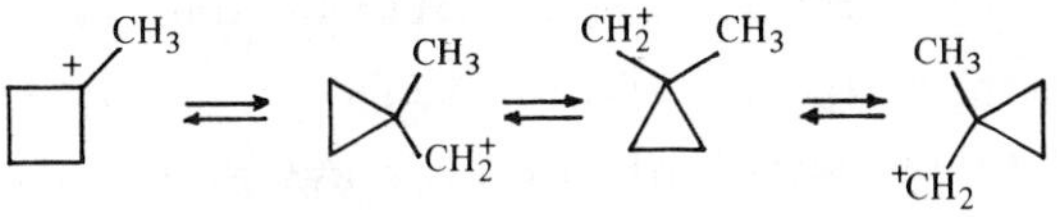

(3)

III. INTERACTION OF CARBENIUM IONS WITH SOLVENT

A. Some Thermodynamic Results

Dynamic calorimetric measurements have been reported[22] of the enthalpy change for the rearrangement of s-butyl to t-butyl cations in SbF_5/SO_2ClF, the identity of the ions being monitored by ^1H NMR spectroscopy. The exothermicity obtained, 14.5 ± 0.5 kcal mol^{-1}, agrees quite closely with values in the range 15 to 17 kcal mol^{-1}, obtained in the gas phase by mass spectrometric methods, and with the value of 18 kcal mol^{-1} calculated using *ab initio* methods.[18] The close correspondence between solution and gas phase indicates that these carbenium ions have a low structural dependence on electrostatic solvation energy; in Grunwald-Leffler symbolism, $\delta_R \Delta H_{solv} \sim 0$. In this respect they differ from ammonium or alkoxide ions, which are capable of entering into hydrogen-bonding interactions with appropriate solvents. The conclusion about carbonium ions is consistent with deductions based on activity coefficient behavior derived from acidity functions (e.g., $\underline{H}_R$) in concentrated aqueous acid.[23] The choice of SO_2ClF as solvent for the calorimetric experiments was perhaps a particularly propitious one because this compound is probably incapable of any specific interactions with solutes.

That $\delta_R \Delta \underline{H}_{solv}$ is not always negligible is evident from determinations of $\Delta \underline{H}_{solv}$ for a series of protonated alkylbenzenes (cyclohexadienyl cations) in "magic acid."[24] By combining heats of ionization in the gas phase[25] with heats of both vaporization and ionization in solution, the results in Table 2 were obtained. These show a small but important structural variation, which parallels changes in ionic size and also explains the Baker-Nathan effect of alkyl groups on protonation in solution.

TABLE 2. Structural Variation of Heats of Ionization $(\Delta \underline{H}_i)$ and Solvation $(\Delta \underline{H}_{solv})$, in kcal mol^{-1}, of Protonated Arenes in Both Gas Phase and Magic-acid Solution

Arene	$\delta_R \Delta \underline{H}_i$ (g)	$\delta_R \Delta \underline{H}_i$ (ma)	$\delta_R \Delta \underline{H}_{solv}$
Ph-Me	(0.00)	(0.00)	(0.00)
Et	-0.85	2.84	2.50 ± 0.75
Pr^n	-1.7	2.13	1.96 ± 0.39
Pr^i	-2.1	2.61	2.86 ± 0.58
Bu^t	-2.3	3.87	3.60 ± 0.96

Another approach to structural effects on the thermodynamics of carbenium ion solvation has been described.[26] Standard free-energy changes $\left(\Delta \underline{G}_i^0\right)$ for proton-transfer reactions of type (4) in the gas phase were obtained using pulsed-ion cyclotron-resonance spectroscopy. The unsaturated carbenium ion precursors ranged from styrene to azulene. It turns out that $\Delta \underline{G}_i^0(g)$ is linearly related to the standard free energies of protonation of the unsaturated compound in aqueous acids as shown in Eq. (5). Thus there appears to be a difference of 35 kcal mol^{-1} in relative free energies of ionization between the gas phase and solution, independent of carbenium ion structure over a range of some 23 kcal mol^{-1} in $\Delta \underline{G}_i^0$. The implication is clear; solvation, in this case by a solvent capable of both nucleophilic solvation and hydrogen bonding, as well as purely electrostatic solvation, greatly favors ionization, but to an extent independent of carbenium ion structure.

$$NH_4^+ + RR'C{=}CR''R''' \rightleftharpoons RR'C^+{-}CR''R''H + NH_3 \qquad (4)$$

$$\delta_R \Delta \underline{G}_i^o(g) = \delta_R \Delta \underline{G}_i^o(aq) + 35 \qquad (5)$$

This insensitivity of carbenium ion solvation to structure may be ascribable to the high degree of charge delocalization in the benzylic and cycloheptatrienyl cations examined,* together with the inability of the ions to act as hydrogen-bond donors. The introduction of heteroatoms such as oxygen into

---

* Significantly, Gold[27] has estimated widely divergent enthalpies of solvation in aqueous solution for carbenium ions showing different degrees of charge delocalization; for t-butyl cation, $\Delta \underline{H}_{solv}$ = - 50 kcal mol^{-1}, but for cyclohexadienyl cation, $\Delta \underline{H}_{solv}$ = - 73.5 kcal mol^{-1}.

the substrates, however, leads to deviations from Eq. (5), suggestive of changes in the acceptance of hydrogen bonds from the solvent when these substrates are protonated. This seems unlikely to be the only factor that operates because it has long been known that formation of the ions NO_2^+ and NO^+ from nitric and nitrous acids, respectively, follows the same acidity function $(\underline{H}_R)$ in aqueous acids as does the generation of triarylmethyl cations from the corresponding alcohol.

A final example illustrates a nonthermodynamic approach to carbenium ion solvation and is included here because the tentative conclusions seem to be in line with the view that $\delta_R \Delta \underline{H}_{solv}$ is small. The wavelength dependences of the photodecomposition of several carbocations in the gas phase [e.g., reactions (6) and (7)], referred to as "photodissociation spectra," are almost identical with the electronic spectra of the ions in solution.[28] Assuming a constant quantum yield for the photochemical decomposition, the authors advance arguments that it is close to unity throughout, with photodissociation spectra corresponding to absorption spectra in the gas phase. The absence of a solvent shift on the transition energies then suggests that the carbocations (in the ground state) are either inappreciably solvated or that solvent effects on the ground and excited states cancel. The authors are, however, cautious about interpreting their results and further experimental data are awaited.

$$\text{Ph–CO}^+ \xrightarrow{h\nu} \text{Ph}^+ + \text{CO} \qquad (6)$$

$$\text{(H,H)-}\text{CH}_3\text{-}\text{CH}_3\text{-}\text{CH}_3\text{ arenium ion} \xrightarrow{h\nu} \text{CH}_3\text{,CH}_3\text{,CH}_3\text{-phenyl cation} + H_2 \qquad (7)$$

From the foregoing discussion it can be seen that results from a variety of physical chemical techniques support the view that, although the solvent provides a large stabilizing influence on carbocations, the effect is often rather insensitive to the structure of the cation. This seems to be particularly true if the ions are of roughly similar size, have a delocalized charge, and do not possess sites (in either the cation or its neutral precursor) capable of acting as H-bond donor or acceptor. This conclusion is of far-reaching importance in organic chemistry; it validates comparisons between structures derived by quantum-mechanical calculation and those inferred from, say NMR spectra, and, as we see later, it throws new light on the interpretation of solvolytic reactions where the role of the solvent is crucial.

B. Solvolytic Reactions

Many of the fundamental ideas of carbenium chemistry have come from studies of the solvolytic reactions of alkyl halides and sulfonate esters. In such reactions, the solvent plays a dual role as solvating medium and reactant (present in excess), and much controversy has surrounded the mechanistic distinction between concerted displacement (S_N2 substitution) and a stepwise mechanism involving rate-determining formation of an intermediate carbenium ion (S_N1 substitution). The problem is particularly acute for so-called borderline cases such as the solvolysis of secondary alkyl halides. One mechanistic proposal, advanced in particular by Sneen, is that all solvolyses can be described in terms of reaction (8), in which the intermediate carbenium ion pair can either be formed in the rate-limiting step $\left(S_N1;\ \underline{k}_2(\text{SOH}] \gg \underline{k}_{-1}\right)$ or react with the solvent in the rate-limiting step $\left(S_N2;\ \underline{k}_{-1} \gg \underline{k}_2[\text{SOH}]\right)$. This interpretation may be rendered even more flexible by considering competitive dissociation of the ion pair and reactions of the dissociated carbenium ions with nucleophiles. Sneen's generalization has been the subject of much adverse criticism, and cogent arguments have been offered for the retention of the concerted bimolecular displacement mechanisms.[29] The problem still remains, however, of distinguishing pathways involving carbenium ion formation from the concerted process, and thence attempting to clarify the role of the solvent as a reactant from its role as a solvating agent towards carbenium ions. Examination of solvent effects on solvolysis rates has been one approach.

$$RX \underset{\underline{k}_{-1}}{\overset{\underline{k}_1}{\rightleftharpoons}} R^+X^- \xrightarrow[\underline{k}_2]{SOH} ROS + H^+X^- \qquad (8)$$

One widely used way of correlating (and so, hopefully, understanding) the often very pronounced effect that solvent changes have on solvolysis rates is through the Winstein-Grunwald relation, Eq. (9), proposed for so-called "limiting" (i.e., S_N1) solvolyses.[30] Taking 80% aqueous ethanol as the standard solvent, Eq. (9) relates the kinetic effect of a solvent change for a given substrate to the effect on the solvolysis of t-butyl chloride, the model ionization process; the parameter $\underline{Y}$ is referred to as a measure of the ionizing power of the solvent. The solvolysis of t-butyl chloride proves to be a poor model for the ionization of sulfonate esters because of the differing solvation requirements of incipient halide and sulfonate ions, and 2-adamantyl toluene-p-sulfonate has been proposed as a better reference substrate for such compounds. This is particularly appropriate because the cage structure of the adamantyl group precludes backside S_N2 displacement. Correlation of solvolysis rates using Eq. (9) has been taken to indicate a rate dependence on solvent ionizing power, the

$$\log\left[\frac{k}{k_0}\right]_{RX} = m \log\left[\frac{k}{k_0}\right]_{\text{reference}} = m\,Y \qquad (9)$$

parameter m then measuring the extent to which the transition state approaches the limiting carbenium ion-like structure of the model reaction system.[31]

Bentley and Schleyer[32] now argue that the long-known dispersion of correlations using Eq. (9), that is to say, the production of lines of different slope (m) from solvolytic data obtained using different families of solvents (e.g., ethanol-water mixtures, carboxylic acids), is related to "nucleophilic solvent participation." Thus a linear correlation using Eq. (9) is not to be regarded as implying a pure ionization mechanism; an element of concerted displacement by the solvent is also present, and this may be assessed in ethanol-water (EW) mixtures from the experimental rate-constant ratio, $\left(k_{EW}/k_{AcOH}\right)_Y$, for solvents of the same ionizing power.* In solvolyses that take place without nucleophilic solvent assistance and do not involve internal return of intermediate ion pairs, the ratio $\left(k_{EW}/k_{AcOH}\right)_Y$ is thought to have a value of 0.3 to 0.5 (Y based on *t*-butyl chloride) or approximately 1.0 (Y based on 2-adamantyl tosylate). Larger values of the ratio then indicate direct attack on the substrate by the more nucleophilic ethanol-water mixture, and, as expected, the magnitude of the ratio is related inversely to the value of m. Because acetic acid is itself appreciably nucleophilic, a better definition of nucleophilic solvent assistance (expressed as the ratio of rate constants for nucleophilically assisted and unassisted reactions

---

* This procedure represents the quantification of a method of analysis suggested by Winstein and Grunwald a quarter of a century ago.[30b,c]

TABLE 3. Minimum Estimates of Nucleophilic Solvent Assistance (k_s/k_c) [a]

	Solvent								
Tosylates	CF_3CO_2H	97 wt % $(CF_3)_2CHOH$	97 wt % CF_3CH_2OH	HCO_2H	70 wt % CF_3CH_2OH	AcOH	50% EtOH	80% EtOH	EtOH
2-Adamantyl	1.0	1.0	1.0	1.0	1.0	1.0	1.0	1.0	1.0
Pinacolyl	1.0		3.6	2.4	3.2	8.6	9.5	12	
Cyclohexyl	1.0	0.62	3.2	5.0	6.6	28	61	104	256
Cyclopentyl	1.0			10		105		455	1680
4-Heptyl	1.0	0.87		3.9		28	46	146	
3-Pentyl	1.0	0.60		6.2		46	103	310	1310
2-Pentyl	1.0			9.6		88	195	616	
2-Butyl	1.0		6.3	13	22	140	292	979	
2-Propyl	1.0	0.57	15.3	32	57	472	1130	4430	23 500

[a] Reprinted from Schadt, Bentley, et al.[33] with permission of the American Chemical Society.

k_s/k_c, obtained from titrimetric rate constants, k_t) is that given for alkyl tosylates in Eq. (10).[33] Values of k_s/k_c for a range of tosylates (ROTs) in a variety of solvents are given in Table 3, the reference solvent being trifluoroacetic acid. The values show an inverse correlation with the magnitude of the kinetic α-deuterium isotope effect observed on solvolysis, which, like the magnitude of m, has often been used as a criterion for the carbenium ion mechanism.[31]

$$\frac{k_s}{k_c} = \left[\frac{k_t^{\text{ROTs}}}{k_t^{\text{AdOTs}}}\right]_{\substack{\text{any}\\\text{solvent}}} \cdot \left[\frac{k_t^{\text{AdOTs}}}{k_t^{\text{ROTs}}}\right]_{CF_3CO_2H} \tag{10}$$

Equation (10) is derived on the assumptions that: (a) m is constant for all S_N1 solvolyses; (b) in trifluoroacetic acid all the substrates in Table 3 are solvolyzed by the S_N1 mechanism; and (c) 2-adamantyl tosylate (AdOTs) is solvolyzed by the S_N1 mechanism in all solvents. The implication of these assumptions is that on the ionization pathway, the response to solvent ionizing power of the free-energy difference between the ground state and the transition state (for which the carbenium ion is a model) is independent of the structure of the alkyl group. The connection with the results described in Section III.A is evident. However, it would be interesting to know whether such assumptions are valid for more delocalized transition states such as those in benzylic systems, where positive charge can build up at sites remote from the site of heterolysis where steric hindrance to solvation might be less than in alkyl systems. Values of m close to unity were in fact observed in bromine additions to styrene and stilbene, as well as to 1-pentene;[34] this evidence was then used to argue that nucleophilic solvation of the intermediate carbenium ion is not a factor in the changes observed in the stereoselectivity of additions to arylalkenes with changing solvent.

The new picture of solvolysis mechanisms has much to commend it, not least that it brings together again operational and theoretical distinctions between concerted and carbenium ion mechanisms in the vast majority of cases.* On this basis, the ratio $(\underline{k}_s/\underline{k}_c) = 1$ corresponds to S_N1 behavior. Bentley and Schleyer[32] suggest that systems for which $(\underline{k}_s/\underline{k}_c) > 10$ should be classified as S_N2. This leaves a relatively small area of uncertainty, $1 < (\underline{k}_s/\underline{k}_c) < 10$, in which no distinction is drawn between nucleophilic interaction of the solvent with the reaction center on the alkyl group and carbenium ion solvation.

Bentley and Schleyer's analysis of solvolytic reactions places 2-adamantyl and methyl tosylates at extremes of a "spectrum" of solvolysis mechanisms; in the former case, an ion pair is believed to be formed in the rate-determining step, whereas in the latter no intermediate is formed at all. Between these extremes, it could be that there exist systems in which a nucleophilically solvated ion pair (16) is an intermediate in the reaction. Such a mechanism, designated S_N2 (intermediate),[32] could explain evidence of the type on which Sneen's generalization was based, that is, a dependence of the solvolysis rate on the nucleophile (solvent), inversion of configuration during substitution, and the detection of a reactive intermediate. In the light of the foregoing discussion, the number of such cases should turn out to be quite small, and special structural features may be needed for this mechanism.

---

* Some of the kinetic evidence used by both sides in the arguments over the reality of σ-bridged (nonclassical) carbocations may be confused by large contributions from nucleophilic solvent assistance.

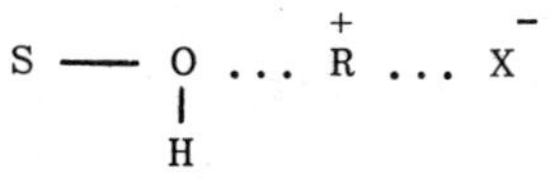

$$\mathrm{S — \underset{\displaystyle H}{\overset{}{O}} \cdots \overset{+}{R} \cdots \overset{-}{X}}$$

(16)

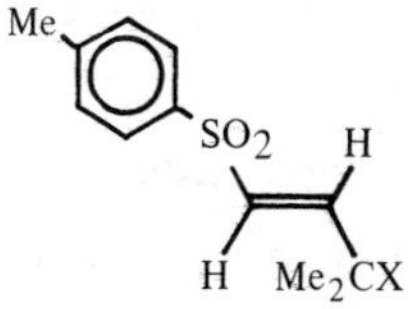

(17)

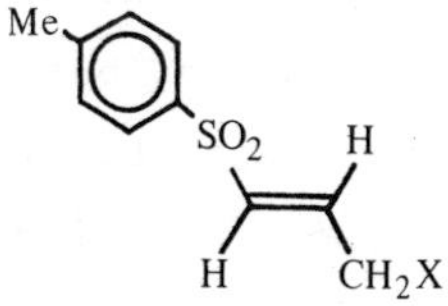

(18)

The special features necessary for the S_N2 (intermediate) mechanism appear to be provided in the highly deactivated tertiary allylic system 17 ($X = Br, OSO_2CH_3$).[35] Lithium azide, for example, reacts with 17 in methanol to give the corresponding allylic azide according to the kinetic law, $v = k[17][LiN_3]$, expected for a bimolecular substitution and, indeed, found for the related primary system 18. Significantly, in the normally less nucleophilic but more polar solvent 60% (v/v) methanol-water solvolysis of 17 competes with azide formation, and the correspondence of kinetic and product data indicates that the rate- and product-controlling steps of the reaction are identical. However, replacement of an α-methyl group by CD_3 retards the rate of the azide reaction at 50° C by a factor of 1.22 and by a factor of 1.38 in the solvolysis; for comparison, the secondary β-deuterium isotope effect is 1.02 for solvolysis of ethyl bromide (in water) and 1.34 for *t*-butyl chloride (in 50% aqueous ethanol). The sequence of reactivity of powerful nucleophiles in methanol toward 17, $N_3^- > SCN^- > PhNH_2 > (H_2N)_2CS$, coincides with that toward triphenylmethyl cations in water-dioxane-acetone. In S_N2-type displacements on primary halides the reactivity sequence is typically $SCN^- > PhNH_2 > N_3^-$, which is similar, but not identical, to that found for 18. The rate of solvolysis of 17 in aqueous alcohol solvents increases with solvent polarity in a manner that fits Eq. (9); values of the slope $\underline{m}$ are about 0.5, suggestive of nucleophilic solvent participation, but $\log\left(\underline{k}_{EW}\ \underline{k}_{AcOH}\right)_Y$ is close to unity. Changing from methanol to dimethylformamide as solvent accelerates the reaction of lithium azide with the primary allylic compound (**18**) about 100 times more than it accelerates the reaction of the tertiary compound (**17**). Thus for **17**, some of the evidence indicates S_N2-type behavior as found for **18**, but the rest strongly suggests the formation of an organic cationic intermediate.

Further evidence for formation of an intermediate with carbenium character from **17** comes from product studies. With buffered sodium thiophenoxide in methanol, **17** gives a mixture of products of direct substitution, elimination, and substitution with allylic rearrangement; **18**, however, gives only the direct substitution product. Furthermore, the ambident nucleophile, thiocyanate ion, reacts with **17** to give a mixture of allylic thiocyanate and isothiocyanate, the ratio RSCN/RNCS being about 1.5 (cf. 2.5 for *t*-butyl chloride under similar conditions); the substitution product from **18** was more than 98% primary allylic thiocyanate.

Bordwell and his co-workers concluded that, despite the similarity of the kinetic forms of their substitution reactions, **17** and **18** react by different mechanisms. The primary system's behavior is consistent throughout with a "classical" S_N2 displacement. The tertiary compound, however, shows a pattern of reactivity suggestive of an intermediate with carbenium ion-like properties formed reversibly from **17**. Reaction (11) represents such a situation, and Bordwell emphasizes that the intermediate (**19**) will be mechanistically significant even if $\underline{k}_{-1} < \underline{k}_2$ (cf. Sneen's interpretation of S_N2 displacement). This intermediate was formulated as an ion "sandwich" formed by entry of the substrate, RX, into the solvation shell of the attacking nucleophile, which then assists in the cleavage of the C—X bond by "solvating" the incipient carbenium ion.

$$Nu^- + RX \underset{k_{-1}}{\overset{k_1}{\rightleftharpoons}} [Nu^- R^+ X^-] \xrightarrow{k_2} NuR + X^- \quad (11)$$

$$\underset{\sim}{19}$$

Clearly, further investigation of this type of reaction is desirable since the vast amount of evidence assembled by Bordwell in favor of 19 is still all circumstantial. The problem (as always) is whether the observed divergence in behavior between 17 and 18 is sufficient to define a different mechanism, the formation of an intermediate, or whether both should be regarded as S_N2 reactions with different degrees of C—X bond breaking in the transition state. The balance of evidence in this instance seems to favor intermediate formation.

Ion association is usually invoked in interpretation of S_N1 solvolyses that take place with net inversion of configuration at the reaction center rather than the complete racemization expected of a free, symmetrically solvated carbenium ion intermediate. When such reactions take place with net retention, however, special explanations are necessary. For example, a nucleophilic component of the solvent (e.g., dioxane in aqueous dioxane) can give rise to an unstable oxonium ion with inversion, the stable final product being formed by hydrolysis, again with net inversion, of this intermediate. Such an explanation seems less likely for the retentive solvolyses of benzylic halides and *p*-nitrobenzoates in phenol solution exhaustively documented by Okamoto, Kinoshita, et al.[36] Compounds of type 20 (R = Me, *t*-Bu, aryl) in phenol containing triethylamine or sodium phenoxide yield predominantly the corresponding phenyl ether with up to 90% retention of configuration. Also formed are *ortho*- and *para*-alkylated phenols in which inversion of configuration is observed, but to a lower degree than retention in the accompanying ether. The ratio of polarimetric to titrimetric rate constants is always greater than unity, indicating the involvement of ion pairs, and the results obtained in the presence of sodium phenoxide suggest the operation of a special salt effect in some instances.

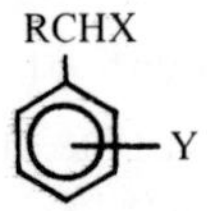

RCHX

Y

(20)

These observations (the opposite of those that led Sneen to propose the ion-pair mechanism for nucleophilic substitution) could presumably be accounted for by a double inversion mechanism, that is, C-alkylation of the phenol by the carbenium ion pair followed by displacement of the phenol molecule by the oxygen atom of a second phenol molecule. However, this would require the second displacement to take place at a rate comparable to proton removal from the intermediate, which would lead to the inverted alkylated phenol. The authors preferred interpretations that place severe constraints on the location of solvent molecules with respect to the carbenium center in the intermediate. Possible structures are four- and six-membered rings (21 and 22), a solvent-separated ion pair 23, and an ion pair shielded at the rear by π-complexing to a phenol molecule (24). Of these, 22 neatly accounts for the observation that small additions of methanol to the solvent phenol yield, in addition to the phenyl ether, some methyl ether formed with retention of configuration. However, at higher methanol concentrations (ca. 5%) in phenol, or with methanol in benzene, the methyl ether shows net inversion.[36b] The preferred interpretation of

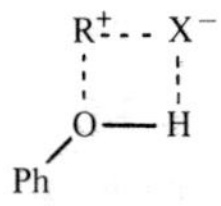

(21) 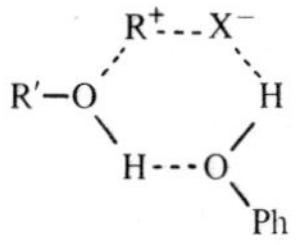 (22) 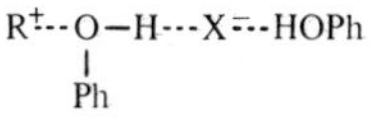 (23)

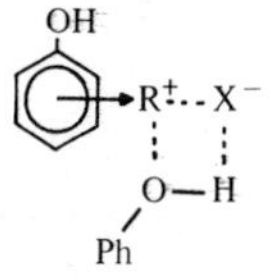

(24) 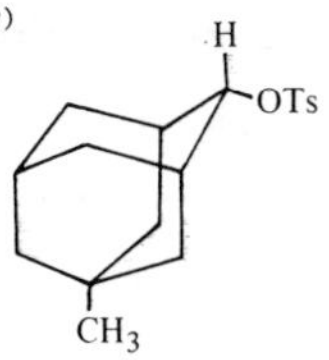 (25) 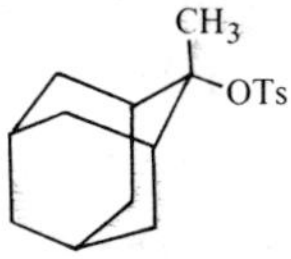 (26)

retention as a result of backside shielding by a phenol molecule (24) is, of course, merely a variant of the double-inversion mechanism.

Another example of solvolysis with excess retention is that of the 2-adamantyl tosylate 25.[37] 2-Methyl-2-adamantyl tosylate (26) shows a much smaller preference for retention, suggesting that the origin of the retention may not be a consequence only of backside shielding by the adamantyl cage.

C. Kinetics of Reactions of Carbenium Ions with Nucleophiles

Even leaving aside the mechanistic complexities, studies of solvolytic reactions provide information about carbenium ions only indirectly. Direct investigation of the kinetic behavior of preformed carbenium ions should provide clearer information on the interaction of the cation with the solvent. Ritchie[38] has determined rate constants for the reaction with nucleophiles of a range of carbenium ions (e.g., Malachite Green, Crystal Violet, arylcycloheptatrienyl, tri-p-anisylmethyl cations) and other organic cations, (e.g., benzenediazonium ion) at concentrations low enough to eliminate complications due to ion association. The relative reactivities of nucleophiles are apparently independent of the structure of the electrophile, although there are such structural effects on equilibrium constants. Second-order rate coefficients ($\underline{k}$) for the combination of carbenium ions and nucleophiles obey the simple relation in Eq. (12), where $\underline{k}_0$ is a constant dependent only on the nature of the cation and $\underline{N}_+$ is a nucleophilic constant, the value of which depends on the nucleophile and solvent, but not on the nature of the electrophile. In practice, $\underline{N}_+$ for a given nucleophile Nu was obtained from the rate constants for reaction of the cation 27 with Nu and with water (a less than perfect choice since the reaction has a general base-catalyzed component)[39] using the relation (13). Ritchie[40] reported a re-evaluation of $\underline{N}_+$ parameters and the remarkable finding that the $\underline{N}_+$ relation in Eq. (12) also governs the relative reactivities of nucleophiles toward carboxylic esters, provided that formation of the tetrahedral intermediate is rate-limiting.

$$\log \underline{k} = \log \underline{k}_0 + \underline{N}_+ \qquad (12)$$

$$\underline{N}_+ = \log \frac{\underline{k}^{Nu}}{\underline{k}^{H_2O}} \qquad (13)$$

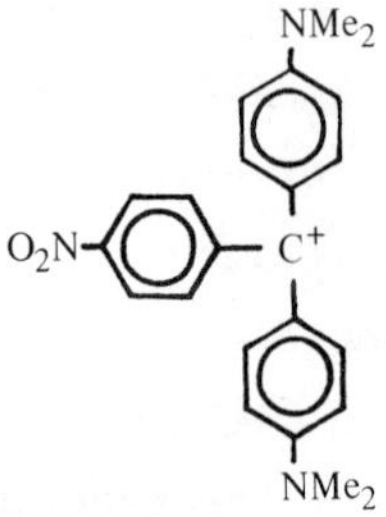

O_2N–C_6H_4–C^+(C_6H_4–NMe_2)$_2$ (27)

The constancy of the relative reactivities of nucleophiles as the structure of the carbenium ion changes is inconsistent with the usual generalization that the more reactive the reagent, the less selective it is between available reaction partners. An early interpretation of the chemistry of systems obeying the

$\underline{N}_+$ relation was that the transition state in the combination of carbenium ions with nucleophiles occurs at an early point on the reaction coordinate when the reactants are still well separated. Variation in the reactivity of the nucleophiles was regarded as arising from partial desolvation, solvation of the carbenium ion being considered unchanged on passing from the ground to transition states. The extension of the $\underline{N}_+$ relation to reactions of esters, however, produced additional results that cast doubt on the generality of this interpretation. Thus the $\underline{N}_+$ values for amines are linearly related to $p\underline{K}_a$, these Brønsted plots having slopes of about 0.5, which is inconsistent with an early transition state. Moreover, amines are much less reactive than anionic nucleophiles of the same $p\underline{K}_a$ value, the reverse of expectation if solvation of the nucleophile is the dominant effect. Finally, it is difficult to interpret the wide variation of absolute reaction rates (10^6-fold in the case of carbenium ion reactions) with electrophile structure in terms only of the solvation of the nucleophile.

A possible solution to the problem has been advanced by Pross,[41] who assumed that the free energy of activation for reaction of a carbenium ion and a nucleophile would involve contributions from desolvation of both the nucleophile and the electrophile and that its magnitude would depend on the "inherent reactivity" of the reactants. On this basis, and making additional simplifying assumptions (perhaps the most dubious of these being that the degree of desolvation of a given electrophile in attaining the transition state is independent of the nature of the solvent), he concluded that the relative reactivity of nucleophiles would be independent of the structure of the carbenium ion if the variation in inherent reactivity of the carbenium ions were balanced by the variation in their solvation. Qualitatively, this seems reasonable. Thus a highly reactive carbenium ion (e.g., one with a localized positive charge) should interact very strongly with the solvent and undergo relatively little desolvation at the transition state for reaction with a nucleophile. On the other hand, a more stable cation would interact more feebly with the solvent but would require greater desolvation to reach the later transition state. Thus the actual contribution that cationic desolvation would make to the free energy of activation could be the same for both carbenium ions.

Such considerations hold out the possibility that the reactivity-selectivity principle may be retained for reactions obeying the $\underline{N}_+$ relation. However, its generality implies that the balance between inherent reactivity and solvation is quantitatively maintained for all systems studied, and these cover a variety of solvents and a diversity of steric situations. The explanation also requires a structural dependence of the solvation energy of carbenium ions, which would theoretically be particularly important if there were heteroatoms in the electrophile[25] (see Section III.A). Of the nine highly delocalized carbenium ions examined by Ritchie, all but two, cycloheptatrienyl cation and its phenyl-substituted analog, contained heteroatoms and might be expected to undergo structure-dependent solvation changes on passing from ground to transition states. Clearly, further data on the kinetics of nucleophilic reactions of carbenium ions less stable than those hitherto studied is desirable. Diffusion control could, of course, limit the range of nucleophiles that could usefully be studied (cf. PhN_2^+). A limited amount of data derived from reactions of the carbenium ions $PhCH_2^+$, Ph_2CH^+, and Ph_3C^+, generated by pulse radiolysis, has already appeared.[42] However, reexamination of the kinetics of reaction of Malachite Green, Crystal Violet, and tri-*p*-anisylmethyl cations with nucleophiles, although confirming the adherence to Eq. (12) in the first two cases, suggests that the slope of the log *k* versus N_+ plot is substantially less than unity (0.81 ± 0.01) for the last.[43] This reexamination should be extended to the more reactive cycloheptatrienyl cations. Further progress in the understanding of reactions of nucleophiles with carbenium ions seems most likely to be made by the development of a more detailed "structural" picture of the approach of the electrophile and nucleophile and their attendant solvent molecules; both Ritchie[39] and Pross[41] have taken the first steps in this direction.

IV. SELECTED TYPES OF CARBOCATIONS

A. Aryl Cations

In recent years great interest has been shown in carbenium ions formed by heterolysis of a bond connecting an sp^2-hybridized carbon atom to a suitable leaving group. The chemistry of vinyl cations has been extensively reviewed, most recently by two notable contributors to the field.[44] Hence this section is devoted to a discussion of recent publications on aryl cations, the usual sources of which are arenediazonium salts.

The decomposition of benzenediazonium bromide in aqueous solution gives only small proportions of bromobenzene, the remainder of the reactant being converted into phenol. The observation that the reaction obeys overall a first-order kinetic law does not distinguish unimolecular and bimolecular mechanisms, but other evidence is more definitive.[45] In the range 80% to 100% H_2SO_4 the reaction rate is insensitive to the acid concentration, even though the water activity changes by more than 100-fold. The entropy of activation of the reaction, +10.5 cal · mol^{-1} deg^{-1}, is very similar to that for the S_N1 solvolysis of *t*-butyl chloride and quite different from that for typical bimolecular nucleophilic substitutions. Changing from water to deuterium oxide as solvent reduces the rate of hydrolysis by less than 4%, and no ring deuteriation occurs, thus ruling out an aryne mechanism.[45a*] The α-nitrogen kinetic isotope effect is 1.0384 ± 0.0010, compared with the β-effect (determined directly) of 1.00106 ± 0.0003.[45c] The theoretical maximum for unimolecular C—N cleavage is in the range of 1.040 to 1.045. The effect of introducing a deuterium atom in place of hydrogen (protium) in the aromatic ring is to reduce the reaction rate substantially; k_H/k_D is 1.22 for ortho-substitution, 1.08 for meta-substitution, and 1.02 for para-substitution. These effects do not arise from relief of steric compressions.[45b] Additional evidence shows that although the reaction rate can be varied by change of solvent, the product proportions are unchanged, indicating separate rate- and product-determining steps.

All of the evidence supports the view that the rate-limiting step in the decomposition of arenediazonium ions is unimolecular cleavage of the C—N bond to give molecular nitrogen and a highly reactive, unselective aryl cation. However, perhaps the most convincing direct evidence for the formation of aryl cations is that of Bergstrom, Zollinger, et al.,[47] who showed that β-^{15}N-benzenediazonium ion undergoes some 8% transformation to the α-labeled ion during thermolysis in trifluoroethanol solution. Moreover, by carrying out the reaction under 300 atmospheres of unlabeled nitrogen, 2.46 ± 0.4% of external nitrogen was incorporated into the diazonium ion at 70% decomposition. Such evidence seems consistent only

---

* Deuterium incorporation and cine substitution have, however, been observed in the decomposition of diazonium salts in pyridinium poly(hydrogen fluoride),[46a] and this was interpreted in terms of charge distribution in the diazonium ions deduced from ^{13}C chemical shifts.[46b]

with intermediate formation of a highly reactive phenyl cation that combines with available nucleophiles (in this instance, the relatively poor nucleophile CF_3CH_2OH, giving $PhOCH_2CF_3$), including eliminated nitrogen or external nitrogen, not usually regarded as being nucleophilic.

Problems remain concerning the structure of aryl cations. Swain, Sheats, et al.[45] could find no evidence that the triplet state (28) suggested by Taft[48] was involved in their reactions.[45a] On the basis of INDO calculations they concluded that the aryl cation is a singlet species with the bonds between C-1 and C-2 and C-1 and C-6 collinear (29), as suggested for vinyl cations. More detailed calculations,[49] using CNDO/2 with configuration interaction, led to the conclusion that 29 is indeed the ground state of the phenyl cation. The singlets with benzene-type geometry (30) and intermediate nuclear arrangements were found to be of higher energy, although these become relatively more stable as the in-plane orbital is populated. Nevertheless, the $\sigma\pi$ triplet (28) had higher energy than 29. It was concluded that 28 is probably not accessible thermally from benzenediazonium ion and that alternative explanations are necessary for the radical- (and carbene-) like behavior that has sometimes been observed for aryl cations.

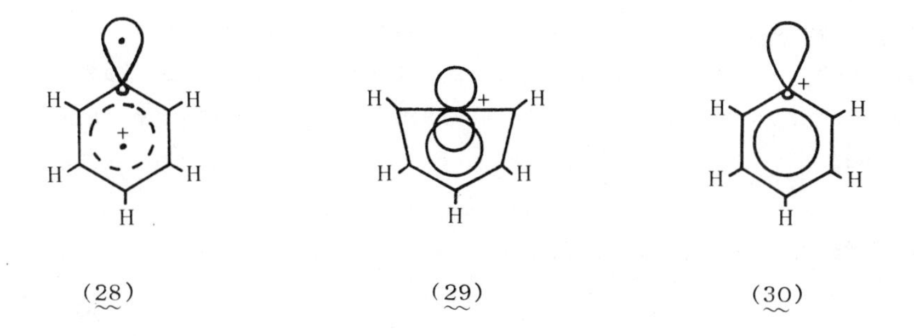

(28) (29) (30)

The triplet state does seem to be obtainable by photolysis. Almost complete triplet ESR spectra have been described[50] for a range of substituted phenyl cations generated by irradiation of diazonium salts of type 31 (cf. the triplet spectrum reported for 32[51]). Although these observations may indicate that the triplet manifold of aryl cations can only be populated from the (triplet) excited diazonium ion, it is worth remembering that earlier calculations[52] suggested that the ground state of p-aminophenyl cation should be a triplet. The present results point the way to further, definitive experimentation.

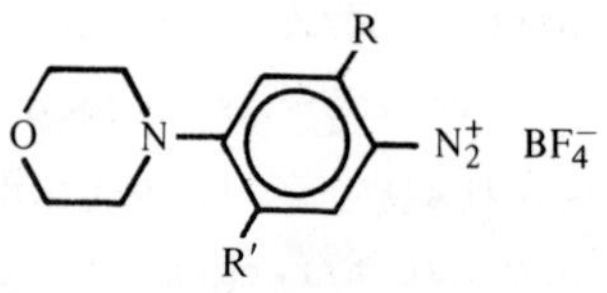

R

O N— —N_2^+ BF_4^-

R′

[R=Me,MeO;R'=H,Me,n-BuO]

(31)

O•

(32)

B. Pentacoordinate Carbocations: Carbonium Ions

Current interest in the role of pentacoordinate carbonium ions in isomerization and fragmentation reactions of alkanes in superacidic media stems from earlier work on the detection of protonated alkanes such as CH_5^+ and $C_2H_7^+$ in mass-spectrometric experiments. Work on the chemistry of such species in the gas phase continues apace, and direct rather than circumstantial evidence for their involvement in reactions in solution is being sought. A selection of studies in both areas is presented here.

Hiraoka and Kebarle have notably advanced the understanding of pentacoordinate carbonium ions using high-pressure mass spectrometry. Not only have they extended the range of protonated aliphatics by the detection of $C_3H_9^+$ (formed from methane and ethyl cation)[53] and $C_4H_{11}^+$, but they have also shown[54] that such species can correspond to structures of two different types. Thus protonated ethane, $C_2H_7^+$, was formed by reaction (14a) in the temperature range -160° to -130° C; the enthalpy change for (14a) was -4 kcal mol^{-1}, corresponding to a heat of formation of 215 kcal mol^{-1} for the ion. Above -130°, the ion was found to be unstable; another ion of the same composition, but with $\Delta H_f = 207$ kcal mol^{-1}, began to appear at temperatures above -100°, and equilibrium (14b) was established in the range 85° to 200°; $\Delta H_{14b} = -11.8$ kcal mol^{-1}. On the basis of calculations of the relative stabilities of isomeric protonated alkanes,[55] the low-temperature ion has been assigned the C—H protonated structure **33** and the high-temperature ion the C—C protonated structure **34**, both formulated with one three-center bond.

$$C_2H_5^+ + H_2 \xrightleftharpoons{-160^0 \text{ to } -130^0} \underset{\mathbf{33}}{C_2H_7^+} \qquad (14a)$$

$$C_2H_5^+ + H_2 \xrightleftharpoons{85^0 \text{ to } 200^0} \underset{\mathbf{34}}{C_2H_7^+} \qquad (14b)$$

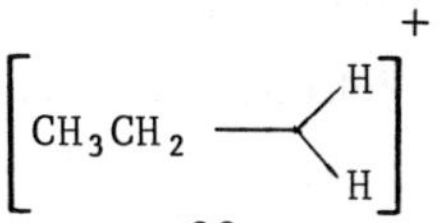

$[CH_3CH_2—<(H)(H)]^+$

33

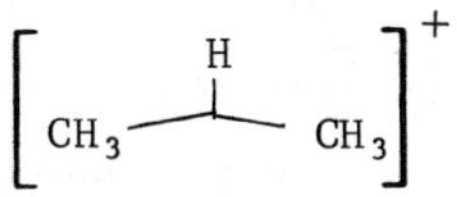

$[CH_3—(H)—CH_3]^+$

34

A form of quasisolvation in the gas phase has been suggested on the basis of studies of the association equilibria between protonated methane and further methane molecules.[56] The association turns out to be exothermic at least up to $CH_5^+(CH_4)_4$ and probably beyond. The interaction is thought to be through three-center bonds between the hydrogens of lowest electron density in CH_5^+ and the C—H bonds of methane as shown in **35**.

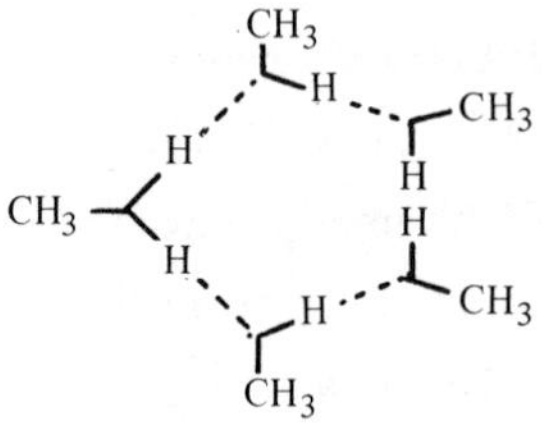

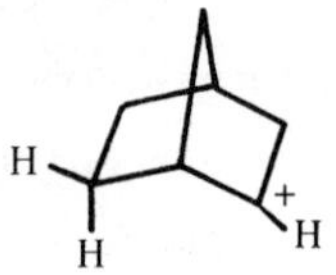

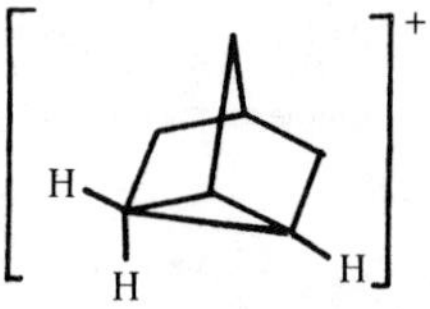

(35) (36) (37)

In solution, spectroscopic evidence for the simple cationated alkanes is absent. Although there have been reports of spectra of ions such as the 2-norbornyl cation (36), which can be formulated as the pentacoordinated carbonium ion 37, their interpretation is not without controversy.[11] Reliance thus tends to be placed on the nature of the reaction products and circumstantial evidence; recent examples concern the apparent hydride transfer from isobutane to acetyl cation giving acetaldehyde[57] and the alkylation of aromatics by paraffins in superacidic media.[58]

The mechanism of the apparent hydride transfer in reaction (15) in H_2SO_4-FSO_3H mixtures has been investigated in detail,[59] and this highlights some of the types of problems involved. Leaving aside the question of whether the reaction as indicated in reaction (15) is thermodynamically feasible, two pieces of evidence make it clear that it is not a simple intermolecular hydride transfer. The first is that the reaction appears to be subject to acid catalysis, since in three different acid mixtures log $\underline{k}$ (where $\underline{k}$ is the observed second-order velocity constant in a given acid mixture) is linearly dependent on $-\underline{H}_0$, the Hammett acidity function. Second, protons in the reaction medium are scrambled with the hydrogen transferred from the isobutane. It was suggested that reactions of this type take place by interaction of an acid (e.g., FSO_3H) with the most basic C—H bond of the isoalkane, this pair then transferring a hydride ion and a proton to the carbenium ion. Depending on the acidity of the catalyzing acid, the acid-alkane interaction could lead to a complex that, in the limit, would have the structure of $\underline{C}_s$ symmetry expected for the protonated alkane. Once again the difficulty of distinguishing a true intermediate of low stability from a transition state appears. In this instance, however, no exchange of isotopic hydrogen with the isobutane was observed in the absence of the benzylic cation, and this suggests that the symmetrical protonated alkane is not an intermediate. A concerted termolecular mechanism via transition state 38 was preferred.

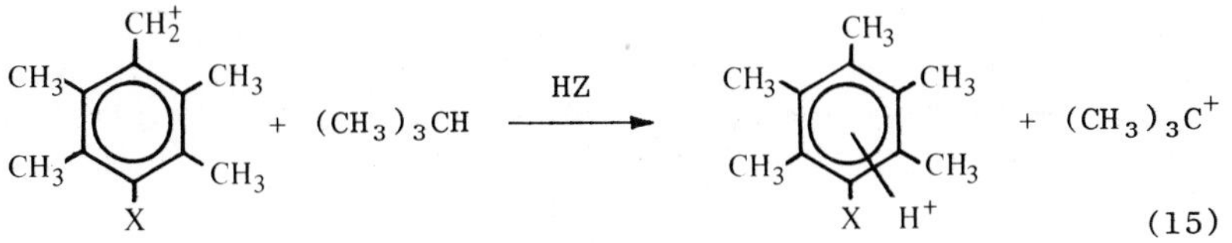

$$X\text{-}C_6(CH_3)_4\text{-}CH_2^+ + (CH_3)_3CH \xrightarrow{HZ} [X\text{-}C_6(CH_3)_5H]^+ + (CH_3)_3C^+ \quad (15)$$

CH3, X, CH3, CH3, CH3, C+, H, H, FSO3---- H---H, C, CH3, CH3, CH3

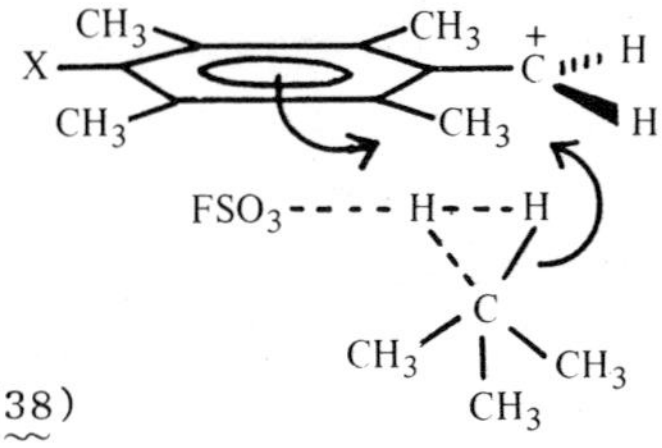

(38)

$$R_3C^+ + H_2 \rightleftharpoons R_3CH + H^+ \quad (16)$$

A somewhat similar state of affairs probably exists in the reduction of carbenium ions with molecular hydrogen in strongly acidic media [see reaction (16)]. The patent literature indicates commercial interest in reactions of this type, and Wristers has described experiments on the hydrogenation of benzene in superacidic media such as $HF\text{-}TaF_5$ ($50^\circ C$) containing pentane.[60] By operating under hydrogen pressures of 35 atmospheres in the presence of some isopentane, hydrogen is consumed and the reaction is catalytic in both the proton and tertiary hydride ion sources. The key steps of the catalytic process are shown in reactions (17) through (19). It should be noted that the delocalized cyclohexadienyl and allylic cations produced in the course of the benzene reduction do not apparently react directly with hydrogen; the _t_-alkyl cation is necessary to mediate the process. The reaction is not very selective for simple hydrogenation and, as well as cyclohexane, alkylated benzenes, methylcyclopentane, and acyclic hexanes are produced.

$$C_6H_6 + H^+ \longrightarrow C_6H_7^+ \qquad (17)$$

$$C_6H_7^+ + CH_3\text{—}C(CH_3)(CH_2CH_3)\text{—}H \longrightarrow C_6H_8 \text{ and/or } C_6H_8 + CH_3\text{—}C^+(CH_3)(CH_2CH_3) \qquad (18)$$

Further protonation and hydride transfer

$$CH_3\text{-}C^+(CH_3)(CH_2CH_3) + H_2 \longrightarrow CH_3\text{-}CH(CH_3)(CH_2CH_3) + H^+ \qquad (19)$$

V. CONCLUSION

Carbocations are important in such diverse fields as the petrochemical industry and biochemistry. Traditionally, studies on carbocations have led the way in developing the physical principles of organic chemistry. Although this chapter has not attempted to be comprehensive, even in its coverage of the current trends in the field it is hoped that it will have shown that the understanding of the behavior of carbocations and their application for practical ends continue to tax the ingenuity of chemists.

VI. REFERENCES

1. D. Bethell and V. Gold, Carbonium Ions. An Introduction, Academic, London, 1967.

2. For a spirited justification of the relevance of studies in superacidic media, see T. S. Sorensen, Acc. Chem. Res., 9, 257 (1976).

3. D. G. Farnum, Adv. Phys. Org. Chem., 11, 123 (1975).

4. G. A. Olah and D. A. Forsyth, J. Am. Chem. Soc., 97, 3137 (1975).

5.* S. Fliszár, Can. J. Chem., 54, 2839 (1976).

6. (a) H. Volz, J.-H. Shin, and H.-J. Streicher, Tetrahedron Lett., 1297 (1975); (b) H. Volz and R. Miess, ibid., 1665 (1975). See also G. A. Olah, G. Liang, and Y. K. Mo, J. Org. Chem., 39, 2394 (1974).

7.* G. A. Olah, P. W. Westerman, and D. A. Forsyth, J. Am. Chem. Soc., 97, 3419 (1975).

8. G. A. Olah, R. J. Spear, and D. A. Forsyth, J. Am. Chem. Soc., 98, 6284 (1976).

9.* D. A. Forsyth, R. J. Spear, and G. A. Olah, J. Am. Chem. Soc., 98, 2512 (1976).

10. G. E. Maciel, Top. Carbon-13 NMR Spectrosc., 1, 53 (1974).

11. See, for example, G. M. Kramer, Adv. Phys. Org. Chem., 11, 177 (1975); G. A. Olah, Acc. Chem. Res., 9, 41 (1976).

12.* D. G. Farnum and R. E. Botto, Tetrahedron Lett., 4012 (1975).

13.* (a) H. C. Brown, E. N. Peters, and M. Ravindranathan, J. Am. Chem. Soc., 97, 2900 (1975); (b) H. C. Brown, M. Ravindranathan, and E. N. Peters, ibid., 96, 7351 (1974). See also H. C. Brown, S. Ikegami, K.-T. Liu, and G. L. Tritle, ibid., 98, 2531 (1976).

14. D. G. Farnum and A. D. Wolf, J. Am. Chem. Soc., 96, 5166 (1974).

15. H. C. Brown, M. Ravindranathan, K. Takeuchi, and E. N. Peters, J. Am. Chem. Soc., 97, 2899 (1975).

16. J. B. Stothers, Carbon-13 NMR Spectroscopy, Academic, New York, 1972.

17.* D. P. Kelly, G. R. Underwood, and P. F. Barron, J. Am. Chem. Soc., 98, 3106 (1976).

18. L. Radom, J. A. Pople, and P. v. R. Schleyer, J. Am. Chem. Soc., 94, 5935 (1972).

19. D. P. Kelly and H. C. Brown, J. Am. Chem. Soc., 97, 3897 (1975).

20. G. A. Olah and G. Liang, J. Am. Chem. Soc., 98, 7026 (1976).

21.* G. A. Olah, R. J. Spear, P. C. Hiberty, and W. J. Hehre, J. Am. Chem. Soc., 98, 7470 (1976).

22.* E. W. Bittner, E. M. Arnett, and M. Saunders, J. Am. Chem. Soc., 98, 3734 (1976).

23. E. M. Arnett and G. Scorrano, Adv. Phys. Org. Chem., 13, 83 (1976).

24.* E. M. Arnett and J.-L. M. Abboud, J. Am. Chem. Soc., 97, 3865 (1975).

25. W. J. Hehre, R. T. McIver, J. A. Pople, and P. v. R. Schleyer, J. Am. Chem. Soc., 96, 7162 (1974).

26.* J. F. Wolf, P. G. Harch, and R. W. Taft, J. Am. Chem. Soc., 97, 2904 (1975).

27. V. Gold, J. Chem. Soc., Faraday Trans. 1, 68, 1611 (1972).

28.* B. S. Freiser and J. L. Beauchamp, J. Am. Chem. Soc., 98, 3136 (1976).

29. For example, see D. J. McLennan, Acc. Chem. Res., 8, 281 (1976).

30. (a) E. Grunwald and S. Winstein, J. Am. Chem. Soc., 70, 846 (1948); (b) S. Winstein, E. Grunwald, and H. W. Jones, ibid., 73, 2700 (1951); (c) S. Winstein, A. H. Fainberg, and E. Grunwald, ibid., 79, 4146 (1957).

31. For a discussion of additional factors that may influence the magnitude of m, see J. S. Lomas and J.-E. Dubois, J. Org. Chem., 40, 3303 (1975).

32.* T. W. Bentley and P. v. R. Schleyer, J. Am. Chem. Soc., 98, 7658 (1976).

33.* F. L. Schadt, T. W. Bentley, and P. v. R. Schleyer, J. Am. Chem. Soc., 98, 7667 (1976).

34. M.-F. Ruasse and J.-E. Dubois, J. Am. Chem. Soc., 97, 1977 (1975).

35.* (a) F. G. Bordwell and T. G. Mecca, J. Am. Chem. Soc., 97, 123 (1975); (b) F. G. Bordwell and T. G. Mecca, ibid., 97, 127 (1975); (c) F. G. Bordwell, P. F. Wiley, and T. G. Mecca, ibid., 97, 132 (1975).

36.* (a) K. Okamoto, T. Kinoshita, and U. Osada, J. Chem. Soc., Perkin Trans. 2, 2834 (1975); (b) K. Okamoto, T. Kinoshita, Y. Takemura, and H. Yoneda, ibid., 1426 (1975); (c) K. Okamoto, T. Kinoshita, T. Oshida, T. Yamamoto, Y. Ito, and M. Dohi, ibid., 1617 (1976).

37. J. A. Bone, J. R. Pritt, and M. C. Whiting, J. Chem. Soc., Perkin Trans.2, 1447 (1975).

38. C. D. Ritchie, Acc. Chem. Res., 5, 348 (1972).

39.* C. D. Ritchie, D. J. Wright, D.-S. Huang, and A. A. Kamego, J. Am. Chem. Soc., 97, 1163 (1975).

40.* C. D. Ritchie, J. Am. Chem. Soc., 97, 1170 (1975).

41.* A. Pross, J. Am. Chem. Soc., 98, 776 (1976); see also A. Pross, Tetrahedron Lett., 1289 (1975).

42. R. J. Sujdak, R. L. Jones, and L. M. Dorfman, J. Am. Chem. Soc., 98, 4875 (1976).

43.* K. Hillier, J. M. W. Scott, D. J. Barnes, and F. J. P. Steele, Can. J. Chem., 54, 3312 (1976).

44. (a) Z. Rappoport, Acc. Chem. Res., 9, 265 (1976); (b) M. Hanack, ibid., 9, 364 (1976).

45.* (a) C. G. Swain, J. E. Sheats, and K. G. Harbison, J. Am. Chem. Soc., 97, 783 (1975); (b) C. G. Swain, J. E. Sheats, D. G. Gorenstein, and K. G. Harbison, ibid., 97, 791 (1975); (c) C. G. Swain, J. E. Sheats, and K. G. Harbison, ibid., 97, 796 (1975); (d) C. G. Swain and R. J. Rogers, ibid., 97, 799 (1975).

46. (a) G. A. Olah and J. Welch, J. Am. Chem. Soc., 97, 208 (1975); (b) G. A. Olah and J. L. Grant, ibid., 97, 1546 (1975).

47.* R. Bergstrom, R. G. M. Landells, G. H. Wahl, Jr., and H. Zollinger, J. Am. Chem. Soc., 98, 3301 (1976).

48. R. W. Taft, J. Am. Chem. Soc., 83, 3350 (1961); see also R. A. Abramovitch and J. G. Saha, Can. J. Chem., 43, 3269 (1965); R. A. Abramovitch and F. F. Gadallah, J. Chem. Soc. (B), 497 (1968); T. Cohen and J. Lipowitz, J. Am. Chem. Soc., 86, 2514 (1964).

49.* H. H. Jaffé and G. F. Koser, J. Org. Chem., 40, 3082 (1975).

50. A. Cox, T. J. Kemp, D. R. Payne, M. C. R. Symons, D. M. Allen, and P. Pinot de Moira, J. Chem. Soc., Chem. Commun., 693 (1976).

51. E. Wasserman and R. W. Murray, J. Am. Chem. Soc., 86, 4203 (1964).

52. E. M. Evleth and P. M. Horowitz, J. Am. Chem. Soc., 93, 5636 (1971).

53. K. Hiraoka and P. Kebarle, J. Chem. Phys., 63, 394, 1689 (1975).

54.* K. Hiraoka and P. Kebarle, J. Am. Chem. Soc., 98, 6119 (1976).

55. W. A. Lathan, W. J. Hehre, and J. A. Pople, J. Am. Chem Soc., 93, 808 (1971).

56.* K. Hiraoka and P. Kebarle, J. Am. Chem. Soc., 97, 4179 (1975); cf., F. H. Field and D. P. Beggs, ibid., 93, 1585 (1971).

57. G. A. Olah, A. Germain, H. C. Lin, and D. A. Forsyth, J. Am. Chem. Soc., 97, 2928 (1975).

58. G. A. Olah, P. Schilling, J. S. Staral, Y. Halpern, and J. A. Olah, J. Am. Chem. Soc., 97, 6807 (1975).

59.* P. van Pelt and H. M. Buck, J. Am. Chem. Soc., 98, 5864 (1976).

60.* J. Wristers, J. Am. Chem. Soc., 97, 4312 (1975).

5

FREE RADICALS

LEONARD KAPLAN

Union Carbide Corp.,
P. O. Box 8361, S. Charleston, W. Va. 25303

I. STRUCTURE OF PROTOTYPAL RADICALS

A. Methyl

The photoelectron spectrum of $CH_3\cdot$, produced by pyrolysis of $CH_3N{=}NCH_3$ at 670°, contained extremely weak bands that were considered to be members of a vibrational progression with a separation of ≃720 cm^{-1}.[1] They were assigned, reasonably, as being associated with the out-of-plane C-H bending of CH_3^+, and it was concluded from that assignment that "there must be a change in the degree of planarity upon going from <u>the</u> [emphasis added] methyl radical to the methyl cation." It was assumed that CH_3^+ is planar, and the vibrational progression was taken to indicate a small degree of nonplanarity, estimated to be ≃5°, for $CH_3\cdot$.

Proof of the nonplanarity of $CH_3\cdot$ would be a most significant development. Although the sequence of interpretations is logically consistent with the results, an indication based on those results, even on granting of the assignment and presumption regarding structure of CH_3^+, is that <u>a</u> methyl radical be nonplanar. In view of the high temperature of the experiments and the relative insensitivity of the energy of $CH_3\cdot$ to its structure,[2a] the presence of methyl radicals of different structure would be anticipated; the species that exhibited the vibrational progression could be thermally populated methyls, not of lowest energy, whose presence would be expected. Indeed, if the preferred geometry of $CH_3\cdot$ is planar, and if its energy increases by as much as 1/2 kcal/mol per degree of deviation from planarity (θ),[2a] the fraction of thermally equilibrated species of $\theta = 5°$ present at 670° would be a significant fraction of that of $\theta = 0°$.[3]

B. Ethyl

The photoirradiation of RCO_2O_2CR' in an inert matrix at low temperature has been found to produce radicals in concentrations sufficient for the convenient observation of their IR spectra.[4] The technique was calibrated by the use of acetyl benzoyl peroxide and produced the well-known IR spectrum of the methyl radical.

A transient molecule, reasonably considered to be $CH_3CH_2\cdot$, was produced from $CH_3CH_2CO_2O_2CR$. A comparison of its absorption bands with those of $CH_2{=}CD_2$ [3113 and 3033 vs. 3095 and 3016 cm^{-1} (assigned to C-H stretching) and 1366 vs. 1384 cm^{-1} (assigned to in-plane scissor deformation), respectively] combined with the reasoning that similarity of force constants indicates similarity of electronic structure, which indicates similarity of bond angles and distances, leads to a consistent assignment to planar $MeCH_2\cdot$ as the carrier. Earlier ESR work is also consistent with the proposition that ethyl is planar.[2b,5]

C. *t*-Butyl

Electron spin resonance work on the structure of $(CH_3)_3C\cdot$ continues.[6] We believe the objections of Symons[6f] to be reasonable and sound, given the results[6a] on which they were based; however, subsequent results[6b,c,e] supersede his comments and do provide substantial support for the proposition that $(CH_3)_3C\cdot$ does not have a structure that is planar at the radical site when time-averaged over all rotamers. Note that, in the limit of a "high" barrier to C-CH_3 rotation. $(CH_3)_3C\cdot$ can be planar at the radical site only accidentally, since the plane of the four carbons need not, in any one rotamer, be a plane of symmetry of the molecule. Also, $(CH_3)_3C\cdot$ was considered to have inversional symmetry, and a barrier to inversion of ≃600 cal/mol was calculated.[6c] On the time scale of such an inversion, $(CH_3)_3C\cdot$ may well not have inversional symmetry.

The photoelectron spectrum of $(CH_3)_3C\cdot$ has been found to be consistent, subject to our discussion in Sections I.A and I.C, with *a* *t*-butyl radical being nonplanar.[1]

II. CHARACTERISTICS OF PROTOTYPAL RADICALS

A. Ultraviolet Spectra of Isopropyl and _t_-Butyl

Photolysis of $(CH_3)_3CN{=}NC(CH_3)_3$[7] and of $(CH_3)_2CHN{=}NCH(CH_3)_2$[7b] in the gas phase produced transient molecules whose UV spectra, reasonably assigned to $(CH_3)_3C\cdot$ and, somewhat less securely, $(CH_3)_2CH\cdot$, respectively, were broad continua from 220 to 270 mμ, consisting of a more intense absorption at ≈230 to 235 mμ and a weaker one at ≈250 to 255 mμ.

B. Infrared Spectrum of Phenyl

Photoirradiation of $PhCO_2O_2CR$ in an inert matrix at low temperature (see Section IB) permitted observation of a transient species exhibiting a band at 710 cm^{-1}. This was reasonably attributed to an absorption (considered to correspond to a C-H out-of-plane bending mode) of the phenyl radical, a molecule not previously detected by IR spectroscopy.[4]

C. A Radical Anion Comprised of Tetracoordinated Carbons Only

The ESR spectrum observed on γ-irradiation of a solid solution of perfluorocyclobutane in neopentane at 77 K was attributed to the radical anion of perfluorocyclobutane.[8] Although electron attachment to perfluorocyclobutane is a known process,[9] the title class of compounds, including _c_-$C_4F_8^{\cdot-}$, was previously unknown. The assignment would be more secure and the possibility of error[10] lessened if systems containing isotopically substituted substrate and matrix were examined. Study of a tetrahedrally symmetric fluorocarbon, such as 1,3,5,7-tetrafluoroadamantane,[11] could be of interest.

D. Near-infrared Spectra of Peroxyl Radicals

Oxygen-containing compounds, the IR absorptions of which appeared in the region 5000 to 10000 cm^{-1}, were generated in the gas phase by means of Hg-photosensitized decomposition of a precursor (in parentheses following each radical listed here) in the presence of O_2.[12] Assignments, somewhat weak, were made to $CH_3OO\cdot$ (Me_2CO), $CH_3CH_2OO\cdot$ (EtN=NEt, Et_2CO), $(CH_3)_2CHOO\cdot$ (C_3H_8), and $CH_3CO_3\cdot$ (Me_2CO, MeCOCOMe).

E. The Isoelectronic Series $H_2CN\cdot$, $H_2CO^{+\cdot}$, $H_2BO\cdot$

The low temperature γ-irradiation of CH_2O in H_2SO_4 or D_2SO_4 was reported to produce a radical the ESR spectrum of which is consistent with one reasonable for $H_2CO^{+\cdot}$, the cation radical of formaldehyde.[13] After warming of the mixture the spectrum disappeared as that of HCO grew to a final intensity equal to that of the initial spectrum of "$CH_2O^{+\cdot}$".

Vaporization at 2000 to 2100° of elemental boron in the presence of H_2 and codeposition with Ar into a matrix at 4 K produced a paramagnetic material considered to be $H_2BO\cdot$;[14] we regard the assignment as weak. These spectra are described, along with that of $H_2CN\cdot$, in Table 1.

TABLE 1
ESR Spectra

	$H_2CN^{\cdot}$	$H_2CO^{+\cdot}$	$H_2BO\cdot$
a_H, gauss	87.4[15a], 87.5[15b], 91.2[15c], 89.0[15d], 88[15e], 87.2[15f], 87.4[15g], 91[15h]	90.3	130.
g	2.002[15a], 2.0031[15c], 2.0028[15f], 2.0025[15g], 2.0034[15h], 2.0025[15b]	2.017	2.008

F. Cation Radical of a Sulfoxide

Low-temperature γ-irradiation of $(CH_3)_2SO$ produced a radical, the ESR spectrum of which is not inconsistent with one reasonable for $(CH_3)_2SO^{+\cdot}$, the cation radical of DMSO.[16]

III. ENERGETICS OF PROTOTYPAL RADICALS

A. D(PhO-H), ΔH_f(PhO•)

The oxygen-hydrogen bond dissociation energy of phenol has been the subject of several highly uncertain determinations and estimates.[17] The problem has recently been approached by combination of K for ($RO_2\cdot$ + PhOH $\rightleftarrows$ RO_2H + PhO•) and D(ROO-H), itself determined by combination (Ar = 2,4,6-tri-*t*-butylphenoxy) of K for ($RO_2\cdot$ + ArOH $\rightleftarrows$ RO_2H + ArO•) and D(ArO-H), to give D(PhO-H) = 88 kcal/mol.[18] Unfortunately, the first-mentioned K and D are both dependent on long, involved, multistep, and hence tenuous, lines of support. It is regrettable that a quantity as basic as the heat of formation of phenoxyl is not known with greater confidence.

B. $D(R_3\overset{+}{N}\text{-}H)$

Proton affinities of amines ($H^+ + R_3N \rightarrow R_3\overset{+}{N}H$), determined by equilibrium ion cyclotron resonance techniques, were combined with ionization protentials of amines ($R_3N \rightarrow R_3N^{+\cdot} + e^-$), determined by photoelectron spectroscopy, to yield $D(R_3\overset{+}{N}\text{-}H)$, the nitrogen-hydrogen bond-dissociation energies ($R_3\overset{+}{N}H \rightarrow R_3N^{+\cdot} + H\cdot$), of many aliphatic and alicyclic ammonium ion.[19] Although the absolute values are uncertain by much more than the error usually associated with the determination of the dissociation energy of a bond in a neutral molecule, the relative values are probably accurate to within a few kcal/mol.

C. The 3-Cyclopropenyl Radical

The 3-cyclopropenyl radical is potentially the simplest fully conjugated cyclic-D_{nh} "resonance-stabilized" radical; an indication of the degree of realization of this potential would thus be of great interest. The reduction potentials of a series of R^+ have been determined and combined with known pK_{R^+} values of ROH to give a measure of D(R-OH).[20] Cyclopropenyl, trityl, and tropylium tetrafluoroborates were among those examined in CH_3CN, with the results shown in Table 2. Possible experimental problems and ambiguities are discussed here[20] and elsewhere.[21]

TABLE 2
Relative "Bond Dissociation Energies"

R	$ROH \rightarrow R\cdot + HO\cdot$, kcal/mol
3-Cyclopropenyl	+ 22
Triphenylmethyl	(0)
7-Cycloheptatrienyl	-5

Unfortunately, $D(Ph_3C\text{-}OH)$ is unavailable because the heat of sublimation of Ph_3COH is unknown. Consequently, D[(3-cyclopropenyl-OH] cannot be compared to D[(alkyl)-OH] or D[(cycloalkyl)-OH]. Instead, a scale of relative changes in "resonance energy" as $ROH \rightarrow R\cdot$ was proposed from which it was concluded that D[(3-cyclopropenyl)-OH] is abnormally high, a result for which there is no unique "explanation." Abnormality *per se* does not imply a particular "origin" of the abnormality; here, "abnormally" high "stability" of ROH and "abnormally" low "stability" of R· cannot be distinguished. As Wasielewski and Breslow state, "In any case, no large amount of stabilization is detectable in the cyclopropenyl radical from these data."[20]

D. Stability and Selectivity

By presenting a temperature-dependent ordering of a series of radicals according to their selectivity of reaction with $CCl_4/BrCCl_3$, the ordering thus being a function of the temperature chosen for comparison, Giese[22] supplies another prima facie case, using an example in free-radical chemistry, for the illegitimacy of determining an order of stability from an order of selectivity without regard to the relation of the operating temperature to an "isokinetic" ("isoselective") temperature, a difficulty encountered not infrequently in studies of the chemistry of ions. This would constitute another indication that, if something is to be determined, it should be measured, rather than something else whose relationship to the desired property is presumed to be understood.

Data were obtained from a study of the reaction of $BrCCl_3/CCl_4$ with R•, generated from RCO_3-*t*-Bu, and with the intermediate, presumed to be R•, produced in the reacting system RHgX-$NaBH_4$-$BrCCl_3$-CCl_4.

> The selectivities of the radicals . . . lie on a straight line . . . in the Eyring plot in a temperature range between -20° and +130°C . . . Above and below [50° to 60°] . . . the sequences of selectivities are reversed. At 0°C . . . *tert*-undecyl radical is considerably more selective than the methyl radical. . . . 2-*tert*-Butylphenyl radical is more selective than the unsubstituted vinyl radical. At 130°C, on the other hand, the methyl radical is the most selective π radical and the vinyl radical the most selective σ radical . . . Furthermore, above 80°C the differences in selectivity do not decrease as expected . . . with rising temperature; instead they become greater.[22]

There are some problems of experimentation and interpretation. The reactions involving RHgX are reported to proceed "instantaneously" at the lowest temperature of the studies. Since the report is a communication, it lacks experimental details. Particularly important regarding the instantaneous reactions would be experiments proving that the reaction was conducted so as to permit a true kinetic competition and provide a true measure of the selectivity of the intermediate; otherwise, the "selectivities" need not be inherent, and could be confounded by factors such as relative rates of diffusion, rates of mixing, local selective depletion of trapping agent, and inhomogeneities in temperature.

The presentation and subsequent interpretation of data is misleading. Key elements of the specific verbal arguments (quoted in the preceding paragraphs) are unreal. The following comparisons of selectivity are made: (a) (t-undecyl·, 0°) > (CH_3·, 0°), (b) (CH_3·, 130°) > (t-undecyl·, 130°), (c) (2-t-Bu-Ph·, 0°) > (vinyl·, 0°), and (d) (vinyl·, 130°) > (2-t-Bu-Ph·, 130°). The quantities underlined were not actually determined experimentally.[22]

In the diagram provided, many of those lines having great visual impact and serving to dramatize inversions in selectivity are actually extrapolations into the key regions. Those radicals whose selectivities actually are reported at both high and low temperatures are presented in Table 3. The only entry significantly out-of-line, that is, the relative selectivity of which is significantly dependent on temperature, is methyl.

TABLE 3
Radical Selectivities

Radical	log(Selectivity, 130°)	log (Selectivity, 0°)[22]
2-Octyl	2.76[24]	3.8
1-Octyl	2.89[24]	3.6
Methyl	3.02[23]	3.5
7-Norbornyl	2.23[24]	3.2
Cyclopropyl	2.4 (extrapolated)[24]	2.8

In summary, the results are stimulating, but should be accepted only conditionally, pending examination of a full paper.

IV. EXPERIMENTAL TECHNIQUES

An ultimately useful experimental or mechanistic (Section V) technique generally passes through several stages of development, including: (a) a preutility stage, during which one is confined to studying the technique, rather than using it to answer questions (i.e., one is a prisoner of the processes and questions that emerge *a posteriori* from application of the technique), (b) a stage of selective or occasional utility, during which it is sometimes applicable to questions and systems of one's choosing but is more generally brought to bear on whatever happens to be going on during its application, and (c) a stage of widespread utility, during which its application results in answering preconceived questions rather than in learning more about the technique itself. For example, CIDNP

(Section V) is now in stage (b), whereas NMR, in less than 20 years, progressed from being a phenomenon of interest only to physicists to having long been used routinely by college undergraduates [i.e., beyond stage (c)]. Now in stage (a) is the work of Clough on intramolecular rotational tunneling in free radicals and its interaction with other processes.[25]

Photo-and γ-irradiation are frequently used stage (b) techniques of generating radicals in matrices for study by the use of ESR spectroscopy. Most of the large amount of such work, although fruitful, has been executed myopically. Lines of reasoning and experimentation have been extended far beyond any secure base of knowledge and understanding. Willard's work[26] has been concerned with the ". . . investigation of the nature, yields, and decay mechanisms of trapped intermediates produced . . . [in] glassy and polycrystalline organic matrices."[26a] While they are now out of the mainstream of organic free-radical chemistry, there will come a time of reckoning during which such studies of the intimate details of processes occurring during and as a result of such irradiation will be very useful as support for radiolysis/photolysis as a technique of generation, ESR spectroscopy as a technique of study, and for whatever is done with a radical once it is generated in a matrix.

The advantages and limitations of optical absorption detection as compared to ESR detection in a flash-photolysis system have been discussed.[27]

V. MECHANISTIC TECHNIQUES

A. Generation of Radicals

The efficient formation of a given carbonium ion has long been a relatively simple matter, the only major requirement being the availability of a practicable synthetic route to a material with a suitable functionality at the future cationic site. Introduction of this material into an appropriate medium usually leads ultimately to the production of the desired carbonium ion. The latter point is almost taken for granted, the major concern being synthesis of a precursor.

The situation with regard to free radicals is very different. Texts on free radicals list many methods of generation; many papers are published in the area of free-radical mechanisms. This might give the impression that, in practice, there are many methods at one's disposal to apply to the particular problem of interest. This is not the case. It may be so for some simple radicals but, for most, especially if they involve complex or functionally substituted molecules, it is not always easy, and sometimes it is very difficult to generate a radical in high yield at a site of one's choosing. Consequently, one often studies what one gets [Section IV, case (b)].

The major drawbacks to the clean and efficient generation of a desired radical are as follows:

1. Some of the more common methods involve the decomposition of a stable compound to a radical, such as acyloxy, acyl, or alkoxy, which subsequently produces the desired radical, R•. In practice, it is quite common for the initially formed radical to enter into reactions other than decomposition to R•, thus decreasing the yield of R•. It is not unusual for these side reactions to predominate.
2. Many methods inherently can lead to extremely complex mixtures. This results, of course, whenever the radical is produced in a chemical environment that does not offer it a clearly preferred reaction path. Decomposition of peroxides, peresters, and azo compounds in the usual "inert" solvents suffers seriously from this probelm; decarbonylation of aldehydes and decomposition of hypohalites do not.
3. Use of a given method almost always forces one to work in a relatively narrow range of temperature, the upper and lower bounds of which are usually determined by the rate of decomposition of a precursor. Whereas if one desires to work in a different range of temperature, he can usually find a method of generation which would allow him to do so, this will usually involve very different chemistry and new difficulties.
4. All methods that generate R• from a molecule producing more than one radical suffer from cage recombination and internal redistribution problems.

What is desired, then, is a method that: (a) cleanly produces the desired radical in high yield without the intermediacy or, ideally, presence of other radicals, (b) results in its conversion to one type of product, and (c) permits a wide choice of operating temperature with the same chemistry and type of reaction system.

Since 1966, when the first papers appeared from our and other laboratories indicating recognition that the use of the reaction of organotin hydrides with organic halides was a means of tactically and cleanly producing a desired radical for the purpose of studying that radical, the use of organotin hydrides as radical-generating agents has increased enormously. This has been with good reason since they do not suffer from drawbacks (1), (2), and (4) and suffer only mildly from (3). [One has a reasonable amount of flexibility in temperature as the nature of the organotin hydride, the starting halide (Cl, Br, or I), and the use of initiation are possible temperature-related variables.]

B. Trapping of Radicals

Trapping is often used if one wants to: (a) simplify the chemistry by converting a system that would give a complex mixture of products in the absence of trapping agent into a simple one, (b) study mechanism and identify intermediates, and (c) produce a desired product in high yield.

One has long been able to "dial a lifetime" for a cleanly produced carbonium ion merely by choice of an appropriate solvent and trapping nucleophile from the large number of available possibilities. Thus one can choose which reactions of a carbonium ion he will permit to occur and which he will suppress. This can be done over an enormous range of lifetimes because of the wide range of available nucleophiles. Also, one deals with the same type of reaction system irrespective of choice of nucleophile.

The situation with regard to free radicals is relatively primitive: (a) a wide range of trapping agents is not available, (b) what range is available is composed of materials of very different chemistry, and one would not always deal with the same type of reaction system, and (c) most important, many of the methods contain the seeds of their own destruction. As an example, suppose that a radical is generated by the decomposition of a peroxide and one attempts to trap one radical by addition of a trapping agent, T:

$$\begin{array}{ccccc} RCO_2\cdot & \xrightarrow{\underline{k}_1} & R\cdot & \xrightarrow{\underline{k}_2} & \text{products} \\ \downarrow \underline{k}_3[T] & & \downarrow \underline{k}_4[T] & & \end{array} \qquad (1)$$

This scheme suffers from difficulties (a) and (b), and there may be an additional one: if almost all radicals R· are to be trapped before they can do anything else, $\underline{k}_4[T]$ must be much greater than $\underline{k}_2$. In practice, if this condition is fulfilled, $\underline{k}_3[T]$ is very often not much less than $\underline{k}_1$. In general, if a trapping agent adequate to trap almost all R· is used, it will also trap the radical precursor of R·, and one may not even be able to observe much of the desired reaction. This problem is intimately related to the general problem of cleanly generating radicals.

Organotin hydrides are useful trapping agents as: (a) they are very efficient,[28] (b) drawbacks (1) and (2) (section V.A) are eliminated together because, in the scheme

$$RX + -\overset{|}{\underset{|}{Sn}}\cdot \rightarrow R\cdot + -\overset{|}{\underset{|}{Sn}}X$$

$$R\cdot + -\overset{|}{\underset{|}{Sn}}H \rightarrow RH + -\overset{|}{\underset{|}{Sn}}\cdot \qquad (2)$$

the species, $-\overset{|}{\underset{|}{Sn}}\cdot$, which cleanly generates R•, is derived from the one that traps it ($-\overset{|}{\underset{|}{Sn}}H$). The addition of a single substance, $-\overset{|}{\underset{|}{Sn}}H$, provides both generating and trapping species, which are interrelated by a simple radical chain process. (c) Drawback (4) does not apply, and (d) difficulties (a) and (b) (Section V.B) are eliminated together, particularly if other Group IV hydrides are included among the trapping agents.

C. Recent Examples of the Generation and Trapping of Radicals

The elimination of drawbacks (1) and (2) (section V.A) is of particular importance in the economical design and choice of radical generation/trapping systems. Another system that both generates and traps radicals by use of the same reagent has recently been described:[29] A variety of halides, RX, was reacted with $Bu_3SnCH_2CH{=}CH_2$ under conditions of thermolysis and photolysis to yield $CH_2{=}CHCH_2R$; here an allyl group in $R_3SnCH_2CH{=}CH_2$ may serve the same function as does H in R_3SnH.

Most of the bases were touched in accumulating the usual assortment of evidence in support of a free-radical mechanism. However, the work is often described imprecisely or vaguely, and experimental details are frequently not provided in this full paper; summary statements substitute for data and detailed procedure. In one experiment, for which details _are_ supplied, production of racemic product from optically active RX, the control experiment to establish the optical stability of RX under the reaction conditions, is inadequate, the all-too-common practice is followed of stating that a material was subjected to the "reaction conditions," although the actual conditions were far from those prevailing during the reaction.

The most direct evidence for the intermediacy of R•, exclusive of its observation and monitoring, would be a trapping experiment. The evidence that is provided is concerned with: (a) the involvement of radicals, not necessarily R•, in the reaction, (b) orders of relative reactivity, and (c) racemization in the course of the formation of product. None of this is as direct as would be a trapping experiment whose outcome is unique to radicals.[30] The reaction of Eq. (3) is taken as

$$\begin{gathered} CH_2{=}CH(CH_2)_4I + Bu_3SnCH_2CH{=}CH_2 \rightarrow \\ Bu_3SnI + (\text{cyclopentyl})CH_2CH_2CH{=}CH_2 \end{gathered} \tag{3}$$

evidence for the trapping of the cyclopentylcarbinyl radical, formed by cyclization of $CH_2{=}CH(CH_2)_3CH_2\cdot$. No support, other than the nature of the product, is provided for this mechanism. A route that combines a C_5-analog of the homo-S_H' reaction,[31] involving an ε-elimination,[32] and a displacement of Sn by I· is not excluded.

The reaction of allylcobaloximes with $BrCCl_3$ and with CCl_4 to give a halocobaloxime plus an allyltrichloromethane have been reported in a communication.[33] "From the influence of initiators, inhibitors, and the nature of the reagent on the rates of the . . . reactions, we believe that they involve chain reactions in which trichloromethyl radicals and cobaloxime(II) complexes are the chain carriers," that is, a mechanism similar to that proposed[29] for the corresponding reactions of allyltributyltin.

Another method of generation is discussed by Danen and Rose:[34] "We have demonstrated . . . that the free radical abstraction of a halogen atom is a convenient and unambiguous method of generating a specific free radical for the purpose of studying homolytic processes . . ."

Another method, the photolysis of halides, had long appeared to be a good method of generation of radicals at preselected sites, particularly useful at lower temperatures where most techniques, being thermal, are inapplicable. Since the findings that photolyses of 1-halonorbornanes,[35a-c] 1-halomethylnorbornanes,[35c] 1,2-diiodo-3,3-dimethylnorbornane,[35d] and 1,2-diiodoadamantane[35d] lead to products derived from the corresponding cations, and reports of similar behavior by 2-_exo_-iodonorbornane,[35a] 3-_endo_-hydroxy-2-iodo-4-_endo_-norbornanecarboxylic acid lactone,[35a] and the corresponding tricyclic deoxo ether,[35a] it has unfortunately become clear that, in general, this technique is of very limited utility in cases where it must be known with confidence that the transient produced is a radical and that the products are radical and that the products are radical-derived. Indeed, ". . . irradiation of alkyl . . . iodides in solution is a convenient and powerful means for the generation of carbocationic intermediates."[35c] Recent work[36] involves also reports of the photolysis of vinyl halides,[37] for which mechanisms involving inter-[37a] and intramolecular[37b] nucleophilic attack on the α-carbon of the ionic-chemistry-prone excited state of the substrate were neither excluded nor mentioned. Photolyses of 1- and 2-halo-,[35c,38] 1,3-dihalo-,[38] and 1,3,5-trihaloadamantanes[38] have also been reported. A factor confounding those experiments involving bridgehead halides and trapping by a nucleophile is the formal possibility that the stereochemical requirements of the S_N process are less severe for the n-σ* excited state of a substrate than for its ground state.

D. Chemically Induced Dynamic Nuclear and Electron Polarization

CIDNP and CIDEP, as both phenomena and techniques, are the most important additions to the methodology of investigation of free radicals and their mechanisms of reaction to appear in a long time. After the usual period of euphoria that follows introduction of new techniques, during which it was felt that a great deal of useful information was obtainable from simplistic interpretation of a few experiments, the new methods are now being applied more cautiously, rigorously, and deliberately as an element in the study of the properties and chemistry of free radicals.

It is becoming more widely recognized that CIDNP is best used in combination with other techniques in order to compensate for its limitations of being able to detect and raise to prominence what are minor reaction pathways, and of being unable, in its common usage, to indicate the relative importance of different pathways.[39]

Relatively few uses of CIDNP have taken it from stage (b) to stage (c) (Section IV). One of the applications to a preconceived mechanistic question (i.e., a question formulated first and then attacked by use of CIDNP, rather than one asked because results of CIDNP studies already available could answer it) is a study of the Stevens rearrangement.[40] Other studies are described by different investigators.[41-46]

VI PERSISTENT RADICALS

Griller and Ingold's work on persistent radicals[47] is important, having provided greatly expanded concepts of the nature and the determinants of the behavior of radicals in solution. Conjugation, a property of the radical and not of its products or the transition states leading to them, previously was thought to be the requirement for persistence; the work illustrates the important role of F-strain in the reference compound in determining "stability" and underscores the distinction between stability [D(R-Y)] and persistence.

Ingold's seminal contributions in this area precede (1973-1974) the period of the present review. Subsequently:

> Once it was appreciated that persistent carbon-centered radicals were produced primarily by steric and not by electronic factors [emphasis deleted], the number and variety of such radicals underwent explosive growth. . . . The reluctance . . . to distinguish between the thermodynamic "stabilization" of a radical and its "persistence" in solution resulted in a very long delay between Gomberg's discovery of triphenylmethyl and the generation of persistent alkyl radicals that were not stabilized by p-π delocalization. However, once it was appreciated that persistence was principally a consequence of steric factors, it became a fairly simple matter to generate almost all types

of carbon-centered radicals in persistent forms. These radicals are so new that their chemical behavior and the bulk of their physical properties are still unknown. The general concept of sterically induced persistence has already been successfully extended to heteroatom-centered radicals.[47]

Recent examples, exclusive of those cited by Griller and Ingold,[47] are reported by other authors.[48]

VII. NEW RADICAL PROCESSES

A. 1,2-Shift of Hydrogen from Carbon to Carbon

"Early claims to the contrary, no bona fide intramolecular vicinal shift of hydrogen in a free radical has been substantiated, particularly in solution. But even in the gas phase, hydrogen atom transfers via transition states smaller than five-membered are not observed."[49]

The history of work directed toward this end, or which purported to achieve it, is involved. Our citation of this reference is not a "cop-out," a means of avoiding responsibility and judgment; although it is somewhat severe on some of the reports of 1,3-shifts, we do accept its substance. The occurrence in a fluid medium of a 1,2-shift of hydrogen, in an alkyl radical in particular, has not been demonstrated. Indeed, the standards against which the credibility of work in this area has come to be judged are such that the claim itself of such an observation is often taken as an indication of experimental or interpretative unreliability of the work.

Gordon, Tardy, et al.,[50] indicating that earlier studies "suffer from chemical complexities," report that the results of the pyrolysis at 520° to 580° of CD_3CH_3 "may be understood in terms of a mechanism involving isomerization of ethyl radical by a 1,2-H(or -D) shift . . ." Unfortunately, their work is similarly afflicted. There are many reaction pathways, each isotopically perturbed; some of them are considered and used selectively as a basis for a necessarily complex treatment and interpretation of data. This is an almost inevitable consequence of studying, at elevated temperature, a reacting system, all options of which are not normally preferred free radical processes. Such a situation cannot lead to a tightly constructed economical mechanism.

B. Disproportionation of Radicals via α-Abstraction

Proton CIDNP was observed during the reaction of "*t*-BuMgCl" with $BrCCl_3$ in $CH_3OCH_2CH_2OCH_3$.[51]

The phase of polarization (enhanced absorption) of the *t*-BuCl produced (total yield = 1.4%; total material balance on *t*-Bu ≃ 50%) was the same as that of the products formulated plausibly as arising from reactions in the cage of the primary radical pair. The same phase of polarization was observed also during the reaction of "*t*-BuMgCl" with CCl_4 in the presence of a large excess of styrene, conditions designed to lead to the trapping of free *t*-Bu· and hence to the suppression of formation of *t*-BuCl outside of the cage; yields of the products of this reaction are not reported.

From these and from the observations that "the yield of *tert*-butyl chloride . . . depends very little on the concentrations and molar ratios of the reagents [presumably, $BrCCl_3$ and "*t*-BuMgCl"]" and that "the presence of styrene in the reaction mixture increases the polarization signal of . . . *tert*-butyl chloride a little like other products from the primary radical pair," it is concluded that *t*-BuCl is a product of the *t*-Bu· $\cdot CCl_3$ cage, considered to be the parent of the other cage products. It is then tacitly assumed that any reaction of a caged *t*-Bu· must be with its partner in the cage. One specific alternative is considered, reaction of caged *t*-Bu· with tetrahalomethane. It is rejected on the grounds that, in the $BrCCl_3$ case, the product "would be" *t*-BuBr, not *t*-BuCl; the questions of the stability of *t*-BuBr under the reaction conditions and the validity of equating relative amounts of products and relative intensities of their polarization were not considered. Although reaction of $BrCCl_3$ with free *t*-Bu· would lead preferentially to *t*-BuBr, it is not obvious that its reaction with $\overline{t\text{-Bu}\cdot\ \cdot CCl_3}$ also would do so, a similar proposal by Kaptein, Verheus, et al. in the case of $\overline{i\text{-Pr}\cdot\ \cdot CCl_3}$[52] notwithstanding. For example, the relative tendencies for spectator interaction of $\cdot CCl_3$ with Br versus Cl of $BrCCl_3$, combined with the relative tendencies for abstraction of Br versus Cl by *t*-Bu· need not, *a priori*, balance out in favor of abstraction of Br and spectator interaction with Cl. An observation that may have to be overcome when defending this particular alternative is that a 10-fold increase in $[BrCCl_3]$/["*t*-BuMgCl"] "does not lead to a substantial increase in the yield of *tert*-butyl chloride."

It is proposed that

$$\overline{t\text{-Bu}\cdot\ \cdot CCl_3}^{S} \rightarrow t\text{-BuCl} + :CCl_2 \qquad (4)$$

This would be a new type of free-radical reaction in solution. Its energetics, although favorable, are far less so than those of the reactions leading to *t*-$BuCCl_3$ and $CHCl_3 + Me_2C{=}CH_2$. Disproportionation of radicals via α-abstraction is potentially a useful method for generation of halocarbenes and does not require light, a polar solvent, or strong base.

An alternative route for the formation of $:CCl_2$ and polarized t-BuCl is presented wherein all the products of the initial reaction of "t-BuMgCl" + tetrahalomethane remain in the cage until the formation of t-BuCl. The discussion[51] of this variant, which allows for caged sources of chlorine other than $\cdot CCl_3$, is subject to our comments in the preceding paragraphs.

The amount of O_2 present during reaction and the relevance of the report[80] that t-BuOOCl → t-BuCl (polarized in absorption) are unknown as, conversely, is that of the above[51] results to the origin of the t-BuCl considered[80] to have been formed from t-BuOOCl.

It should be noted that other results have been discussed in terms of $\cdot CCl_3$ paired with radicals even more reactive than t-Bu·[52,53] and with radicals of similar or lower reactivity[54] without invoking α-abstraction. Also, formation of $:CCl_2$ was not invoked in discussions of CIDNP observed during the decomposition of pivaloyl peroxide in CCl_4[55] and during the reaction of "t-BuMgBr" with $BrCCl_3$, formulated in terms of $\overline{\text{t-Bu}\cdot\ \cdot CCl_3}$.[56]

The gas-phase reaction in Eqs. (reactions) (5) through (9) have been put forth.

$$2CF_2Cl\cdot \rightarrow :CF_2 + CF_2Cl_2 \qquad (5)^{57}$$

$$2HCF_2\cdot \rightarrow :CF_2 + CF_2H_2 \qquad (6)^{57f,58}$$

$$H_2CF\cdot + HCF_2\cdot \rightarrow :CF_2 + CHF_3 \qquad (7)^{58b}$$

$$CF_3\cdot + HCF_2\cdot \rightarrow :CF_2 + CHF_3 \qquad (8)^{58e,f,59}$$

$$ClCF_2\cdot + FCCl_2\cdot \rightarrow :CF_2 + CFCl_3 \qquad (9)^{57d}$$

VIII. CHARACTERISTICS OF PROTOTYPAL REACTIONS

Now that free-radical chemistry is no longer confined to the back of the book, it need not be whispered that the preferred stereochemistry of the S_H reaction at carbon is as fundamental to organic chemistry as is that of the S_N reaction.

Upton and Incremona have studied the reaction of 1,1-dichlorocyclopropane with halogens.[60] Their work is discussed here in some detail, both because of its potential importance and for didactic purposes. Notes relating to this discussion appear in Section VIII.D.

A. The Reactions

$$\text{1,1-dichlorocyclopropane} + Cl_2 \xrightarrow[CCl_4]{0\text{-}5^\circ,\ h\nu} ClCH_2CH_2CCl_3 + Cl_2CHCH_2CCl_3 +$$

3.1M 0.5M[a] 32-36%[b] 41-45%

$$ClCH_2CHClCCl_3^{c} + \text{1,1-dichlorocyclopropane} \qquad (10)$$

$$\text{1,1-dichlorocyclopropane} + Br_2 \xrightarrow[\text{neat}]{35^\circ,\ h\nu} BrCH_2CH_2CCl_2Br \qquad (11)$$

6.0 mmol, 0.64 mmol, 95%[d]

B. Their Mechanism(s?)

a. Origin of $Cl_2CHCH_2CCl_3$ and $ClCH_2CHClCCl_3$[f]. It is stated that these compounds arose via chlorination of $ClCH_2CH_2CCl_3$, and not via ring opening of 1,1,2-trichlorocyclopropane, based on the reports that: (a) reaction of neat 1,1,2-trichlorocyclopropane with SO_2Cl_2/Bz_2O_2 at 78° gave 1,1,2,2-tetrachlorocyclopropane[e] in unspecified yield[61] and (b) $ClCH_2CH_2CCl_3$ reacted with $Cl_2/h\nu$ to give $Cl_2CHCH_2CCl_3$ and $ClCH_2CHClCCl_3$ in a ratio of 7:1.[g]

b. Origin of $ClCH_2CH_2CCl_3$ and $BrCH_2CH_2CCl_2Br$. It is proposed that

$$X\cdot + \text{1,1-dichlorocyclopropane} \longrightarrow XCH_2CH_2CCl_2\cdot \xrightarrow{X_2} XCH_2CH_2CCl_2X + X\cdot \qquad (12)$$

i. Evidence that Radicals are Involved. Evidence cited for X = Cl consists of the "known"[h] stability of 1,1-dichlorocyclopropane to ionic conditions,[60] that no reaction occurred in the absence of light in either the presence or absence of HCl and that the reaction was strongly retarded by O_2 and benzoquinone, and moderately accelerated by AIBN and Bz_2O_2.

For X = Br, very little reaction occurred in the absence of light at 45°. Bz_2O_2 and O_2 accelerated the reaction; AIBN did so only slightly. Benzoquinone inhibited the reaction slightly.

ii. Are Radicals Involved in that Manner Shown in Eq. (12)?. Might the observed, preferred stereochemistry of the reaction not be that of a homolytic substitution of X· on carbon? For example, if initial attack of X· is at the 1-carbon and the resulting open radical reacts with X_2 faster than it rotates and loses its "stereochemistry," the preferred direction of attack of X_2 on the open radical would be observed, rather than the preferred stereochemistry of initial attack of X· on the 2-carbon of the starting material. Also, if X_2 reacts with a complex of X• and starting material, and not with open $XCH_2CH_2CCl_2•$, the stereochemical fate of which would already have been determined, the S_H2 displacement would not be occurring at all [i] Mechanisms involving reaction of X_2 with a complex of X• and 1,1-dichlorocyclopropane are dismissed "particularly based on the questionable existence of [the cyclopropane, X• complex],"[j] and a weak argument from which it is concluded that such a complex would be attacked by X_2 at the CCl_2 carbon.[60a] It is conceded, however, that "attack by . . . chlorine and bromine fails to [exclude] . . . exclusive operation of . . . [the complexation] mechanism."[60a] Such a mechanism is eliminated from their consideration, based on the report that reaction of 1,1-dichlorocyclopropane with a Br_2/Cl_2 mixture gave $ClCH_2CH_2CCl_2Br$ and not $BrCH_2CH_2CCl_3$. We do not agree that this report excludes the complexation mechanism.

C. Their Stereochemistry

Reference is made to the following scheme and to Eqs. (13) through (16).[60a]

Scheme

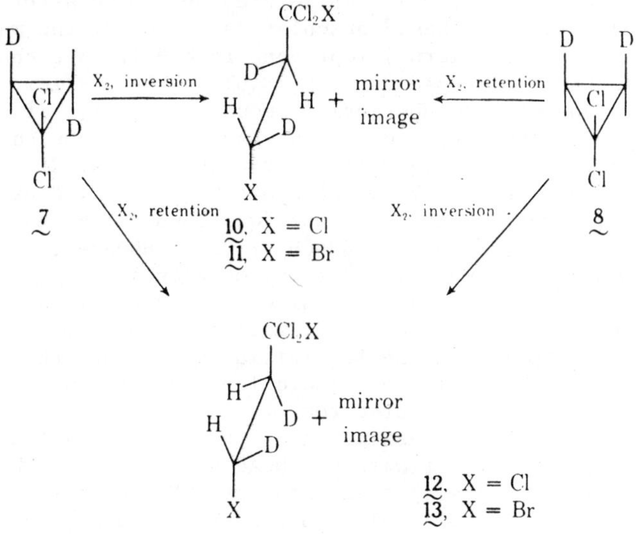

From Upton and Incremona,[60a] reproduced with the permission of the American Chemical Society.

$$\underset{\sim}{7} \xrightarrow[(k)]{Cl_2} 16\% \ \underset{\sim}{10} + 38\% \text{ 1,1,1,3,3-pentachloropropane} \qquad (13)$$

$$\underset{\sim}{10}/\underset{\sim}{12} = 24$$

$$\underset{\sim}{8} \xrightarrow[(k)]{Cl_2} 17\% \; \underset{\sim}{12} + 35\% \text{ 1,1,1,3,3-pentachloropropane} \quad \underset{\sim}{12}/\underset{\sim}{10} = 24 \tag{14}$$

$$\underset{\sim}{7} \xrightarrow[(k)]{Br_2} 67\% \; \underset{\sim}{11}, \; \underset{\sim}{11}/\underset{\sim}{13} = 24 \tag{15}$$

$$\underset{\sim}{8} \xrightarrow[(k)]{Br_2} 55\% \; \underset{\sim}{13}, \; \underset{\sim}{13}/\underset{\sim}{11} = 24 \tag{16}$$

Inversional stereospecificity is high.[1] A *prima facie* case has been presented in support of the proposition that an S_H2 reaction proceeds preferentially with inversion. However, the preferred stereochemistry of an unknown elementary process may have been determined and attention should now turn to the demonstration of mechanism, particularly the number of halogens present in the transition state of the step which seals the stereochemical fate of the product(s).

D. Notes

(a) This was not determined. Amount taken to be the maximum soluble in CCl_4; the solution is presumed to be saturated.

(b) Yields are based on chlorine consumed. The material balance of 1,1-dichlorocyclopropane is ≥ 94%. Note that the remaining ≤ 6% correpsonds to ≤ ~ 40% of the Cl_2 initially present, that is, a substantial amount of the 1,1-dichlorocyclopropane that reacted may not have been accounted for.

(c) Basis of assignment of structure: A. N. Nesmeyanov, R. Kh. Freidlina, and V. I. Firstov, *Dokl. Akad. Nauk SSSR*, *78*, 717 (1951), wherein the structure was assumed, based on the method of preparation ($Cl_3CCH{=}CH_2 + Cl_2$). However, the basis of the comparison of materials is not mentioned.

(d) Yield is based on bromine. The material balance of 1,1-dichlorocyclopropane = 98%; note that the remaining 2% corresponds to 19% of the Br_2 initially present, that is, a substantial amount of the 1,1-dichlorocyclopropane that reacted may not have been accounted for.

(e) The basis of assignment of structure was that the compound did not react with Br_2/HOAc and reacted with aqueous MnO_4^- only very slowly; a satisfactory CH analysis was given.

(f) The basis of assignment of structure was comparison of its properties with those of material described[62] as "probably" $CCl_3CHClCH_2Cl$; see note (c).

(g) Experimental details of this work are not provided. It is stated that the same "photochlorination" reaction was also reported earlier[63] to give the same results; however, the work cited was carried out under the conditions Cl_2/AIBN/90° and gave $Cl_2CHCH_2CCl_3$ and $ClCH_2CHClCCl_3$[c] in the ratio 4.5:1.

(h) What this means is that 1,1-dichlorocyclopropane was inert to concentrated HCl (108°, 9 days)[61] and survived to the extent of 67% when treated with $AcCl{\cdot}AlCl_3$ (refluxing $CHCl_3$, 1 h).[64]

(i) Walling and Fredricks stated that " . . . the possibility that complexes [of Cl• with the cyclopropane] play some role in the low temperature liquid phase reactions of chlorine atoms with cyclopropane cannot be ruled out."[65] Maynes and Applequist characterized, simply by definition, as "fanciful" "a variety of rationales" involving complexing.[60b] Shea and Skell pointed out the possibility of "formation of a complex of cyclopropane and chlorine atom which reacts with Cl_2, inverting one of the methylenes"[60c] and apparently hold, at least in the present context, the view that stereospecificity of a radical addition reaction is not uniquely indicative of mechanism or of the structure of any intermediate.

(j) The references cited in support of this statement are, in our opinion, not relevant.

(k) Isolated yields, based on $Cl_2(Br_2)$. See also notes (b) and (d).

(l) This outcome renders moot serious questions that validly could be raised regarding the isomeric purity of **7** and **8**.

IX. RATES OF PROTOTYPAL REACTIONS

A. Dimerization of Radicals

Dimerization is a basic reaction of radicals. In addition; "Absolute rate constants for many radical-molecule reactions are critically dependent on the rates of combination of the radicals, since many radical-molecule rate constants have been measured relative to the radical combination rates constants. Thermochemical parameters of free radicals can be directly obtained from the Arrhenius parameters of free-radical combination reactions."[66]

Over the last several years, indications that rate constants for the dimerization of hydrocarbon radicals may be different than what was once thought have been emerging with increasing strength and frequency. Recent reports have been confusing, and the outside reader has been faced with a situation much like the one that 8 years ago, prompted Professor Donald Farnum, when asked his "position" on the structure of the 2-norbornyl cation, to remark that it depended on whose JACS communication he read most recently.[67] "Unfortunately . . . problems have only become clearly apparent very recently and all that can be said at present is that the 'correct' values for alkyl recombinations . . . are in a current state of serious uncertainty."[68c]

The situation is now approaching stabilization.

Ethyl,[7b,68,69] isopropyl,[7b,68c,70] and t-butyl[7,66,71,72] radicals dimerize at rates of about $10^{8.6-9.4}$ $M^{-1}s^{-1}$, significantly slower than the collisional frequency, with a small dependence on temperature. The discussions in these papers strikingly illustrate how the mutual interdependence of arguments and conclusions regarding structure, thermodynamics, and kinetics in free-radical chemistry can lead to the revelation of what would otherwise be unapparent problems or errors in one area, as a result of attempts to reconcile differences in another.

B. Rearrangement of Cyclopropylcarbinyl to Allylcarbinyl Radical

The parameters log A (s^{-1}) = 12.5 and E = 5.9 kcal/mol (k = 1.3 x 10^8 s^{-1} at 25°) have been reported for the title reaction, studied by use of ESR spectroscopy.[73]

X. SYNTHETIC APPLICATIONS. INTERCONVERSIONS OF FUNCTIONAL GROUPS

The conversion of alcohols to halides under homolytic conditions may be illustrated as follows:

$$\mathrm{ROH} \longrightarrow \mathrm{RO\overset{O}{\overset{\|}{C}}\overset{O}{\overset{\|}{C}}Cl} \xrightarrow{t\text{-BuOOH}} \mathrm{RO\overset{O}{\overset{\|}{C}}\overset{O}{\overset{\|}{C}}OO\text{-}}t\text{-Bu}$$

$$\xrightarrow[95^\circ]{CCl_4 \text{ or } BrCCl_3 \text{ solvent}} \mathrm{RCl(Br)} \qquad (17)^{74}$$

Examples[74] of this conversion sequence appear in Eqs. (18) through (20).

$$\text{2,2-dimethylcyclohexanol} \longrightarrow \text{1-chloro-2,2-dimethylcyclohexane (44\%)} \qquad (18)$$

$$\mathrm{CH_3CH_2\underset{}{CH}(OH)CH_3} \longrightarrow \mathrm{CH_3CH_2CH(Br)CH_3} \; (25\%) \qquad (19)$$

$$(CH_3)_3C\overset{OH}{\overset{|}{C}}HCH_3 \rightarrow \underset{47\%}{(CH_3)_3C\overset{Cl}{\overset{|}{C}}HCH_3} \qquad (20)$$

(no significant formation of $(CH_3)_2\overset{Cl}{\overset{|}{C}}CH(CH_3)_2$)

The mechanism proposed for these transformations appears in Eq. (21).

$$RO\overset{O}{\overset{\|}{C}}\overset{O}{\overset{\|}{C}}OO\text{-}\underline{t}\text{-Bu} \rightarrow RO\overset{O}{\overset{\|}{C}}\cdot + CO_2 + \underline{t}\text{-BuO}\cdot$$

$$RO\overset{O}{\overset{\|}{C}}\cdot \rightarrow R\cdot + CO_2 \qquad (21)^{74}$$

$$R\cdot \rightarrow RX$$

This mechanism, supported only by the nature of the reaction, and the absence of rearrangements expected to result from the intermediacy of carbonium ions, has as its key element the fragmentation of $RO\underset{O}{\underset{\|}{C}}\cdot$ to $R\cdot + CO_2$, rather than to $RO\cdot + CO$. This would be the first synthetic application of such a reaction.[75] We use the present work as a springboard to expand broadly on the fragmentation step, with considerable documentation, because work on that step *per se* has generally been presented without attention to other approaches to the same intermediates, to the fates of $RO\underset{O}{\underset{\|}{C}}\cdot$ as a general question, or to more recent work involving similar dichotomies in analogous systems.

Reactions formulatable in terms of the fragmentation of $RO\underset{O}{\underset{\|}{C}}\cdot$ to $R\cdot + CO_2$ are divisible into those involving entry into the system via $RO\underset{O}{\underset{\|}{C}}\text{-}X$[75] or via $RO\cdot + CO$;[76] other reactions may be formulated in terms of $RO\overset{O}{\overset{\|}{C}}\text{-}X \rightarrow RO\overset{O}{\overset{\|}{C}}\cdot \rightarrow RO\cdot + CO$.[75c,q]

The reaction $RO\cdot + CO \rightarrow [RO\overset{O}{\overset{\|}{C}}\cdot?] \rightarrow R\cdot + CO_2$ and the well-known reaction, $HO\cdot + CO \rightarrow CO_2 + H\cdot$, have their sulfur analog in the reaction, $RS\cdot + CO \rightarrow [RS\underset{O}{\underset{\|}{C}}\cdot?] \rightarrow R\cdot + COS$, a plausible pathway for the briefly reported reaction, $RSH + CO \rightarrow RH + COS$

[R = $PhCH_2$, C_8H_{17}, C_6H_{13}; 80-300°, 30-150 atm of CO. The reaction is inhibited by ionol, benzoquinone, or transition-metal carbonyls, and accelerated by AIBN or cumyl peroxide. No reaction is observed if CO is replaced by H_2].[77] Possibly related are the reported reactions, R· + COS → RS· + CO; see Eqs. (reactions) (22) through (25):

$$CD_3N{=}NCD_3 \xrightarrow{h\nu} [CD_3\cdot] \xrightarrow{COS} CD_3S\cdot + CO \qquad (22)^{78a}$$

$$H_2 \xrightarrow{h\nu(Hg)} [H\cdot] \xrightarrow{COS} HS\cdot + CO \rightarrow CO + H_2S + S + HgS \qquad (23)^{78b}$$

$$HI \xrightarrow{h\nu} [H\cdot] \xrightarrow{COS} HS\cdot + CO \qquad (24)^{78c}$$

$$H_2 \xrightarrow[\text{discharge}]{\text{microwave}} [H\cdot] \xrightarrow{COS} HS\cdot + CO \rightarrow CO + H_2S \qquad (25)^{78d}$$

Equation (reaction) (26) is described[79a] in terms of Eq. (reaction) (27), and it is also proposed[79b] that the reaction in Eq. (28) occurs during the 2537-Å photolysis of neat CS_2/EtI.

$$ArN_2^+BF_4^- + CS_2 \xrightarrow{NaI} ArSSAr \qquad (26)$$

$$Ar\cdot + CS_2 \rightarrow ArC(=S)S\cdot \rightarrow ArS\cdot + CS \qquad (27)$$

$$CS_2(\text{excited}) + Et\cdot \rightarrow EtS\cdot + CS \qquad (28)$$

In summary, a body of chemistry can be assembled that unifies and generalizes the various routes to and from ROC(=O)· and its sulfur analogs.

XI. ADDENDA

For "honesty in billing," the following article should be commended:

F. C. James and J. P. Simons, "Yet Another Direct Measurement of the Rate Constant for the Recombination of Methyl Radicals," *Int. J. Chem. Kinet.*, *6*, 887 (1974).

For candor in discussion of experimental problems, ambiguities, and sources of error regarding isotope effects in hydrogen atom transfers, the following papers deserve mention:

1. S. Kozuka and E. S. Lewis, "Alkyl Halides with Trialkyltin Hydrides," *J. Am. Chem. Soc.*, *98*, 2254 (1976).

2. E. S. Lewis and M. M. Butler, "The Reactions of Olefins with Mercaptans," J. Am. Chem. Soc., 98, 2257 (1976).
3. E. S. Lewis and K. Ogino, "Radicals from Azo Compounds in Thiophenol," J. Am. Chem. Soc., 98, 2260 (1976), and "Benzylic Hydrogen Abstraction by tert-Butoxy and Other Radicals," J. Am. Chem. Soc., 98, 2264 (1976).

The following papers became available to us too late to be discussed in this volume, but may interest the reader.

1. (a) A. L. Buchachenko, E. M. Galimov, V. V. Ershov, G. A. Nikiforov, and A. D. Pershin, "Isotope Enrichment Induced by Magnetic Interactions in Chemical Reactions," Dokl. Akad. Nauk SSSR (Phys. Chem.), 228, 379 (1976); (b) A. L. Buchachenko, "Enrichment of Magnetic Isotopes in Chemical Reactions," in "Magnetic Effects in Chemical Reactions," Usp. Khim., 45, 761 (1976).
2. P. Toffel and A. Henglein, "The Temperature Dependence of The Electrolytic Dissociation of Some 1-Hydroxy-alkyl Radicals in Aqueous Solution (A Pulse Radiolysis Study)," Ber. Bunsenges. Phys. Chem., 80, 525 (1976).
3. J. A. Den Hollander and R. Kaptein, "Radical Pair Substitution in CIDNP, Spin-Uncorrelated Germinate Radical Pairs," Chem. Phys. Lett., 41, 257 (1976) [cf. H. C. Box and E. E. Budzinsk, "Primary Radiation Damage in Thymidine," J. Chem. Phys., 62, 197 (1975); H. C. Box and E. E. Budzinski, "Electron Spin Resonance Detection of Radicals RS•, RO•, and $R'SR^{++}$ in Irradiated Solids," JCS Perkin II, 553 (1976); S. Schlick and L. Kevan, "Spin Trapping of Radicals Formed in Gamma-Irradiated Methanol:Effect of the Irradiation Temperature from 77K to 300K," Chem. Phys. Lett., 38, 505 (1976)].
4. T. Koenig, T. Balle, and J. C. Chang, "Franck Condon factors for the He(I) Photoelectron Spectrum of Methyl Radical," Spectrosc. Lett., 9, 755 (1976). See section 1A.
5. T. A. Claxton, E. Platt, and M. C. R. Symons, "The geometry of t-butyl radicals: an ab initio study," Mol. Phys., 32, 1321 (1976). See section 1C.

XII. REFERENCES

1.* T. Koenig, T. Balle, and W. Snell, J. Am. Chem. Soc., 97, 662 (1975).

2. L. Kaplan, in "Free Radicals," Vol. 2, J. K. Kochi (ed.), Wiley, New York, 1973: (a) p. 369; (b) p. 377.

3.* For more recent and conventionally interpreted work, see:

J. F. Ogilvie, Spectrosc. Lett., 9, 203 (1976); (b) J. Dyke, N. Jonathan, E. Lee, and A. Morris, J. Chem. Soc. Faraday Trans. II, 1385 (1976).

4.* (a) J. Pacansky and J. Bargon, J. Am. Chem. Soc., 97, 6896 (1975); (b) J. Pacansky, G. P. Gardini, and J. Bargon, ibid., 98, 2665 (1976).

5. Y. Ellinger, R. Subra, B. Levy, P. Millie, and G. Berthier, J. Chem. Phys., 62, 10 (1975).

6. (a) D. E. Wood, L. F. Williams, R. F. Sprecher, and W. A. Lathan, J. Am. Chem. Soc., 94, 6241 (1972); *(b) J. B. Lisle, L. F. Williams, and D. E. Wood, ibid., 98, 227 (1976); *(c) P. J. Krusic and P. Meakin, ibid., 98, 228 (1976); (d) M. C. R. Symons, Molec. Phys., 24, 461 (1972); (e) D. E. Wood and R. F. Sprecher, ibid., 26, 1311 (1973); (f) M. C. R. Symons, Tetrahedron Lett., 207 (1973); (g) H. Paul and H. Fischer, Helv. Chim. Acta, 56, 1575 (1973).

7.* (a) D. A. Parkes and C. P. Quinn, Chem. Phys. Lett., 33, 483 (1975); (b) D. A. Parkes and C. P. Quinn, J. Chem. Soc. Faraday Trans. I, 1952 (1976).

8.* M. Shiotani and F. Williams, J. Am. Chem. Soc., 98, 4006 (1976).

9. See, for example, K. M. Bansal and R. W. Fessenden, J. Chem. Phys., 59, 1760 (1973) and F. J. Davis, R. N. Compton, and D. R. Nelson, ibid., 59, 2324 (1973), and references cited therein.

10. See, for instance, M. T. Jones, J. Am. Chem. Soc., 88, 174 (1966).

11. K. S. Bhandari and R. E. Pincock, Synthesis, 655 (1974).

12.* H. E. Hunziker and H. R. Wendt, J. Chem. Phys., 64, 3488 (1976).

13.* S. P. Mishra and M. C. R. Symons, Chem. Commun., 909 (1975).

14.* W. R. M. Graham and W. Weltner, Jr., J. Chem. Phys., 65, 1516 (1976).

15. (a) E. L. Cochran, F. J. Adrian, and V. A. Bowers, J. Chem. Phys., 36, 1938 (1962); (b) I. S. Ginns and M. C.

R. Symons, J. Chem. Soc. Dalton Trans., 185 (1972); (c) J. A. Brivati, K. D. J. Root, M. C. R. Symons, and D. J. A. Tinling, J. Chem. Soc. (A), 1942 (1969); (d) K. V. S. Rao and M. C. R. Symons, ibid., 2163 (1971); (e) A. Forchioni and C. Chachaty, C. R. Acad. Sci. (Paris), (C), 264, 637 (1967); (f) D. Behar and R. W. Fessenden, J. Phys. Chem., 76, 3945 (1972); (g) M. Fujiwara, N. Tamura, and H. Hirai, Bull. Chem. Soc. Jap., 46, 701 (1973); (h) D. Banks and W. Gordy, Molec. Phys., 26, 1555 (1973).

16.* M. C. R. Symons, J. Chem. Soc. Perkin Trans. II, 908 (1976).

17. (a) N. S. Hush, J. Chem. Soc., 2375 (1953); (b) D. H. Fine and J. B. Westmore, Chem. Commun., 273 (1969) and references cited therein; (c) S. Paul and M. H. Back, Can. J. Chem., 53, 3330 (1975).

18.* L. R. Mahoney and M. A. DaRooge, J. Am. Chem. Soc., 97, 4722 (1975).

19.* D. H. Aue, H. M. Webb, and M. T. Bowers, J. Am. Chem. Soc., 98, 311 (1976).

20.* M. R. Wasielewski and R. Breslow, J. Am. Chem. Soc., 98, 4222 (1976).

21. A. M. Bond and O. E. Smith, Anal. Chem., 46, 1946 (1974) and references cited therein.

22.* B. Giese, Angew. Chem. Int. Ed. Engl., 15, 173 (1976); see also ibid., 15, 688 (1976).

23. From decomposition of t-butyl peracetate: B. Giese, private communication.

24. K. Herwig, P. Lorenz, and C. Rüchardt, Chem. Ber., 108, 1421 (1975).

25.* For example, see (a) S. Clough and J. R. Hill, "Tunnelling Resonance in Electron Spin-Lattice Relaxation," J. Phys. C, 8, 2274 (1975); (b) S. Clough, T. Hobson, and S. M. Nugent, "Change of Frequency of Methyl Group Tunneling Rotation Following Deuteration," ibid., 8, L95 (1975); (c) S. Clough and T. Hobson, "Double-resonance Detection of Tunnelling Sidebands of Electron Spin Resonance Spectrum," ibid., 8, 1745 (1975); (d) S. Clough and J. W.

Hennel, "Evidence for Thermal Quenching of Methyl-group Tunnelling Rotation," ibid., 8, 3115 (1975); (e) S. Clough, "Nuclear Spin-lattice Relaxation of Tunnelling Molecular Groups," ibid., 9, 1553 (1976); (f) S. Clough, J. R. Hill, and M. Punkkinen, "The Measurement of Methyl Group Tunnelling Frequencies by ENDOR," Proc. 18th (1974) Ampere Congr., Magnet. Reson. Relat. Phenom., 2, 389 (1975); (g) S. Clough and J. R. Hill, "The Induction of Dipolar Polarization by Spin Symmetry Conversion of Tunnelling Methyl Groups," ibid., 2, 379 (1975); (h) S. Clough and T. Hobson, "Tunnelling Magnetic Resonances," ibid., 2, 387 (1975); (i) C. Mottley, L. D. Kispert, and S. Clough, "Electron-Electron Double Resonance Study of Coherent and Random Rotational Motion of Methyl Groups," J. Chem. Phys., 63, 4405 (1975); (j) S. Clough, "Tunnelling Effects in Molecular Solids," Proc. 19th (1976) Ampere Congr. Magn. Reson. Relat. Phenom., p. 85, see also R. Srinivasan, in Magnetic Resonance, C. A. McDowell (ed.), International Review of Science, Physical Chemistry, Series Two, Vol. 4, Butterworths, 1975, p. 209.

26.* For example, see: (a) G. C. Dismukes and J. E. WIllard, "Radiolytic and Photolytic Production and Decay of Radicals in Adamantane and Solutions of 2-Methyltetrahydrofuran, 2-Methyltetrahydrothiophene, and Tetrahydrothiophene in Adamantane. Conformation Equilibrium of the 2-Methyltetrahydrothiophene Radical" J. Phys. Chem., 80, 1435 (1976); (b) G. C. Dismukes and J. E. Willard, "Reaction Intermediates Produced in 2-Methyltetrahydrothiophene and its Solutions in 2-Methyltetrahydrofuran by γ Radiolysis and Photolysis at 77°K," ibid. 80, 2072 (1976); (c) M. A. Neiss and J. E. Willard, "Effects of Solutes, Deuteration, and Annealing on the Production and Decay of Radicals in γ-Irradiated 3-Methylpentane Glasses," ibid., 79, 783 (1975); (d) J. E. Willard, "Trapped Electrons in Organic Glasses," ibid., 79, 2966 (1975); (e) N. Bremer, B. J. Brown, G. H. Morine, and J. E. Willard, "Mercury-Photosensitized Production of Free Radicals in Organic Glasses," ibid., 79, 2187 (1975); (f) E. D. Sprague and J. E. Willard, "Photostimulated Reactions of Isolated and Paired Radicals in Solid Hydrocarbons," J. Chem. Phys., 63, 2603 (1975); (g) M. A. Neiss, E. D. Sprague, and J. E. Willard, "Decay Mechanisms of CH_3 in 3-Methylpentane and 3MP-d_{14} Glasses at 77°K as Indicated by Methane Yields," ibid., 63, 1118 (1975); (h) S. L. Hager and J. E. Willard, "Heats of Reaction of Trapped Intermediates in γ-Irradiated Organic Glasses and Relaxation Processes in Unirradiated Glasses Measured by Low Temperature Differential Thermal Analysis," ibid.,

63, 942 (1975); (i) D. D. Wilkey and J. E. Willard, "Search for Low Activation Energy Abstraction of H from C-H Bonds by D Atoms in Hydrocarbons at 77°K," ibid., 64, 3976 (1976).

27. J. R. Bolton and J. T. Warden, in Creation and Detection of the Excited State, Vol. 2, W. R. Ware (ed.), Dekker, New York, 1974, p. 63.

28. (a) L. Kaplan, J. Am. Chem. Soc., 88, 4531 (1966); (b) D. J. Carlsson and K. U. Ingold, ibid., 90, 1055 (1968); (c) F. D. Greene and N. N. Lowry, J. Org. Chem., 32, 882 (1967); (d) W. P. Neumann, R. Sommer, and H. Lind, Annalen, 688, 14 (1965) (we apologize for overlooking this work and omitting it from earlier discussions).

29.* J. Grignon, C. Servens, and M. Pereyre, J. Organomet. Chem., 96, 225 (1975) and earlier reports cited therein. See also M. Kosugi, K. Kurino, K. Takayama, and T. Migita, ibid., 56, C11 (1973).

30. M. Julia, Pure Appl. Chem., 40, 553 (1974), particularly p. 561.

31. L. Kaplan Chem. Commun., 754 (1968).

32. L. Kaplan, J. Org. Chem., 32, 4059 (1967).

33.* B. D. Gupta, T. Funabiki, and M. D. Johnson, J. Am. Chem. Soc., 98, 6697 (1976).

34. W. C. Danen and K. A. Rose, J. Org. Chem., 40, 619 (1975). See also L. Kaplan, J. Am. Chem. Soc., 89, 1753 (1967), and subsequent papers.

35.* (a) P. J. Kropp, T. H. Jones, and G. S. Poindexter, J. Am. Chem. Soc., 95, 5420 (1973); (b) G. S. Poindexter and and P. J. Kropp, ibid., 96, 7142 (1974); (c) P. J. Kropp, G. S. Poindexter, N. J. Pienta, and D. C. Hamilton, ibid., 98, 8135 (1976); (d) B. W. Kaplan, Ph.D. thesis, University of Chicago, 1973.

36. See also P. D. Gokhale, A. P. Joshi, R. Sahni, V. G. Naik, N. P. Damodaran, V. R. Nayak, and S. Dev, Tetrahedron, 32, 1391 (1976).

37.* (a) S. A. McNeely and P. J. Kropp, J. Am. Chem. Soc., 98, 4319 (1976); (b) T. Suzuki, T. Sonoda, S. Kobayashi, and H. Taniguchi, Chem. Commun., 180 (1976).

38.* R. R. Perkins and R. E. Pincock, Tetrahedron Lett., 943 (1975).

39. For discussions, see (a) J. F. Garst and C. D. Smith, J. Am. Chem. Soc., 98, 1526 (1976); (b) P. R. Bowers, K. A. McLauchlan, and R. C. Sealy, J. Chem. Soc. Perkin Trans. II, 915 (1976).

40.* (a) W. D. Ollis, M. Rey, I. O. Sutherland, and G. L. Closs, Chem. Commun., 543 (1975); (b) U.-H. Dolling, G. L. Closs, A. H. Cohen, and W. D. Ollis, ibid., 545 (1975).

41. P. W. N. M. Van Leeuwen, R. Kaptein, R. Huis, and C. F. Roobeek, "Photolysis of Methyl(triphenylphosphine)gold(I)," J. Organomet. Chem., 104, C44 (1976).

42. (a) N. A. Porter, J. G. Green, and G. R. Dubay, "^{15}N CIDNP in Azo Compound Decomposition," Tetrahedron Lett., 3363 (1975); (b) N. A. Porter and J. G. Green, "^{15}N Study of the Phenyldiazenyl Radical. A Search for Phenyl Migration, ibid., 2667 (1975); (c) E. Lippmaa, T. Saluvere, T. Pehk, and A. Olivson, "Chemical Polarization of ^{13}C and ^{15}N Nuclei in the Thermal Decomposition of Diazoaminobenzene (1,3-Diphenyltriazene)," Org. Magnet. Res., 5, 429 (1973).

43. J. A. Den Hollander, "Radical Pair Substitution in Chemically Induced Dynamic Nuclear Polarization. Cooperative Effects," Chem. Phys., 10, 167 (1975).

44. (a) B. Blank, A. Henne, G. P. Laroff, and H. Fischer, "Enol Intermediates in Photoreduction and Type I Cleavage Reactions of Aliphatic Aldehydes and Ketones," Pure Appl. Chem., 41, 475 (1975); (b) A. Henne and H. Fischer, "Low Temperature Photochemistry of the Acetone/2-Propanol System," Helv. Chim. Acta, 58, 1598 (1975).

45. C. F. Poranski, Jr., W. B. Moniz, and S. A. Sojka, "Carbon-13 Fourier Transform CIDNP. Kinetics and Mechanism of the Photochemical Decomposition of Benzoyl Peroxide in Chloroform," J. Am. Chem. Soc., 97, 4275 (1975).

46. J. F. Garst, J. A. Pacifici, V. D. Singleton, M. F. Ezzel, and J. I. Morris, "Dehalogenations of 2,3-Dihalobutanes by Alkali Naphthalenes. A CIDNP and Stereochemical Study," J. Am. Chem. Soc., 97, 5242 (1975).

47.* D. Griller and K. U. Ingold, Acc. Chem. Res., 9, 13 (1976).

48. (a) J. W. Cooper, D. Griller, and K. U. Ingold, "Electron Paramagnetic Resonance Spectra and Structure of Some Vinyl Radicals in Solution," J. Am. Chem. Soc., 97, 233 (1975); (b) G. Brunton, D. Griller, L. R. C. Barclay, and K. U. Ingold, "Kinetic Applications of Electron Paramagnetic Resonance Spectroscopy. 26. Quantum-Mechanical Tunneling in the Isomerization of Sterically Hindered Aryl Radicals," ibid., 98, 6803 (1976); (c) K. S. Colle, P. S. Glaspie, and E. S. Lewis, "Equilibrium Dissociation of Triphenylmethyl Dimer," Chem. Commun., 266 (1975); (d) P. G. Cookson, A. G. Davies, and B. P. Roberts, "Electron Spin Resonance Spectra of Cyclohexadienyl Radicals," ibid., 289 (1976); (e) W. Ahrens, K. Schreiner, H. Regenstein, and A. Berndt, "Carbon-13 ESR Coupling Constants of Persistent and Stabilized Allyl Radicals," Tetrahedron Lett., 4511 (1975); (f) H. Hillgärtner, W. P. Neumann, and B. Schroeder, "Bis(trimethyltin)Benzopinacolate, Its Reversable Radical Dissociation and Reactions," Annalen, 586 (1975); (g) W. P. Neumann, B. Schroeder, and M. Ziebarth, "Preparation, Stability and ESR Spectroscopy of Ketyl Radicals $RR'\dot{C}OM(CH_3)_3$," ibid., 2279 (1975); (h) C. Hacquard and A. Rassat, "Study of the 2,4,6-Tri-t-butylpyrylyl Radical by Electron Paramagnetic Resonance. Calculation of the Couplings," Molec. Phys., 30, 1935 (1975); (i) H. Sakurai, I. Nozue, and A. Hosomi, "Silyl Radicals. XVI. An Electron Spin Resonance Study on the Homolytic Aromatic Silylation," Chem. Lett., 129 (1976). (j) K. Schreiner and A. Berndt, "Electronic Structure of Tri-tert-butylcyclopropenyl," Angew. Chem. Int. Ed. Engl., 15, 698 (1976); (k) A. Hudson, M. F. Lappert, and P. W. Lednor, "Subvalent Group 4B Metal Alkyls and Amides, Part 4. An Electron Spin Resonance Study of some Long-lived Photochemically Synthesised Trisubstituted Silyl, Germyl, and Stannyl Radicals," J. Chem. Soc. Dalton Trans, 2369 (1976).

49. J. W. Wilt, in Free Radicals, Vol. I, J. K. Kochi (ed.), Wiley, New York, 1973, p. 378.

50.* A. S. Gordon, D. C. Tardy, and R. Ireton, J. Phys. Chem., 80, 1400 (1976).

51.* V. I. Savin, A. G. Abul'khanov, and Yu. P. Kitaev, Zh. Org. Khim., 12, 484 (1976). All quotations are taken from J. Org. Chem. USSR, 12, 479 (1976).

52. R. Kaptein, F. W. Verheus, and L. J. Oosterhoff, Chem. Commun., 877 (1971).

53. (a) R. Kaptein, M. Fráter-Schröder, and L. J. Oosterhoff, Chem. Phys. Lett., 12, 16 (1971); (b) J. A. Den Hollander, R. Kaptein, and P. A. T. M. Brand, ibid., 10, 430 (1971); (c) J. A. Den Hollander, Ind. Chim. Belg., 36, 1083 (1971); (d) H. D. Roth, ibid., 36, 1068 (1971); (e) H. D. Roth, J. Am. Chem. Soc., 93, 1527, 4935 (1971).

54. (a) C. W. Funke, J. L. M. de Boer, J. A. J. Geenevasen, and H Cerfontain, J. Chem. Soc. Perkin Trans. II, 1083 (1976); (b) C. Walling and A. R. Lepley, J. Am. Chem. Soc., 94, 2007 (1972); (c) M. L. Kaplan and H. D. Roth, Chem. Commun., 970 (1972); (d) G. A. Nikiforov, Sh. A. Markaryan, L. G. Plekhanova, B. D. Sviridov, S. V. Rykov, V. V. Ershov, A. L. Buchachenko, T. Pehk, T. Saluvere, and E. Lippmaa, Org. Mag. Res., 5, 339 (1973).

55. R. Kaptein, in Chemically Induced Magnetic Polarization, A. R. Lepley and G. L. Closs (eds), Wiley, New York, 1973, p. 163.

56. R. G. Lawler, in Progress in Nuclear Magnetic Resonance Spectroscopy," Vol. 9, J. W. Emsley, J. Feeney, and L. H. Sutcliffe (eds.), Pergamon, Oxford, 1973, p. 186.

57. (a) J. R. Majer, C. Olavesen, and J. C. Robb, Trans. Faraday Soc., 65, 2988 (1969); (b) J. R. Majer, D. Phillips, and J. C. Robb, ibid., 61, 110 (1965); (c) R. Bowles, J. R. Majer, and J. C. Robb, ibid., 58, 1541; (d) 2394 (1962); (e) M. G. Bellas, O. P. Strausz, and H. E. Gunning, Can. J. Chem., 43, 1022 (1965); (f) G. O. Pritchard and M. J. Perona, J. Phys. Chem., 73, 2944 (1969).

58. (a) G. O. Pritchard and M. J. Perona, Int. J. Chem. Kinet., 1, 413; (b) 509 (1969); (c) J. A. Kerr and D. M. Timlin, ibid., 3, 1 (1971); (d) G. O. Pritchard and J. T. Bryant, J. Phys. Chem., 72, 1603 (1968); (e) G. O. Pritchard and J. T. Bryant, ibid., 70, 1441 (1966); (f) G. T. Bryant and G. O. Pritchard, ibid., 71, 3439 (1967); (g) J. B. Hynes, R. C. Price, W. S. Brey, Jr., M. J. Perona, and G. O. Pritchard, Can. J. Chem., 45, 2278 (1967).

59. M. J. Perona, J. T. Bryant, and G. O. Pritchard, unpublished data cited in reference 58d.

60.* (a) C. J. Upton and J. H. Incremona, J. Org. Chem., 41, 523 (1976); (b) G. G. Maynes and D. E. Applequist, J. Am. Chem. Soc., 95, 856 (1973) and (c) K. J. Shea and P.

S. Skell, ibid., 95, 6728 (1973).

61. P. G. Stevens, J. Am. Chem. Soc., 68, 620 (1946).

62. H. Gerding and H. G. Haring, Rec. Trav. Chim., 74, 841 (1955).

63. Ya. I. Rotshtein and N. G. Sokolovskaya, Zh. Org. Khim., 7, 26 (1971).

64. H. Hart and G. Levitt, J. Org. Chem., 24, 1261 (1959).

65. C. Walling and P. S. Fredricks, J. Am. Chem. Soc., 84, 3326 (1962).

66. K. Y. Choo, P. C. Beadle, L. W. Piszkiewicz, and D. M. Golden, Int. J. Chem. Kinet., 8, 45 (1976).

67. D. G. Farnum, 21st National Organic Chemistry Symposium, American Chemical Society, Salt Lake City, June 1969.

68.* (a) B. Hickel, J. Phys. Chem., 79, 1054 (1975); (b) D. G. Hughes and R. M. Marshall, J. Chem. Soc. Faraday Trans. I, 413 (1975) and references cited therein; (c) D. L. Allara and D. Edelson, Int. J. Chem. Kinet., 7, 479 (1975).

69.* D. M. Golden, K. Y. Choo, M. J. Perona, and L. W. Piszkiewicz, Int. J. Chem. Kinet., 8, 381 (1976).

70.* K. R. Bull, R. M. Marshall, and J. H. Purnell, Proc. Roy, Soc., A342, 259 (1975).

71.* R. M. Marshall, H. Purnell, and P. D. Storey, J. Chem. Soc. Faraday Trans. I, 85 (1976) and references cited therein.

72.* H. Schuh and H. Fischer, Int. J. Chem. Kinet., 8, 341 (1976).

73.* B. Maillard, D. Forrest, and K. U. Ingold, J. Am. Chem. Soc., 98, 7024 (1976).

74.* F. R. Jensen and T. I. Moder, J. Am. Chem. Soc., 97, 2281 (1975).

75. (a) M. J. Perkins and B. P. Roberts, J. Chem. Soc. Perkin Trans. II, 297 (1974); (b) D. Griller and B. P. Roberts, ibid., 747 (1972); (c) R. Louw, M. van den Brink, and

H. P. W. Vermeeren, ibid., 1327 (1973); (d) P. Beak and S. W. Mojé, J. Org. Chem., 39, 1320 (1974); (e) W. S. Trahanovsky, J. A. Lawson, and D. E. Zabel, ibid., 32, 2287 (1967); (f) P. Cadman, A. J. White, and A. F. Trotman-Dickenson, J. Chem. Soc. Faraday Trans. I, 506 (1972); (g) R. K. Solly and S. W. Benson, Int. J. Chem. Kinet., 1, 427 (1969); (h) D. Griller and B. P. Roberts, Chem. Commun., 1035 (1971); (i) H. G. Kuivila and E. J. Walsh, Jr., J. Am. Chem. Soc., 88, 571 (1966); (j) P. D. Bartlett, B. A. Gontarev, and H. Sakurai, ibid., 84, 3101 (1962); (k) J. C. J. Thynne and P. Gray, Proc. Chem Soc., 295 (1962); (l) J. C. J. Thynne and P. Gray, Trans. Faraday Soc., 59, 1149 (1963); (m) J. C. J. Thynne and P. Gray, ibid., 58, 2403 (1962); (n) J. C. J. Thynne, ibid., 58, 676, 1394, 1533 (1962); (o) J. Warkentin and D. M. Singleton, Can. J. Chem., 45, 3035 (1967); (p) A. Goosen and A. Scheffer, J. Chem. Soc. Perkin Trans. I, 369 (1972); (q) K. Bartel, A. Goosen, and A. Scheffer, J. Chem. Soc. (C), 3766 (1971); (r) M. J. YeeQuee and J. C. J. Thynne, Ber. Bunsenges, Phys. Chem., 72, 211 (1968); (s) see also reference 75a.

76. (a) H. A. Wiebe and J. Heicklen, J. Am. Chem. Soc., 95, 1 (1973); (b) E. A. Lissi, J. C. Scaiano, and A. E. Villa, Chem. Commun., 457 (1971); (c) E. A. Lissi, G. Massiff, and A. E. Villa, J. Chem. Soc. Faraday Trans. I, 346 (1973).

77.* A. S. Berenblyum, N. Ya. Usachev, G. A. Kovtun, V. G. Sorokin, E. D. Radchenko, and I. I. Moiseev, Izv. Akad. Nauk SSSR, Ser. Khim., 2133 (1975).

78. (a) E. Jakubowski, M. G. Ahmed, E. M. Lown, H. S. Sandhu, R. K. Gosavi, and O. P. Strausz, J. Am. Chem. Soc., 94, 4094 (1972); (b) S. Tsunashima, T. Yokota, I. Safarik, H. E. Gunning, and O. P. Strausz, J. Phys. Chem., 79, 775 (1975); (c) G. A. Oldershaw and D. A. Porter, J. Chem. Soc. Faraday Trans. I, 709 (1972); (d) H. Rommel and H. I. Schiff, Int. J. Chem. Kinet., 4, 547 (1972).

79.* (a) L. Benati and P. C. Montevecchi, J. Org. Chem., 41, 2639 (1976); (b) M. Elbanowski, Rocz. Chem., 46, 661 (1972).

80. A. D. Peeshin, N. M. Lapshin, and A. L. Buchachenko, Izv. Akad. Nauk SSSR, Ser. Chim., 1001 (1976).

6
NITRENES

WALTER LWOWSKI

New Mexico State University, Las Cruces, New Mexico 88003

In part, this chapter is intended to update a 1970 monograph.[1] Limits of space make it impossible to mention all substantial nitrene work of the last 6 years, and only those topics are discussed that exemplify major trends or progress in the field. The particular examples used had to be selected somewhat arbitrarily, and it is hoped that those authors whose work was not mentioned will generously forgive, realizing that the length of the chapter had to be limited.

I. ALKYLNITRENES

A. Thermolysis of Alkyl Azides

Thermolysis of alkyl azides results in the formation of alkylnitrenes and imines,[2] and much or all of the latter might be produced by rearrangement of the nitrenes. However, the data on hand do not rule out a concerted thermal rearrangement of some alkyl azides, and it is conceivable that some systems produce nitrenes and imines in parallel reactions, with some or all of the nitrene rearranging to imine in turn. Intermolecular alkylnitrene reactions are very inefficient, and only traces of their products are observed. Intramolecular reactions are much more facile; the pyrolysis of ethyl azide gives aziridine in 35% yield.[3] Intramolecular addition to a C=C double bond has been observed,[4] as well as intramolecular attack on aromatic rings.[2] However, the major isolated products are usually imines (see reaction (1)]. There is little difference between the migratory

$$R^1C(R^2)(R^3)\text{—}N_3 \longrightarrow R^1C(R^2)(R^3)\text{—}N + R^1R^2C{=}NR^3 + R^1R^3C{=}NR^2 + R^2R^3C{=}NR^1 \qquad (1)$$

aptitudes of the groups R (even between those of alkyl and aryl), indicating that the major rearrangement pathway (in the examples studied) goes through the nitrene.[2]

B. Photolysis of Alkyl Azides

The photolysis of alkyl azides usually yields imines, apparently in a concerted rearrangement.[5-7] Nitrenes may also be produced, but in some systems their presence is difficult to ascertain, especially if their rearrangement to imines is rapid.

Intramolecular C—H insertion occurred on photolysis of 6β-azido-5α-pregnane, with a 6% yield of insertion into the C-10 β-methyl group.[8] Intermolecular nitrene products were formed in over 30% yield in photolyses of 2H-hexafluoropropyl azide in cyclohexene.[9] The same azide, when photolyzed neat, gave a 73% yield of the imine produced by migration of the CF_3CHF group. This indicates that it is the nitrene that rearranges, unless it is first intercepted by cyclohexene or cyclohexane [see reaction (2)].[9]

$$F_3C\text{-}CFH\text{-}CF_2\text{-}N_3 \longrightarrow F_3C\text{-}CFH\text{-}CF_2\text{-}N \longrightarrow F_2C{=}N\text{-}CFH\text{-}CF_3 \quad (2)$$

In the majority of alkyl azide photolyses, only imines are isolated. Different authors, measuring migratory aptitudes in different sets of azides, have proposed different mechanisms. One[5] concludes that the migrating alkyl group displaces nitrogen from N_α by backside attack in a geometry akin to that of the Beckmann rearrangement. Abramovitch and Kyba[6] propose a transition state [reaction (3)] in which the migrating alkyl group moves to an electron-depleted orbital in the photoexcited azide, in a plane orthogonal to the C—C—N_α plane. In conformationally flexible (e.g., open-chain) systems, this leads to the migration of the medium-size group: the conformer most populous (in the ground state or perhaps the excited state) rearranges much faster than it comes to conformational equilibrium. Systems **1** (of Montgomery and Saunders[7]), bearing an aryl group at C_α, follow Abramovitch's rule, whereas the all-alkyl systems **2** showed virtually no preference of ethyl over methyl migration. Significant steric hindrance might be needed to cause a measurable preference of one conformation over the

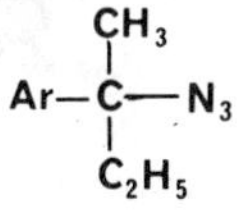

$$\mathrm{Ar{-}C(CH_3)(C_2H_5){-}N_3} \quad \mathbf{1} \qquad\qquad \mathrm{Alkyl{-}C(CH_3)(C_2H_5){-}N_3} \quad \mathbf{2}$$

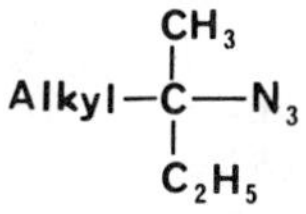

other[7] (in terms of Abramovitch's model), or perhaps the concerted mechanism changes to a stepwise (nitrene) mechanism in going from the α-aryl to the all-alkyl system. The Abramovitch-Kyba model is as follows:[6]

predominant conformation → (hν) excited state with depleted y,z-orbital → predominant product (3)

$S(L)C{=}N{-}M + N_2$

ABRAMOVITCH - KYBA MODEL

Photolyses of steroidal azides gave product distributions consistent with Abramovitch's model.[10] These experiments also show large differences between the product distributions from photolyses and thermolyses of the same azides.

Whatever the detailed mechanism of imine formation may be in a given case, the imine-forming rearrangements are capable of producing highly strained rings. The thermolysis of cyclopropyl azides gave 1-azetines in good yields.[11,12] Photolysis of 1-azidoadamantane gave azahomoadamantene (4-azatricyclo-[4.3.1.1^{3,8}]-4-undecene), isolated as its dimer or in the form of alcohol adducts.[13] The photolysis of 1-azidonorbornane[14] in alcoholic solvents must be reinterpreted in the light of Quast and Eckert's[13] results; the one- and two-membered bridges migrate (in nearly the statistically expected ratio) to the nitrene, forming bridgehead imines that then add alcohol. In methanol the total yield of 1-methoxy-2-azabicyclo[3.2.1]-octane and 1-methoxy-2-azabicyclo[2.2.2]-octane was 78% [see reaction (4)].

Pyrolysis, but not photolysis, of 3-azidoazetidinones induced acyl migration to form imidazolin-2-ones [see reaction (5)].[15] Some steroidal α-azidoketones, on photolysis (rather than thermolysis), gave ring expansion of a six- to a seven-membered ring, with apparent acyl migration.[16]

It seems that the formation of imines from azides can follow a multitude of mechanistic paths.

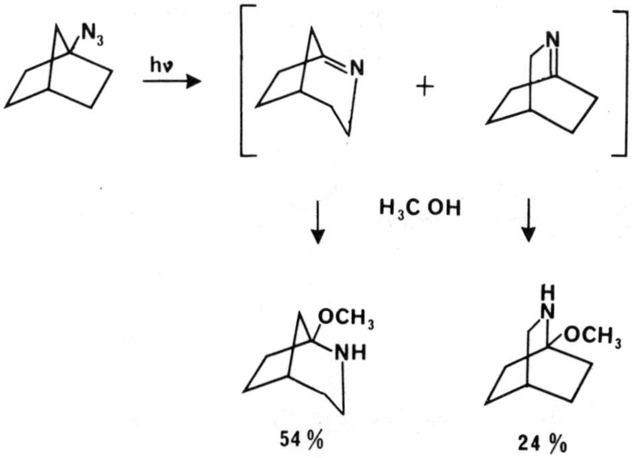

(4)

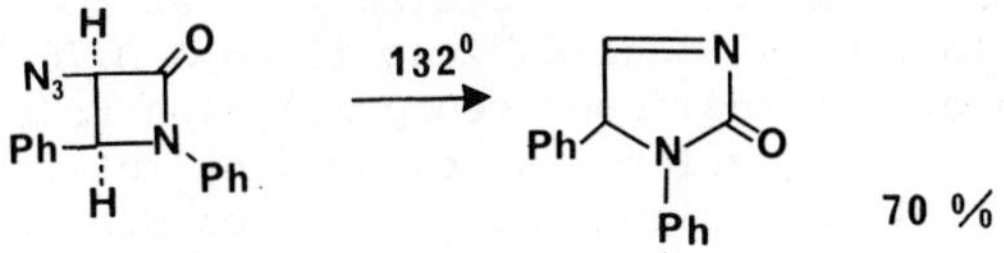

(5)

II. ARYLNITRENES

A. Generation

Arylhydroxylamines were converted to their N,O-bis(trimethylsilyl) derivatives, which gave the nitrenes on heating to 100° [see reaction (6)].[17] Phenylnitrene, produced by this method, was observed to add to cyclohexene, giving 7-phenyl-7-azabicyclo[4.1.0]heptane in 2% yield (see reaction (7)]. N-Arylhydroxylamines can also be dehydrated by phosphorus pentoxide to give nitrenes. The 2-nitrenobiphenyl thus obtained gave carbazole in 20% yield.[18]

The nitrene generated by photolysis of phenyl azide in low-temperature matrices is predominantly in the singlet state; only 12% to 13% is generated directly in the triplet state.[19]

$$Ar\text{-}NHOH + 2\,Me_3SiCl \longrightarrow Ar\text{-}N(SiMe_3)\text{-}OSiMe_3 \xrightarrow{100^o} ArN + Me_3SiOSiMe_3 \quad (6)$$

$$\text{azepine (NEt}_2\text{, H}_2\text{) } 95\% \xleftarrow{HNEt_2} PhN \xrightarrow{\text{cyclohexene}} PhN\text{(aziridine) } 2\% + PhNH_2\ 58\% \quad (7)$$

B. Intramolecular Cyclizations

The cyclization of 2-nitrenobiphenyls to carbazoles has been studied in the gas phase[20] and in solution.[21] In the gas phase the reaction had a rate constant of at least $1.4 \times 10^6\ s^{-1}$, too fast to measure exactly. In cyclohexane at 300^oK the rate constant was $2.18 \times 10^3\ s^{-1}$ and $\Delta H^{\ddagger}$ was 11.46 ± 0.76 kcal/mol.[20] The competition of internal bond reorganization (see reaction (8)] with carbazole formation was measured by intercepting the 7-azabicyclo[4.1.0]heptatriene reorganization product with diethylamine to give the azepine 5.[22] The results indicate a rapid equilibrium between the nitrene 3 and the azabicycloheptatriene 4. However, the carbazoles might be formed from more than one precursor, as indicated by Sundberg's flash-photolytic studies,[21,23] which allow one to write an approximate mechanism [see reaction (8)]. A flash-photolytic investigation of the

(Reaction scheme (8): nitrene 3 ⇌ azabicycloheptatriene 4; 3 → NH intermediate → carbazole; 4 ⇢ intermediate; 4 + Et_2NH → azepine 5 bearing NEt_2)

(8)

reaction of phenylnitrene with secondary amines agrees with this mechanism for the azepine formation, and rate constants for the reaction of the closed-shell bicyclic azirine with secondary amines are reported to be in the range 10^4 to 10^9 liter · mol^{-1} s^{-1}.[23]

Because of attack of the nitrene at the electron-rich 1-position, the cyclization of arylnitrenes of type 6 usually occurs with rearrangement to form a spiro intermediate [see reactions (9) and (10)].[24,25] The formation of the spiro

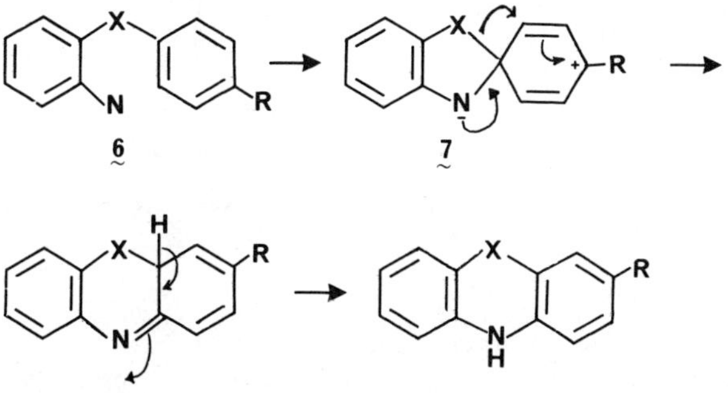

X = O; S

(9)

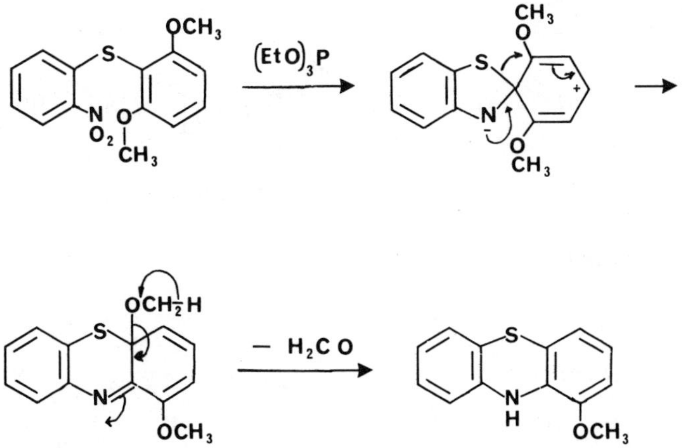

(10)

intermediate 7 is facilitated by electron availability at C-1. If electron-withdrawing groups are attached to C-1, attack on C-2 and C-6 becomes competitive. Thus 2'-azidophenyl-3-methylphenylsulfone, on thermolysis, gave (in 84% yield) a mixture of 29% of rearranged and 71% of unrearranged products [see reaction (11)].[26] Other complex cyclizations of arylnitrenes bearing

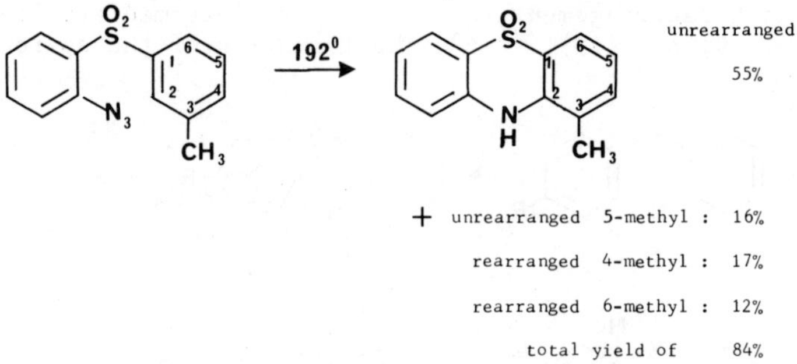

192^0 unrearranged 55%

(11)

+ unrearranged 5-methyl : 16%

rearranged 4-methyl : 17%

rearranged 6-methyl : 12%

total yield of 84%

heterocyclic substituents have been described.[27,28]

Suschitzky's group[29] provided a number of elegant examples for the dependence of the reaction course of arylnitrene cyclizations on the spin state of the nitrene. One example is shown in reaction (12), and it is noteworthy that the thermolysis of the

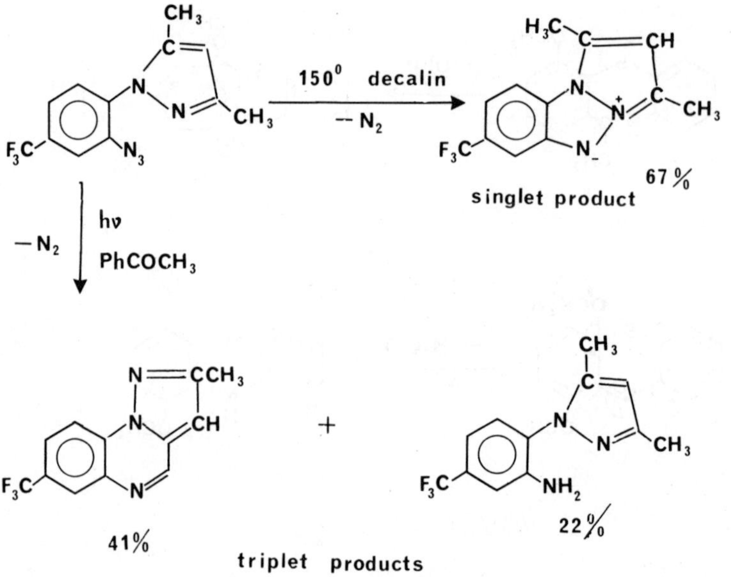

150^0 decalin $-N_2$

67% singlet product

$-N_2$ hν $PhCOCH_3$

(12)

41% 22%

triplet products

azide in decalin yielded none of the two triplet products, whereas the photolysis in acetophenone solution gave none of the singlet product. The two papers[29] contain further examples, with much information on the yield of products from the singlet and from the triplet sets.

C. Intermolecular Reactions

Cadogan and Gosney[30] reported that triarylarsines act as highly efficient nitrene traps for aryl- as well as for many other types of nitrenes. Non-nitrene routes to the resulting AsAsAs-triarylarsine imides are also available;[30] thus their mere formation cannot suffice as evidence for the formation of a nitrene in a given reaction.

Although arylnitrenes are usually inefficient in intermolecular reactions, one method[17] for their generation allows one to trap at least a few percent using olefins. Attaching electron-withdrawing substituents to the phenyl ring greatly increases the yields in a variety of intermolecular reactions.[31,32] Stereospecific addition to olefins of C_6F_5-N in 60% yield has been observed;[33] the nitrenes can be made either from the azides or by deoxygenation of nitrosoaryls.[34] Pentafluorophenylnitrene converted acrylonitrile to the cyanoaziridine in 88% yield.[32]

D. Bond Reorganizations

On decomposition of aryl azides in the presence of nucleophiles, especially amines, azepines have long since been known to form,[35-37] often in high yield. An equilibrium between phenylnitrene and the azirine **8** was demonstrated,[23,38-40] [see reaction (13)], and structural effects were studied.[41,42]

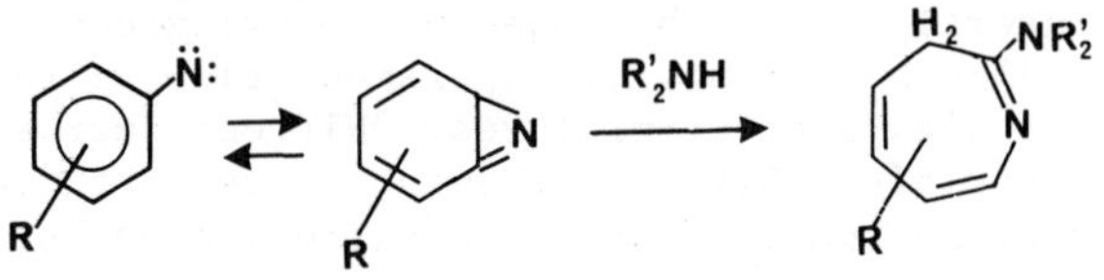

(13)

A new family of bond reorganizations was discovered in gas-phase thermolyses of aryl- and heteroaryl azides.[43] Aside from ring-contraction products, pyridine derivatives were found in the pyrolyses of aryl azides,[44] and the flash pyrolysis of 1,2,3-triazolo[1,5-a]pyridine gave the same products as did that of phenyl azide at lower temperatures.[45,46] The bond reorganizations are similar to those encountered in carbene-carbene reorganizations (see Chapter 3), but variety is introduced by: (a) thermodynamic and kinetic differences due to the replacement of carbon by nitrogen and (b) ability of most such systems to produce alternatively carbenes or nitrenes.[45,47,48] Written for phenylnitrene, the two basic reorganization mechanisms, namely, insertion/extrusion and cycloperambulation, are shown in reactions (14) and (15).

Insertion - Extrusion (14)

Cycloperambulation (15)

The insertion/extrusion mechanism has been shown to operate in phenylnitrene itself, whereas the cycloperambulation process has been demonstrated only for carbenes.* This mechanism readily explains the conversion of pyridylcarbenes into phenylnitrenes,[49] a process studied systematically by Crow, Paddon-Row, et al.[47,50,51] Pyrolysis of 2-picolyltetrazoles at 600^0 and 0.05 torr gave the 2-picolylcarbene, which became stabilized largely by insertion into the adjacent methyl group. In the absence of a 2-methyl group, cycloperambulation of the carbene led to more complex product mixtures. Nitrene extrusion was largely irreversible, whereas carbenes entered both into the cycloperambulation and the insertion/extrusion paths. The reaction of 6-methyl-3-pyridylcarbene is shown as an example in reaction (16).

---

* Recent work of Wentrup establishes an alternative mechanism for the migration of the carbene function in phenylcarbene, C. Wentrup, E. Wentrup-Byrne, and P. Müller, _J. Chem. Soc., Chem. Commun._, 210 (1977).

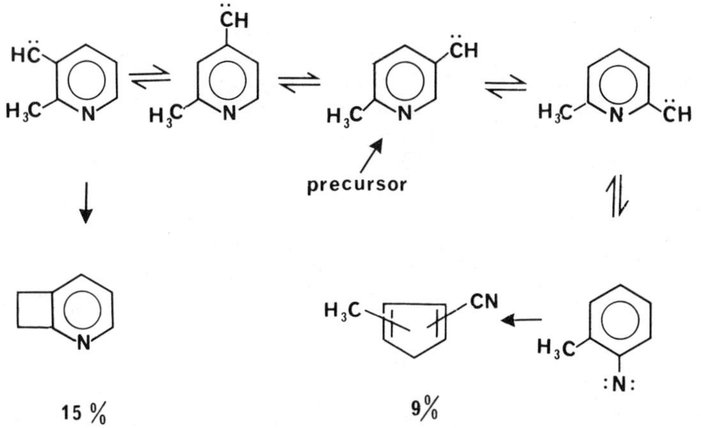

(16)

Pyrolysis of 5-(2'-, 3'-, and 4'-pyridyl)-tetrazoles-5-^{13}C and also ^{13}C NMR analysis of the aniline produced (by hydrogen abstraction by the phenylnitrene) show that both the cycloperambulation and the insertion/extrusion mechanisms operate, with the latter predominating. The double labeling with deuterium in the tetrazole ring showed the ^{13}C and the deuterium labels to stay together. The aniline produced from 2-pyridyltetrazole showed 14% cycloperambulation [see reaction (17)].[51]

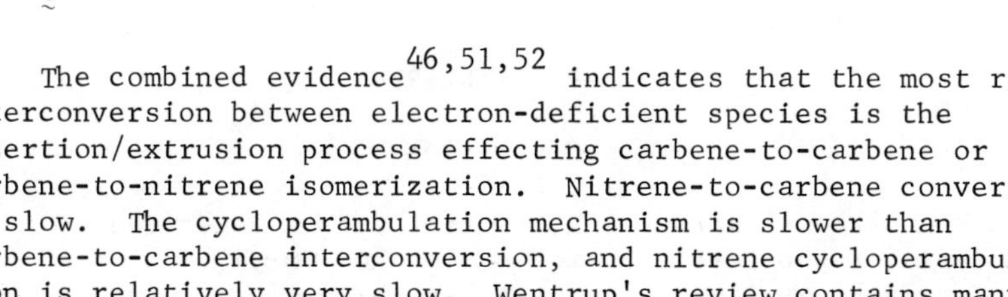

(17)

The combined evidence[46,51,52] indicates that the most rapid interconversion between electron-deficient species is the insertion/extrusion process effecting carbene-to-carbene or carbene-to-nitrene isomerization. Nitrene-to-carbene conversion is slow. The cycloperambulation mechanism is slower than carbene-to-carbene interconversion, and nitrene cycloperambulation is relatively very slow. Wentrup's review contains many other examples of carbene-to-nitrene interconversions.[45]

Aryl- and heteroarylnitrenes undergo ring-contractions[44,53,54] in addition to the reactions mentioned above. On the basis of labeling studies, Wentrup[55] proposed a mechanism that accounted for the label distribution in the products from phenylnitrene-2-^{13}C. The label was retained in the direct nitrene products, aniline and azobenzene,[52] but in the contraction product, cyano-cyclopentadiene, each ring carbon carried 15%, and the cyano-carbon, 25% of the label [see reaction (18)].[55] It is reasonable

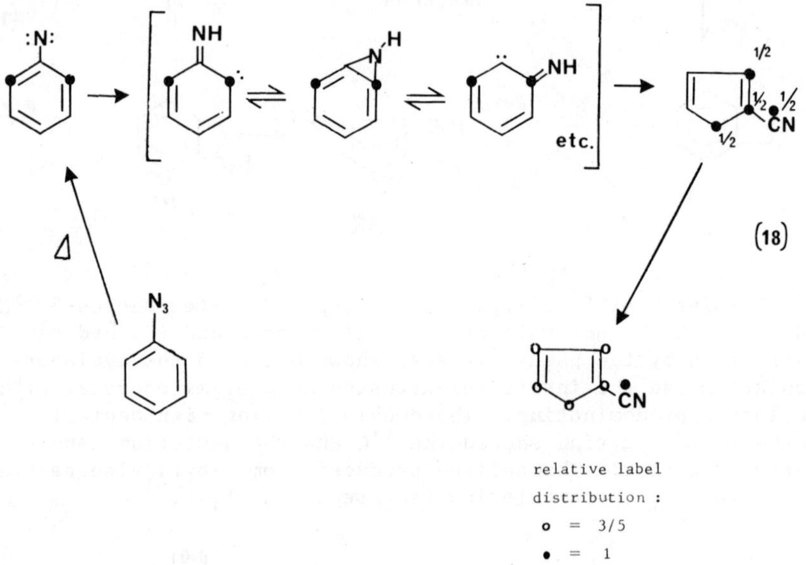

(18)

relative label distribution :
o = 3/5
• = 1

to assume that the rearrangements discussed here are of the singlet species. Wentrup[56,57] has calculated their heats of formation by CNDO/2 and extended Hückel methods, and finds the nitrenes always to be more stable than the isomeric carbenes, in good agreement with the experiments.

III. CARBONYLNITRENES

A. Generation

Alkoxycarbonylnitrenes, RO–CO–N, can be generated by thermolysis or photolysis of the azides RO–CO–N_3, and also conveniently by α-elimination.[1] Alkanoyl- and aroylnitrenes, however, are much more cumbersome to generate, the main route being the photolysis of the corresponding azides, R–CO–N_3. Therefore, several alternative methods for making R–CO–N have been studied. Eibler and Sauer[58] found that the photolysis, but not the thermolysis, of 5-phenyl-1,3,4-dioxazolin-2-one in 2-methylbutane gave a 23% yield of the C—H insertion products of benzoylnitrene [see reaction (19)]. The nitrene showed the same selectivity (between tertiary, secondary, and primary C—H bonds) as did benzoylnitrene generated by the photolysis of benzoyl azide. Like benzoyl azide, the heterocycle gave about equal yields of benzoylnitrene and phenyl isocyanate on photolysis. The sulfur analog of the dioxazolinone, 5-phenyl-1,3,2,4-dioxathiazolin-2-oxide, produced much smaller yields of benzoylnitrene C—H insertion products [see reaction (19)].[58]

$$H_5C_6\text{-(dioxazolinone ring: N, O, CO, O)} \xrightarrow[-CO_2]{h\nu} C_6H_5NCO + \underline{\underline{C_6H_5CO\text{-}N}} \xleftarrow[-COS]{h\nu} \text{(O, N, OS, O ring)}C{-}C_6H_5 \qquad (19)$$

Swern[59] studied the photolysis of N-benzoyliminodimethylsulfurane, which produced benzoylnitrene. However, he regards benzoyl azide as a better source for the nitrene.

Aminimides, such as pyridinium N-carbonyl- and alkoxycarbonylimides, also form carbonylnitrenes on photolysis, albeit in often modest yields.[60] Becker, Beyer, et al. obtained relatively high yields of acetyl- and benzoylnitrene by photolyzing 4-acylimino-1,2,4-triazolium aminimides [reaction (20)].[61]

$$\text{1-benzyl-1,2,4-triazolium-4-}\bar{N}{-}COCH_3 \xrightarrow{h\nu} \text{1-}NCH_2C_6H_5\text{-1,2,4-triazole } (91\%) + N{-}COCH_3 + CH_3NCO \qquad (20)$$

$$N{-}COCH_3 \xrightarrow{CH_3OH} CH_3ONH{-}COCH_3 \; (41\%) \qquad CH_3NCO \xrightarrow{HOCH_3} CH_3NHCOOCH_3 \; (21\%)$$

This is the most advantageous route to acylnitrenes yet reported, provided it proves to be generally applicable. Certain azimines also generated carbonylnitrenes on photolysis or thermolysis, but the reaction did not succeed for all potential carbonylnitrenes, with loss of nitrogen from the azimine as a major alternative reaction path.[62] Most interestingly, Gait, Rees, et al.[62] obtained a 29% yield of the C—H insertion product of EtOCO-N into cyclohexane using benzophenone sensitized photolysis [see reaction (21)]. Direct photolysis of the benzo[c]cinnoline N-ethoxycarbonylimide in cyclohexane gave only a trace of ethyl N-cyclohexylcarbamate. Thermolysis in decalin at 165° gave a 19% yield of a mixture of C—H insertion products, together with benzocinnoline. Benzoyl- and acetylnitrene could not be obtained from the corresponding benzocinnoline N-acylimines.

$N-CO_2C_2H_5$ ⇌ $CO_2C_2H_5$ → $NCO_2C_2H_5$ + N_2

thermolysis or triplet sensitized photolysis (21)

+ $:\ddot{N}-CO_2C_2H_5$ trapped by C-H insertion

B. Reactions

As one might expect, singlet carbonylnitrenes add readily to unshared electron pairs, if a stable bond can be formed in this manner. Simple dialkylsulfides form iminosulfuranes[63] if the sulfides have special structural features, such as allylic sulfides[64] or disulfides.[65] The selectivities of the additions of various carbenes and nitrenes (including carbonylnitrenes) to sulfides have been studied and found to be very low, in accordance with the presumption of a low energy of activation for the addition of singlet nitrenes to unshared electron pairs.[66]

C. Solvent Effects

Interaction of electron-deficient species, especially singlet sextet species, with molecules bearing unshared electron pairs should be expected. Gleiter and Hoffmann[67] have calculated such interaction for singlet nitrenes. Aside from the case of formation of a permanent bond (e.g., with amines), they found a marked stabilization of singlet nitrenes for a situation in which the nitrene forms a linear array with two solvating molecules symmetrically arranged, X:-----$\overset{R}{N}$-----:X. Other (and as yet uncalculated) geometries might conceivably also be stabilizing. The chemical consequences to be expected from such a nucleophilic solvation are: (a) increased singlet nitrene lifetime (singlet dissociation and intersystem crossing being retarded, (b) decreased reactivity and increased selectivity due to potential loss of the nucleophilic solvation in the transition state, and (c) increased bulk and effects of steric hindrance in reactions of the solvated nitrene.

Experience confirms the expected effects of solvents containing atoms with unshared electron pairs. Alewood, Benn, et al. studied the thermolysis of neopentyl azidoformate in 11 solvents.[68] The yield of the intramolecular C—H insertion product, 5,5-dimethyl-tetrahydro-1,3-oxazin-2-one, was nil in cyclohexane or benzene solutions, but 90% in dichloromethane, and intermediate in the other solvents. Stabilization by halogenated solvents of the singlet states of carboalkoxynitrenes, RO—CO—N,[69] and of alkanoylnitrenes[70,71] has been observed. Hexafluorobenzene[69,72] increased the yields of C—H insertion products from ethoxycarbonylnitrene by a factor of 1.5 over those observed in pure hydrocarbon solvents. The C—H insertion yields in *n*-octadecyloxycarbonylnitrene-cyclohexane-hexafluorobenzene systems showed a dependence on hexafluorobenzene concentration[69] similar to the dependence on dichloromethane concentration observed[73] in the system ethoxycarbonylnitrene-dimethylcyclohexane-dichloromethane.

The question of a solvent effect on nitrene selectivity in C—H insertion reactions has been studied by Tardella.[74,75] Thermolyses of ethyl azidoformate in solutions of hydrocarbons containing sterically hindered tertiary C—H groups favored insertion into the hydrocarbon's secondary C—H bonds when dichloromethane was present in the solution, compared to the same reactions run without dichloromethane. One plausible explanation for this decreased ratio of insertion into tertiary versus secondary C—H groups is the possibility of greater bulk of the dichloromethane solvated singlet nitrene.

Singlet pivaloylnitrene adds stereospecifically to trans olefins, such as *trans*-4-methyl-2-pentene.[76] The yield of the trans-aziridine is 7.4% in an olefin-dichloromethane mixture containing 78 mol % of olefin. Decreasing the olefin concentration to 3 mol % increases the aziridine yield to 20.5%, just the opposite of what would be expected if the solvent (dichloromethane) were truly inert. An increase in yield by a factor of 2 was observed[71] in the intramolecular C—H insertion reactions of some alkanoylnitrenes when the solvent was changed from cyclohexane to dichloromethane. For example, n-hexanoylazide yielded 17% of lactams and 2% of N-cyclohexylhexanamide on photolysis in cyclohexane, but 33.5% of γ- and δ-lactams in dichloromethane solution.

The apparently greater susceptibility to the "dichloromethane effect" of alkanoylnitrenes, compared with alkoxycarbonylnitrenes, may be due to an intramolecular competition for solvation by the ether oxygen in the alkoxycarbonylnitrene [see reaction (22)].

RO(C=O)N: + CH_2Cl_2 ⇌ solvate A   vs.   R(C=O)N: + CH_2Cl_2 ⇌ solvate B   (22)

IV. CYANONITRENE

Little use seems to have been made of cyanonitrene, although the preparation of its precursor, cyanogen azide, has now been published in detail.[77] A more convenient alternate precursor would no doubt rekindle the interest in the versatile and intriguing species NCN. Its isomer, CNN, has again been observed.[78]

V. SULFONYLNITRENES

A. Generation

Thermolysis of azides ($-SO_2N_3$) remains the commonly used path to sulfonylnitrenes. Temperatures of at least 120° are needed, rendering alternative methods desirable. Photolysis of sulfonyl azides is being used, but is often unsatisfactory because of the formation of complex product mixtures and copious amounts of tar. Thus non-azide sources for sulfonylnitrenes have been sought by many investigators, but the search has been frustrating. Because the same holds true for many other nitrenes, a brief summary of the attempts over the last 5 years follows. Photolyses of N-arenesulfonylpyridinium imines[79] and of aryl isocyanates[80] did not give significant yields of nitrenes, nor did the dechlorination of N,N-dichloroarenesulfonamides,[80] the dehydrochlorination of N-chloroarenesulfonamides,[81] or the lead tetraacetate oxidation of arenesulfonamides.[80,82] The treatment of N-arenesulfonylhydroxylamines with phosphorus pentoxide gave a few percent of the sulfonylnitrenes. Thermolysis of methanesulfonylimidophenyliodonium ylide gave traces of the nitrene,[80] but better may be the decomposition of the ylides, $Ar-SO_2-N^{-}-I^{+}-Ph$.[83] Attempts to generate sulfamoylnitrenes, R_2N-SO_2-N, by thermolysis of the corresponding azides were unsuccessful.[84]

B. Reactions

The relative reactivities of the tertiary, secondary, and primary C—H bonds of 2-methylbutane toward methanesulfonylnitrene (generated photolytically)[85] were found to be 9.6:4.2:1, a selectivity less than that of ethoxycarbonylnitrene (34:9:1). Breslow[86] found the insertion of alkanesulfonylnitrenes (generated by azide thermolysis) to select between the tertiary, secondary, and primary C—H bonds of 2,4-dimethylpentane in the (statistically corrected) ratios of 6.0:2.3:1. The insertion into the tertiary C—H bonds of cis- and trans-1,2-dimethylcyclohexane was completely stereospecific, leading to the conclusion that only the singlet sulfonylnitrene inserts.[86]

The reactivities, per X—H bond, of the secondary C—H bonds of 2-methylbutane and cyclohexane have been compared with that of the OH bond of ethanol.[85] For methanesulfonylnitrene the ratios were 0.95:1:29, respectively, whereas ethoxycarbonylnitrene gave the ratios 0.8:1:23. This agrees with the contention that the first step in the reaction of either nitrene with the OH group is addition of the nitrene to an unshared electron pair of the alcohol oxygen. The additions of p-toluenesulfonylnitrene (generated photolytically from the azide) to various amines[87] and sulfides[88] are quite unselective. Thus iminosulfuranes were formed from dimethylsulfide and di-tert-butylsulfide at about the same rate. This again points to the great ease with which singlet nitrenes add to unshared electron pairs.

Singlet sulfonylnitrenes attack aryl rings with the initial formation of benzoaziridines, which are in equilibrium with

N-sulfonylazepines.[80,89] The latter can be trapped with tetracyanoethylene, failing which they rearrange to N-sulfonylanilines with great ease.[80,89] Triplet sulfonylnitrenes attack aromatic systems as one would expect of highly electrophilic 1,1-diradicals. Thus methyl benzoate and triplet methanesulfonylnitrene give predominantly methyl o-methanesulfonylamidobenzoate.[89]

The thermolysis of ferrocene-1,1'-disulfonyl azide gave intermolecular nitrene products with aliphatic as well as with aromatic substrates.[90]

VI. PHOSPHORYLNITRENES

A "well-behaved" class of phosphorylnitrenes has finally been discovered.[91] On photolysis, diethyl- and diphenylphosphoryl azides gave nitrenes that insert readily into C—H bonds. Diethylphosphoryl azide, photolyzed in cyclohexane, gave diethyl cyclohexylphosphoramidate in 88% yield and diethylphosphoramide in 12% yield, a high-yield performance unequaled by other nitrenes. The relative reactivities of the tertiary, secondary, and primary C—H bonds of 2-methylbutane were 6.0:4.3:1 for diethylphosphorylnitrene, and 3.4:1.2:1 for diphenylphosphorylnitrene.[91] Taking selectivity as an inverse measure of reactivity, this makes the latter nitrene the most reactive known. Extreme lack of selectivity was also shown in the reaction of diethylphosphorylnitrene with tert-butanol; a 35% yield of the C—H insertion product into the nine C—H bonds of the methyl groups was obtained, along with a 33% yield of the t-butoxamino compound, $(EtO)_2PO-NH-Ot\text{-}Bu$, and a 27% yield of the amide, $(EtO)_2PO-NH_2$. On photolysis in cyclohexane, bis(dimethylamino)-phosphoryl azide gave only a 7% yield of the C—H insertion product;[91] the decomposing azide reacted predominantly by dimethylamino migration, giving $Me_2N-P(O)=N-NMe_2$, which subsequently polymerized.

Diphenylphosphinylnitrene is probably formed by thermolysis or photolysis of trimethylammonium(diphenylphosphinyl)imide, $Me_3N^+-N^-POPh_2$ in dimethylsulfoxide. Thermolysis at 180° gave a 29% yield of N-diphenylphosphinyldimethylsulfoximine.[92] This sulfoximine could not be obtained by thermolysis in dimethylsulfoxide of diphenylphosphinyl azide, Ph_2PO-N_3.[93]

VII. AMINONITRENES

A. Introduction

The knowledge and use of aminonitrenes have been advanced greatly over the last 7 years, especially in the area dealing with intermolecular reactions. Reviews are available covering the literature into 1971.[94,95]

Unsubstituted aminonitrene, H_2N-N, has been trapped at low temperatures[96] after the vacuum pyrolysis of cesium tosylhydrazide, Cs^+ p-$MeC_6H_4SO_2N^- - NH_2$, and its mass spectrum has been recorded. The reaction of H_2N-N with ammonia at 72°K is faster than that of trans-diazene, $HN=NH$.

Most of the work on intermolecular aminonitrene reactions became possible due to the substitution of electron-withdrawing groups on the amino-nitrogen, N_β. This decreases the importance of the resonance structure **10**, relative to that of contributor **9**, and increases the reactivity of aminonitrenes to the point where they will add to $C=C$ double bonds (but, according to present information, not enough to make them insert into $C-H$ bonds).

$$R_2\ddot{N}_\beta-\ddot{N}_\alpha{:} \longleftrightarrow R_2\overset{+}{N}_\beta=\ddot{N}_\alpha^{-} \qquad (23)$$

9 **10**

The more popular aminonitrenes with reduced contributions by the resonance structure **10** fall into two classes, namely, the acyl- and diacylaminonitrenes, in which the formally unshared electron pair on N_β is involved in amide resonance, and those aminonitrenes in which this electron pair is part of a hetero-aromatic system. A good example of the latter class has been studied in some mechanistic detail by Mayer, Sauer, et al.[97]

B. Generation

The most important route to the "electron-depleted" amino-nitrenes is Rees's method of lead tetraacetate oxidation of the corresponding amines, R_2N-NH_2.[98] The reaction produces acetic acid, the removal of which (e.g., by added calcium oxide) is often essential to avoid product decomposition [see reaction (24)].[99]

Lead tetraacetate has been employed most often, but other oxidants can serve, and produce an intermediate identical with that generated by lead tetraacetate. Sauer and colleagues[97,100]

$$\text{C}_6\text{H}_4(\text{O})(\text{N-NH}_2)\text{CO} \xrightarrow{Pb(CH_3COO)_4} \text{C}_6\text{H}_4(\text{O})(\text{N-}\ddot{N}:)\text{CO} + 2\ CH_3COOH + Pb(CH_3COO)_2 \quad (24)$$

oxidized 1-amino-2,5-diphenyl-1,3,4-triazole with four different oxidants [$Pb(OAc)_4$, $Pb(OCOPh)_4$, $Ph-I(OAc)_2$, $Ph-I(OCOCHCl_2)_2$] and trapped the nitrene with six different olefins [styrene, p-methoxy-, p-chloro-, p-methylstyrene, cis- and trans-β-methylstyrene]. The reactions produced both aziridines and also benzonitrile and nitrogen, the decomposition products of the nitrene. The product ratios were evaluated in terms of two rate constants, k_1 for the nitrene fragmentation and k_2 for the olefin addition [cf. reaction (25)] by means of Eq. (25a). For each olefin the ratio k_1/k_2 was found to be independent of olefin concentration and of the oxidant used. Thus the existence of a discrete nitrene as the partitioning intermediate, rather than an intermediate containing both the aminotriazole and the oxidant moieties, is virtually certain.

Reaction (25): 1-amino-2,5-diphenyl-1,3,4-triazole (PhC, CPh, N—N, N–NH₂) $\xrightarrow{Ox}$ triazolyl nitrene (N–:N:); $\xrightarrow{k_1}$ $N_2 + 2\ PhCN$; $\xrightarrow[k_2]{C=C}$ N-(2,5-diphenyl-1,3,4-triazol-1-yl)aziridine (N–N< C—C) (25)

$$\frac{k_1}{k_2} = \frac{\frac{1}{2}[PhCN][C=C]}{[aziridine]} \quad (mol \cdot liter^{-1}) \qquad (25a)$$

Oxidation of 1,1-dimethylhydrazine with cupric chloride in aqueous solution gave a complex, $[Me_2N{=}N]_2Cu_3Cl_3$, which released dimethylaminonitrene on treatment with acid, followed by neutralization.[101] Tetramethyl-2-tetrazene was produced. A mixture of the complexes from 1,1-dimethyl- and 1,1-diethylhydrazine gave 1,1-dimethyl-4,4-diethyl-2-tetrazene, indicating that $R_2N{-}N$ had indeed been formed. The regeneration of a nitrene from its metal complexes is most unusual; such complexes are normally too stable to release nitrenes under tolerable reaction conditions.

The photolysis[102] and thermolysis[103] of substituted N-aminosulfoximines is another new route to aminonitrenes. They can also be generated by alkaline treatment of 2,2-disubstituted sulfonylhydrazides,[104,105] by base-induced decomposition of a methansulfinylhydrazine, PhMeN − NH − SOMe,[106] or by simple deprotonation of diazenium salts.[107]

The thermolysis of certain phthalimidylaziridines is important both for synthetic purposes (generation of phthalimidonitrene in the absence of an oxidant and under thermally mild conditions) and mechanistically because it leaves little doubt as to the existence of a free phthalimidonitrene [see reaction (26)].[108]

NPht, COCH$_3$ —benzene, reflux→ N—NPht

Me$_2$SO → O=S(Me)(Me)=N−NPht 34 %

→ PhtN−N (aziridine) 52 % (26)

NPht = phthalimidyl

C. Reactions

The most common products of aminonitrene-forming reactions are tetrazenes, $R_2N-N=N-NR_2$.[94,95] They might be formed by dimerization of aminonitrenes, but that explanation is cumbersome where the aminonitrenes are short-lived or are present only in low concentration. An alternative mechanism involves attack of an aminonitrene, R_2N-N, on a molecule of starting material, R_2N-NH_2. Addition to the unshared electron pair of the amine function, followed by tautomerization, should give a tetrazane, $R_2N-NH-NH-NR_2$. Rees et al.[109] were able to isolate several such tetrazanes by oxidizing some N-amino heterocycles with lead tetraacetate. N-Aminophthalimide gave a 90% yield of the tetrazane $PhtN-NH-NH-NPht$ (dec. 210° to 212°), which could be oxidized further to give the tetrazene $PhtN-N=N-N-Pht$. The tetrazane mechanism has been supported by the findings of other workers;[110] it is summarized in reaction (27).

$$R_2N\text{-}NH_2 \xrightarrow{Ox} R_2N\text{-}\ddot{N} \xrightarrow{R_2N\text{-}NH_2} R_2NNHNHNR_2 \xrightarrow{Ox} R_2NN{=}NNR_2 \quad (27)$$

Other modes of stabilization of aminonitrenes include sigmatropic shifts in the nitrene, if its structure is suitable, such as in **11**.[111] Alternatively, some charge-separated adducts of aminonitrenes can undergo sigmatropic rearrangements, for example, **12** in reaction (29).[112] Ring expansion has been

$$\underset{\mathbf{11}}{:\bar{N}{=}\overset{+}{N}(\text{—})\text{—}CH_2\text{—}C{=}C} \longrightarrow \text{—}N{=}N\text{—}C\text{—}C{=} \quad (28)$$

$$PhtN\text{—}N + Ar\text{—}S\text{—}CH_2\text{—}CH{=}CHR \longrightarrow \underset{\mathbf{12}}{Ar\text{—}\overset{+}{S}(\text{—}\bar{N}\text{—}NPht)\text{—}CH_2\text{—}CH{=}CHR} \longrightarrow Ar\text{—}S\text{—}N(NPht)\text{—}CHR\text{—}CH{=}CH_2 \quad (29)$$

observed to be the most facile reaction path in an example due to Rees [see reaction (30)].[99]

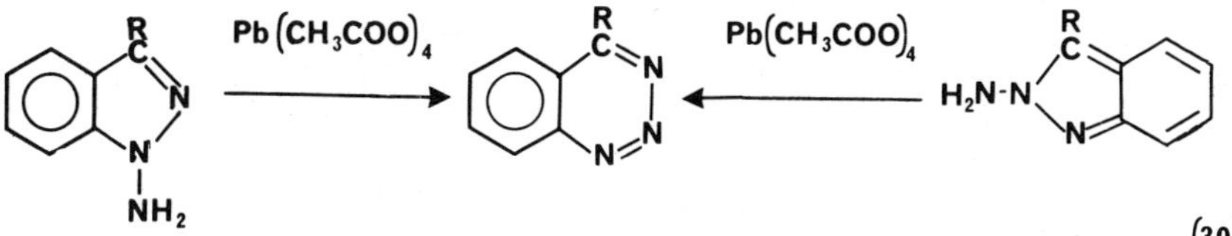

(30)

The intermolecular addition of aminonitrenes to olefins is stereospecific, with retention of the geometric configuration of the olefin,[100,113,114] even in highly diluted solutions.[113] Strained[115] and highly substituted[116] olefins react readily. The additions of phthalimidonitrene[117] and 3-nitrenobenzoxazole[118] to 1,3-dienes[118] (or to α,β-unsaturated esters[117]) places the substituent on the aziridine nitrogen in the <u>syn</u> position relative to the olefinic double bond (or ester function) [see reaction (31)]. The <u>syn</u>-<u>N</u>-aminoaziridines

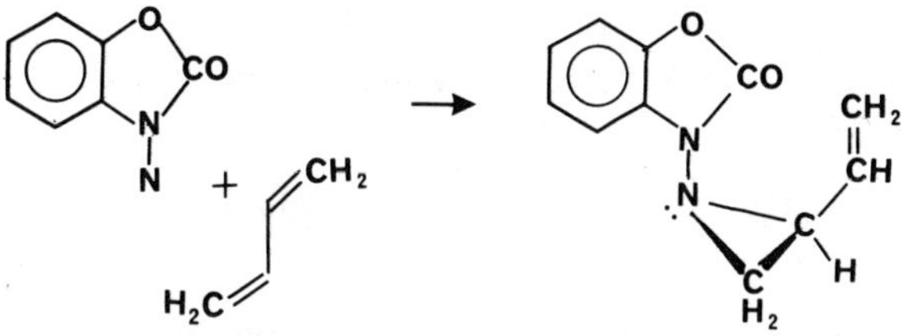

(31)

equilibrate on warming by <u>N</u>-inversion, demonstrating the greater stability of the <u>anti</u>-<u>N</u>-invertomers. Foucaud and co-workers[119,120] have explained the initial formation of the <u>syn</u> isomer by considering MO interactions in the transition state.

Rees[121] added phthalimidonitrene to acetylenes and observed the isomerization of the initially formed 1-H-azirines to 2-H-azirines. The addition of phthalimidonitrene to a sulfoxide was found to be stereospecific, most likely proceeding with retention of configuration at the asymmetric sulfur.[122]

VIII. REFERENCES

1. W. Lwowski (ed.), Nitrenes, Wiley-Interscience, New York, 1970.

2.* R. A. Abramovitch and E. P. Kyba, J. Am. Chem. Soc., 96, 480 (1973).

3. W. Pritzkow and D. Timm, J. Prakt. Chem., 32, 178 (1966).

4. J. J. Looker, J. Org. Chem., 36, 1045 (1971).

5. R. M. Moriarty and R. C. Reardon, Tetrahedron, 26, 1379 (1970).

6. R. A. Abramovitch and E. P. Kyba, J. Am. Chem. Soc., 93, 1537 (1971).

7. F. C. Montgomery and W. H. Saunders, Jr., J. Org. Chem., 41, 2368 (1976).

8. A. Pancrazi, Q. Khuong-Huu, and R. Goutarel, Tetrahedron Lett., 5015 (1972).

9.* R. E. Banks, D. Barry, M. J. McGlinchey, and G. J. Moore, J. Chem. Soc. (C), 1017 (1970).

10.* A. Pancrazi and Q. Khuong-Huu, Tetrahedron, 31, 2041, 2049 (1975).

11. A. B. Levy and A. Hassner, J. Am. Chem. Soc., 93, 2051 (1971); A. Hassner, A. B. Levy, E. F. McEntire, and J. E. Galle, J. Org. Chem., 39, 585 (1974).

12. G. Szeimies, U. Siefken, and R. Rink, Angew. Chem., 85, 173 (1973).

13.* H. Quast and P. Eckert, Justus Liebigs Ann. Chem., 1227 (1974).

14. J. O. Reed and W. Lwowski, J. Org. Chem., 36, 2864 (1971).

15. J. S. Millership and H. Suschitzky, J. Chem. Soc., Chem. Commun., 1496 (1971).

16. W. A. Court, O. E. Edwards, C. Greico, and W. Rank, Can. J. Chem., 53, 463 (1975).

17.* F. P. Tsui, Y. H. Chang, T. M. Vogel, and G. Zon, J. Org. Chem., 41, 3381 (1976).

18. K. T. Potts, A. A. Kutz, and F. C. Nachod, Tetrahedron, 31, 2171 (1974).

19. A. Reiser and L. J. Leyshon, J. Am. Chem. Soc., 93, 4051 (1971).

20. P. A. Lehman and R. S. Berry, J. Am. Chem. Soc., 95, 8614 (1973).

21.* R. J. Sundberg, D. W. Gillespie, and B. A. DeGraff, J. Am. Chem. Soc., 97, 6193 (1975).

22. R. J. Sundberg and R. W. Heintzelman, J. Org. Chem., 39, 2547 (1974).

23. B. A. DeGraff, D. W. Gillespie, and R. J. Sundberg, J. Am. Chem. Soc., 96, 7491 (1974).

24. J. I. G. Cadogan and S. Kulik, J. Chem. Soc., Chem. Commun., 233 (1970).

25.* J. I. G. Cadogan, Acc. Chem. Res., 5, 303 (1972).

26. J. I. G. Cadogan, J. N. Done, G. Lunn, and P. K. K. Lin, J. Chem. Soc., Perkin Trans. 1, 1749 (1976).

27. G. R. Cliff, G. Jones, and J. M. Woolard, J. Chem. Soc., Perkin Trans. 1, 2072 (1974).

28. T. Kametani, F. F. Ebetino, and K. Fukimoto, J. Chem. Soc., Perkin Trans. 1, 861 (1974).

29.* I. M. McRobbie, O. Meth-Cohn, and H. Suschitzky, Tetrahedron Lett., 925, 929 (1976).

30. J. I. G. Cadogan and I. Gosney, J. Chem. Soc., Perkin Trans. 1, 460, 466 (1974).

31. R. A. Abramovitch and S. R. Challand, J. Chem. Soc., Chem. Commun., 1160 (1972).

32. R. E. Banks and A. Prakash, Tetrahedron Lett., 99 (1973).

33.* R. A. Abramovitch, S. R. Challand, and Y. Yamada, J. Org. Chem., 40, 1541 (1975).

34. R. A. Abramovitch, S. R. Challand, and E. F. Scriven, J. Am. Chem. Soc., 94, 1374 (1972); J. Org. Chem., 37, 2705 (1972).

35. L. Wolff, Justus Liebigs Ann. Chem., 394, 59 (1912).

36. R. Huisgen, D. Vossius, and M. Appl, Chem. Ber., 91, 1, 12 (1958).

37. W. v. E. Doering and R. A. Odum, Tetrahedron, 22, 81 (1966).

38. R. A. Odum and G. Wolf, J. Chem. Soc., Chem. Commun., 360 (1973).

39. R. J. Sundberg, S. R. Suter, and M. Brenner, J. Am. Chem. Soc., 94, 513 (1972).

40. T. DeBoer, J. I. G. Cadogan, H. H. McWilliam, and A. G. Rowley, J. Chem. Soc., Perkin Trans. 2, 554 (1975).

41. J. Rigaudy, C. Igier, and J. Barcelo, Tetrahedron Lett., 3845 (1975).

42. B. Iddon, M. W. Pickering, and H. Suschitzky, J. Chem. Soc., Chem. Commun., 759 (1974).

43. P. A. S. Smith, in reference 1, pp. 155 ff.

44.* W. D. Crow and C. Wentrup, Tetrahedron Lett., 5569 (1968).

45. C. Wentrup, Topics in Current Chemistry, 62, 173 (1976).

46. See reference 45, pp. 227, 228, and Table 19.

47. W. D. Crow and M. N. Paddon-Row, J. Am. Chem. Soc., 94, 4746 (1972); W. D. Crow, A. N. Khan, and M. N. Paddon-Row, Aust. J. Chem., 28, 1741 (1975).

48. W. D. Crow and M. N. Paddon-Row, Aust. J. Chem., 26, 1705 (1973).

49. W. D. Crow and C. Wentrup, Tetrahedron Lett., 6149 (1968).

50. W. D. Crow, M. N. Paddon-Row, and D. S. Sutherland, Tetrahedron Lett., 2239 (1972).

51.* W. D. Crow, A. N. Khan, M. N. Paddon-Row, and D. S. Sutherland, Aust. J. Chem., 28, 1763 (1975).

52.* W. D. Crow and M. N. Paddon-Row, Aust. J. Chem., 28, 1755 (1975).

53. W. D. Crow and C. Wentrup, Tetrahedron Lett., 4379 (1967).

54. E. Hedeya, M. E. Kent, D. W. McNeil, F. P. Lossing, and T. McAllister, Tetrahedron Lett., 3415 (1968).

55. C. Wentrup, reference 45, p. 236.

56. C. Wentrup, Tetrahedron, 30, 1301 (1974).

57. C. Wentrup, reference 45, p. 244, Table 27.

58.* E. Eibler and J. Sauer, Tetrahedron Lett., 2565 (1974).

59. Y. Hayashi and D. Swern, J. Am. Chem. Soc., 95, 5205 (1973).

60. T. Sasaki, K. Kanematsu, A. Kakehi, I. Ichikawa, and K. Hayakawa, J. Org. Chem., 35, 427 (1970).

61. H. G. O. Becker, D. Beyer, and H.-J. Timpe, Z. Chem., 10, 264 (1970).

62.* S. F. Gait, M. E. Peek, and C. W. Rees, J. Chem. Soc., Perkin Trans. 1, 19 (1975).

63. W. Ando, N. Ogino, and T. Migita, Bull. Chem. Soc. Jpn., 44, 2278 (1971).

64. W. Ando, H. Fuji, I. Nakamura, and T. Migita, Int. J. Sulfur Chem., 8, 13 (1973).

65. W. Ando, H. Fuji, and T. Migita, Int. J. Sulfur Chem., A2, 143 (1972).

66. D. C. Appleton, D. C. Bull, J. McKenna, J. M. McKenna, and A. R. Walley, J. Chem. Soc., Chem. Commun., 140 (1974).

67. R. Gleiter and R. Hoffmann, Tetrahedron, 24, 5899 (1968).

68. P. F. Alewood, M. Benn, and R. Reinfried, Can. J. Chem., 52, 4083 (1974).

69. D. S. Breslow and E. I. Edwards, Tetrahedron Lett., 2041 (1972).

70.* G. R. Felt, S. Linke, and W. Lwowski, Tetrahedron Lett., 2037 (1972).

71.* W. Lwowski and S. Linke, Justus Liebigs Ann. Chem., 8 (1977).

72. R. C. Belloli and V. A. LaBahn, J. Org. Chem., 40, 1972 (1975).

73. R. C. Belloli, M. A. Whitehead, R. H. Wollenberg, and V. A. LaBahn, J. Org. Chem., 39, 2128 (1974).

74. P. Tardella, Atti Acad. Naz. Lincei, Cl. Sci. Fis. Mat. Nat. Rend., 48, 443 (1970); Chem. Abstr., 73, 120174p (1970).

75. P. A. Tardella and L. Pellacani, J. Org. Chem., 41, 2034 (1976).

76. G. R. Felt and W. Lwowski, J. Org. Chem., 41, 96 (1976).

77. F. D. Marsh, J. Org. Chem., 37, 2966 (1972).

78. R. L. DeKock and W. Weltner, Jr., J. Am. Chem. Soc., 93, 7106 (1971).

79. R. A. Abramovitch and T. Takaya, J. Org. Chem., 38, 3311 (1973).

80. R. A. Abramovitch, T. D. Bailey, T. Takaya, and V. Uma, J. Org. Chem., 39, 340 (1974).

81. F. Ruff and A. Kuczsman, J. Chem. Soc., Perkin Trans. 2, 509 (1975).

82. J. I. G. Cadogan and I. Gosney, J. Chem. Soc., Chem. Commun., 586 (1973).

83. Y. Yamada, T. Yamamoto, and M. Okawara, Chem. Lett., 361 (1975).

84. R. A. Abramovitch and K. Miyashita, J. Chem. Soc., Perkin Trans. 1, 2413 (1975).

85. T. Shingaki, M. Inagaki, N. Torimoto, and M. Takebayashi, Chem. Lett., 1181 (1972).

86.* D. S. Breslow, E. I. Edwards, E. C. Linsay, and H. Omura, J. Am. Chem. Soc., 98, 4268 (1976).

87. D. C. Appleton, J. McKenna, J. M. McKenna, L. B. Sims, and A. R. Walley, J. Am. Chem. Soc., 98, 292 (1976).

88. J. McKenna, Tetrahedron, 30, 1555 (1974).

89.* R. A. Abramovitch, G. N. Knaus, and V. Uma, J. Org. Chem., 39, 1101 (1974).

90. R. A. Abramovitch and W. D. Holcomb, J. Org. Chem., 41, 491 (1976).

91.* R. Breslow, A. Feiring, and F. Herman, J. Am. Chem. Soc., 96, 5937 (1974).

92. E. Kameyama, S. Inokuma, and T. Kurwamura, Bull. Chem. Soc. Jpn., 49, 1439 (1976).

93. F. Weissbach and W. Jugelt, J. Prakt. Chem., 317, 394 (1975).

94. D. M. Lemal, reference 1, pp. 361 ff.

95. B. V. Ioffe and M. A. Kuznetsov, Russ. Chem. Rev., 41, 131 (1972).

96.* N. Wiberg, G. Fischer, and H. Bachhuber, Angew. Chem., 88, 386 (1976).

97.* K. K. Mayer, F. Schroeppel, and J. Sauer, Tetrahedron Lett., 2899 (1972).

98.* R. S. Atkinson and C. W. Rees, J. Chem. Soc., Chem. Commun., 1230 (1967); D. J. Anderson, T. L. Gilchrist, D. C. Horwell, and C. W. Rees, J. Chem. Soc. (C), 576 (1970).

99. D. J. C. Adams, S. Bradbury, D. C. Horwell, M. Keating, C. W. Rees, and R. C. Storr, J. Chem. Soc., Chem. Commun., 828 (1971).

100. F. Schroeppel and J. Sauer, Tetrahedron Lett., 2945 (1974).

101.* J. R. Boehm, A. L. Balch, K. F. Bizot, and J. H. Enemark, J. Am. Chem. Soc., 97, 501 (1975).

102. D. J. Anderson, T. L. Gilchrist, D. C. Horwell, and C. W. Rees, J. Chem. Soc., Chem. Commun., 146 (1969).

103. C. W. Rees and M. Yelland, J. Chem. Soc., Chem. Commun., 377 (1969).

104. S. B. Matin, J. C. Craig, and R. P. K. Chan, J. Org. Chem., 39, 2285 (1974).

105. B. V. Ioffe and L. A. Kartsova, Zh. Org. Khim., 10, 989 (1974).

106. S. Mataka and J.-P. Amselme, J. Chem. Soc., Chem. Commun., 554 (1974).

107. M. A. Kuznetsov and B. V. Ioffe, Zh. Org. Khim., 11, 1420 (1975); Chem. Abstr., 83, 147049v (1975).

108.* D. W. Jones, J. Chem. Soc., Chem. Commun., 884 (1972).

109.* D. J. Anderson, T. L. Gilchrist, and C. W. Rees., J. Chem. Soc., Chem. Commun., 800 (1971).

110. L. Hoesch and A. S. Dreiding, Helv. Chim. Acta, 58, 980 (1975).

111. J. E. Baldwin, J. E. Brown, and G. Höfle, J. Am. Chem. Soc., 93, 788 (1971).

112. R. S. Atkinson and S. B. Awad, J. Chem. Soc., Chem. Commun., 651 (1975).

113. R. S. Atkinson and C. W. Rees, J. Chem. Soc. (C), 772 (1969).

114. L. Hoesch and A. S. Dreiding, Helv. Chim. Acta, 58, 1995 (1975).

115. G. R. Meyer and J. Stavinoha, Jr., J. Heterocycl. Chem., 12, 1085 (1975).

116. L. Hoesch, Chimia, 29, 531 (1975).

117. R. S. Atkinson and R. Martin, J. Chem. Soc., Chem. Commun., 386 (1974).

118. R. S. Atkinson and J. R. Malpass, J. Chem. Soc., Chem. Commun., 555 (1975).

119. H. Person, F. Tonnard, A. Foucaud, and C. Fayat, Tetrahedron Lett., 2495 (1973).

120. H. Person, C. Fayat, F. Tonnard, and A. Foucaud, Bull. Soc. Chim. France, 635 (1974).

121.* D. J. Anderson, T. L. Gilchrist, and C. W. Rees, J. Chem. Soc., Chem. Commun., 147 (1969).

122. S. Colonna and C. J. M. Stirling, J. Chem. Soc., Chem. Commun., 1591 (1971).

7

SILYLENES

PETER P. GASPAR

Washington University Saint Louis, Missouri 63130

---

*This publication was prepared with financial assistance from the United States Energy Research and Development Administration. This is technical report COO-1713-71.

I. INTRODUCTION

When this author, together with Professor B. J. Herold, last attempted to survey the chemistry of silylenes in 1970,[1] a great deal of activity in the field was apparent. The attention of organic and physical chemists had been captured by carbenes almost two decades earlier. The continuous stream of interesting results that flowed from investigations of carbene chemistry led quite naturally to a widespread curiosity about the behavior of higher Group IV carbene analogs. Indeed, the general acceptance of silylenes within the canon of reactive intermediates (as opposed to mere structural and spectroscopic curiosities) occurred only when insertion by silylenes into the silicon-hydrogen bond was firmly established in 1970. This is, of course, an analog of the carbene insertion reaction, and led us to warn that the distinctive features of the chemistry of divalent Group IV compounds might be overlooked if the interpretation of their reactions were colored by the expectation that their behavior would resemble that of carbenes.

Our present task is to review the developments in silylene chemistry in the years 1975 and 1976, but most of the work of the 1970s is mentioned, and some earlier background material is presented. The last 2 years have seen tremendous strides in the development of new routes to the generation of silylenes, the discovery of new reactions, and the growth of our knowledge about how silylene reactions occur. Many of the most exciting recent discoveries related to silylene reactions, which closely resemble carbene reactions. Are we to conclude from our present knowledge that nature has designed silylenes to humbly mimic carbenes, or is the situation different? Chemists make the discoveries for which their minds are prepared. Perhaps if we were cleverer we would have found the novel reactions of silylenes, and perhaps we still shall.

II. THE GENERATION OF SILYLENES

A. Thermal Silylene Extrusions

Five years ago the established methods for the generation of silylenes required rather harsh conditions, such as the temperatures above 800°C needed for the <u>anti</u>-disproportionation of tetrahalosilanes over silicon, a general method for the preparation of dihalosilylenes:[1]

$$SiX_4 + Si \rightleftharpoons 2\ SiX_2 \quad (X = F, Cl, Br, I)$$

Recently, much milder methods, both thermal and photochemical, have been developed for silylene generation. A great deal of effective synthetic and mechanistic work is still carried out employing the high-temperature extrusion of silylenes from disilanes, a reaction whose mechanism is being intensively studied. We briefly review these and other thermal silylene extrusions and silylene-producing silicon atom insertions before turning our attention to photochemical silylene extrusion and the thermolysis of silacyclopropanes, themselves unknown in 1970.[2] The persistent question as to whether silylene arises in the pyrolysis of monosilane has also continued to receive attention, and it is treated here.

A general method for the production of silylenes is the unsymmetrical thermolysis, with rearrangement, of disilanes:[3]

$$XYZSi\text{-}SiXYZ \rightarrow XYSi: + SiXYZ_2$$

The migrating group Z can be hydrogen, halogen, or alkoxy, whereas X and Y can be hydrogen, halogen, alkoxy, alkyl, or aryl. Competition between hydrogen and halogen shifts in the thermolysis of appropriately substituted disilanes has been studied.[4]

$$RSihal + Si(hal)RH_2 \xleftarrow[\text{shift}]{\text{1,2-hydrogen}} R(H)(hal)Si\text{—}Si(hal)(H)R \xrightarrow[\text{shift}]{\text{1,2-halogen}} RSiH + Si(hal)_2RH$$

The relative rate of 1,2-hydrogen shift compared to 1,2-chlorine shift was 4.4 ± 0.4 for MeHClSiSiHClMe. Only hydrogen shifts were observed for FH_2SiSiH_2F and ClH_2SiSiH_2Cl. Thus these latter compounds are clean sources of HSiF and HSiCl, respectively.

If 10- to 20-s. contact times in the hot zone are employed for gas-phase disilane pyrolyses, temperatures of 325° to 350°C are required for the decomposition of a 1,2-dimethoxytetraalkyldisilane. Even higher temperatures (500° to 525°C) are needed for a 1,2-dichloro compound. Dihydro- and difluorodisilanes require intermediate temperatures.[3] These reactions have half-lives of a few hours in the liquid phase at temperatures about 150° lower than those employed in the gas phase. When the flash-vacuum-pyrolysis technique of Hedaya[5] is applied to silylene extrusion from disilanes, milder conditions are achieved because contact times are less than 0.1 s, although temperatures above 600°C are employed.[6]

The clear demonstration by Atwell and Weyenberg of the generality of the high-temperature extrusion of silylenes from disilanes[7] has given perhaps the greatest impetus to the study and application of silylene chemistry in the past decade. The careful kinetic studies of Atwell, Weyenberg, and colleagues,[8,9]

Purnell and coworkers,[10-12] and Ring and coworkers,[13] and the trapping experiments carried out by these workers[8,14-16] dispelled any doubts that free silylenes were formed. Particularly convincing was the finding that copyrolysis of disilane in the presence of alkylsilanes containing Si-H bonds led to the formation of the alkyldisilanes that would result from Si-H insertion:[14,15]

$$Si_2H_6 + H_{\underline{n}}SiMe_{(4-\underline{n})} \rightarrow SiH_4 + H_3SiSiH_{(\underline{n}-1)}Me_{(4-\underline{n})}$$

$$(\underline{n} = 1,2,3,4)$$

$$Si_2D_6 + H_3SiMe \rightarrow SiD_4 + HD_2SiSiMe$$

The absence of radical coupling products $Me_{(4-\underline{n})}H_{(\underline{n}-1)}SiSiH_{(\underline{n}-1)}Me_{(4-\underline{n})}$ eliminated from serious consideration an abstraction-recombination mechanism.

B. Silicon Atom Insertion

The reactions of atomic silicon afford another rather general route to the formation of silylenes. A decade ago Skell and Owen reported the formation of 1,1,1,3,3,3-hexamethyltrisilane from the cocondensation of silicon vapor and trimethylsilane and suggested that insertion of a silicon atom into the Si-H bond of trimethylsilane produces a singlet silylene that can attack another trimethylsilane molecule.[17]

$$Si + HSiMe_3 \rightarrow HSiSiMe_3$$

$$Me_3SiSiH + HSiMe_3 \rightarrow Me_3SiSiH_2SiMe_3$$

Similar consecutive insertions have been reported by Skell for evaporated silicon atoms and disilane, trimethylsilane, dimethylsilane, methylsilane, methanol, hydrogen chloride, and hydrogen bromide.[18,19] Trisilane yields of approximately 30% have been obtained from such reactions.

Recoiling silicon atoms also give rise to trisilanes on reaction with silanes, but much higher yields of disilanes are obtained, for instance:

$$^{31}Si + SiH_4 \rightarrow {}^{31}SiH_4 + H_3{}^{31}SiSiH_3 + H_3Si{}^{31}SiH_2SiH_3$$

$$13 \pm 2\% \quad 48 \pm 5\% \quad 7 \pm 2\% \text{(refs. 20,21)}$$

$$^{31}Si + HSiMe_3 \rightarrow H_3{}^{31}SiSiMe_3 + Me_3Si{}^{31}SiH_2SiMe_3$$

$$30\% \quad 15\% \quad \text{(ref. 22)}$$

It has been suggested that the disilane (and silane) products in the reactions of recoiling silicon atoms arise from silylene intermediates formed by primary abstraction or insertion/dissociation processes.[21]

$$^{31}Si + H\text{-}\overset{|}{\underset{|}{Si}}\text{-} \rightarrow \begin{cases} ^{31}SiH \rightarrow \rightarrow {}^{31}SiH_2 \rightarrow \text{products} \\ \uparrow \\ [H\text{-}^{31}Si\text{-}Si \leftarrow]^* \end{cases}$$

An indeed elegant experiment based on the premise of silylene formation in the reactions of recoiling silicon atoms has been employed to assign a singlet electronic state to nucleogenic silylene.[23] It has recently been found, however, that the products that arise from the reactions of recoiling silicon atoms with organic 1,3-dienes, even in the presence of silane, differ from the products of thermally generated SiH_2 with the same substrates.[24] This discrepency may be due to differences in reaction conditions, but it does reopen the question as to whether silylene is really an important intermediate in recoil systems.

The extrusion of silylenes from 7-silanorbornadienes occurs over a wide temperature range, depending on the thermal stability of the bicyclic compound. This method for silylene generation was pioneered by Gilman and coworkers.[25,26] For the liquid-phase pyrolysis of 2,3-dibenzo-1,2,3,4-tetraphenyl-7,7-dimethyl-7-silanorbornadiene, a temperature of 300°C was required.[25] In contrast, the 1,2-diphenyl analog apparently eliminates dimethylsilylene at temperatures below 0°C.[27]

Me Me Si ϕ ϕ ϕ ϕ —Δ, 300°→ ϕ ϕ ϕ ϕ (1,2,3,4-tetraphenylnaphthalene) + $SiMe_2$

A 7-silanorbornadiene that extrudes dimethylsilylene at the temperature of refluxing carbon tetrachloride has been reported.[28] The use of silanorbornadienes as silylene sources must be regarded with caution in view of Barton's evidence that silylene extrusion may be stepwise,[29] and thus silylene transfer may occur without the formation of free silylene.

C. Photochemical Silylene Extrusions

Silylenes have been directly detected in flash-photolysis experiments and implicated in other photochemical studies,[1] but it is only in the last few years that the photolysis of polysilanes has been recognized as a useful source of silylenes for reaction studies.

Extrusion of dimethylsilylene from dodecamethylcyclohexasilane under irradiation with a low-pressure mercury lamp, first reported in 1970,[30] was deduced from the formation of Si-H insertion products from the silylene as well as the lower homolog of the cyclosilane. Small quantities of other products were also formed.

$$(Me_2Si)_6 \xrightarrow[HSiMeEt_2]{h\nu} (Me_2Si)_5 + HSiMe_2SiMeEt_2$$

Linear and branched-chain polysilanes $Me(SiMe_2)_nMe$ (n = 3-6) and $Me_3Si[(Me_3Si)SiMe]_nSiMe_3$ (n = 1,2) were also formed under photolysis with loss of $SiMe_2$, which could be trapped by insertion into $HSiMeEt_2$.[31] When cyclic organotrisilanes were subjected to photolysis, it was found that the central silicon is extruded exclusively. A variety of silylenes has thus been generated, as indicated by trapping experiments with H-Si bonds.[32]

$$\text{cyclo-}(CH_2)_4(SiMe_2)SiRR'(SiMe_2) \xrightarrow[HSiMeEt_2]{h\nu} \text{cyclo-}(CH_2)_4(SiMe_2)_2 + HSiMe_2SiMeEt_2$$

R	R′
Me	i-Pr
Me	Mr
$(CH_2)_4$	

The short wavelengths required for photolysis of saturated polysilanes can also cause secondary reactions, a disadvantage shared with high-temperature pyrolysis. The longer wavelength radiation from a high-pressure mercury arc suffices for the extrusion of methylphenylsilylene from 2,3-diphenyloctamethyltetrasilane.[33] Insertion into $HSiMeEt_2$ was again observed, as well as addition of methylphenylsilylene to cyclohexene, which is discussed in the text that follows.

Extrusion of silylene from the trisilacycloheptane system has been found to be highly stereospecific, with complete retention of the *cis-trans* stereochemistry for both *cis-* and *trans-*starting compounds.[34] This result is consistent with a concerted elimination of the silylene, but Si-Si homolysis as the initial step is not regorously excluded, since chiral silyl radicals have been found to be configurationally stable.

On the *assumption* that silylene extrusion is concerted in

$$\text{cyclo-}[(CH_2)_4\text{-SiMePh-SiMe}_2\text{-SiMePh}] \xrightarrow[HSiMeEt_2]{h\nu} \text{cyclo-}[(CH_2)_4\text{-SiMePh-SiMePh}] + HSiMe_2SiMeEt_2$$

polysilane photolysis, Ramsey has deduced from application of the Woodward-Hoffmann rules that a $\sigma \rightarrow \sigma^*$ transition of the polysilane leads to extrusion of a silylene in an excited state, $^3B_1$ or $^1B_1$.[35] Since it is believed that thermal methods produce ground-state $^1A_1$ silylenes, it would be very useful if photochemical methods yielded higher electronic states. No experimental evidence in support of this interesting prediction has yet been presented.

Whether silylene extrusion in polysilane photolyses is concerted does not seem to have been clearly established, and some evidence indicates otherwise. Careful study of dodecamethylcyclohexasilane photolysis in cyclohexane revealed that the photolysis rate depends strongly on the concentration of the cyclic polysilane.[36] When the cyclohexasilane is irradiated in the presence of a large excess of $HSiMeEt_2$, the apparent silylene insertion product $HSiMe_2SiMeEt_2$ is the major product, formed in yields as high as 46%, but $H(SiMe_2)_2SiMeEt_2$ is also formed, in yields as high as 18%. When Me_2SiCl_2 is the trapping agent, the yield of $Cl(SiMe_2)_2Cl$ is smaller than the yields of $Cl(SiMe_2)_3Cl$ and $Cl(SiMe_2)_4Cl$. It is apparent that processes other than simple silylene extrusion occur in these photochemical reactions.

To account for the formation of silanes of the type $Me(SiMe_2)_nH$ in the photolysis of permethylated linear polysilanes, Kumada and coworkers have postulated that homolysis of Si-Si bonds, followed by hydrogen abstraction, accompanies the extrusion of dimethylsilylene.[37] It is to be hoped that the tools of modern mechanistic photochemistry will soon be brought to bear on the elucidation of the primary processes in polysilane photolysis. A surer knowledge of what actually transpires will make these photochemical reactions much more useful for the study of silylene reactions. Sakurai has himself stated that homolytic fission of a Si-Si bond, followed by S_H2 attack of a silyl radical center on the remaining Si-Si bond, cannot at present be excluded as an alternative to concerted extrusion:[34]

$$\text{cyclo-}[(CH_2)_4\text{-Si-Si-Si}] \xrightarrow{h\nu} \cdot Si\text{-}(CH_2)_4\text{-Si-Si}\cdot \longrightarrow \text{cyclo-}[(CH_2)_4\text{-Si-Si}] + :Si$$

Unfortunately, this biradical primary intermediate could also act as a silylenoid, a silylene transfer agent. Thus until the primary photochemical processes are established, the intermediacy of free silylene must be regarded as very probable but not proven. As we see in our discussion of the insertion reaction, chemical criteria for the formation of free silylenes are still rather vague. One reaction has been removed from the canon of silylene-generating reactions, and another remains in limbo.

D. Nonformation of Silylenes in Reactions of Alkali Metals with Halosilane-Diene Mixtures

The reductive dehalogenation of dihalosilanes by alkali metals has been claimed to be a source of silylenes in the liquid phase on the basis of the products obtained.[7,38,39]

$$Me_2SiCl_2 + \text{(2,3-dimethylbutadiene)} \xrightarrow[THF]{Na} Me_2Si\text{(3,4-dimethylsilacyclopent-3-ene)} \quad \text{(ref. 38)}$$

$$Me_2ClSiSiClMe_2 + \text{(2,3-dimethylbutadiene)} \xrightarrow[THF]{Li} [Me_2Si] \longrightarrow Me_2Si\text{(3,4-dimethylsilacyclopent-3-ene)} \quad \text{(ref. 39)}$$

It has, however, been demonstrated that the diene rather than the halosilane reacts initially with the alkali metal.[40,41] No firm evidence for the involvement of silylenes in these liquid-phase reactions has been produced. In the gas-phase however the reaction of dichlorodimethylsilane with potassium-sodium vapor in the presence of trimethylsilane led to the formation of the silylene insertion product.[42] No reaction of a halosilane with

$$Me_2SiCl_2 + HSiMe_3 \xrightarrow{K/Na} Me_2HSiSiMe_3$$

alkali metal in solution has produced an intermediate that has given a silylene insertion product on reaction with a Si-H bond. Using the Si-H insertion reaction, perhaps naively, as the touchstone for silylene formation, it seems likely that the gas-phase experiment of Skell and Goldstein[42] did generate dimethylsilylene.

E. Monosilane Pyrolysis

The question as to whether pyrolysis of monosilane SiH_4 releases silylene has remained controversial for over 40 years.[1] In a careful kinetic study of the gas-phase static pyrolysis of SiH_4, Purnell and Walsh found that the initial reaction is of

3/2 order and concluded that it is a unimolecular reaction outside of its first-order region of pressure dependence at the pressures studied (35 to 230 torr).[43] On thermochemical grounds, silylene formation $SiH_4 \rightarrow SiH_2 + H_2$ was preferred over simple homolysis $SiH_4 \rightarrow SiH_3 + H$ as the initial step, but the thermochemistry of silicon compounds is itself an area in which much is still being learned.[44] When Ring, Puentes, et al. studied the pyrolysis of SiH_4-SiD_4 mixtures under both static and flow conditions, they fuond large amounts of HD in addition to D_2 and H_2.[45] The formation of HD was explained in terms of primary homolysis followed by abstraction of hydrogen by the hydrogen atoms liberated.

$$SiH_4 \rightarrow SiH_3 + H$$

$$SiD_4 \rightarrow SiD_3 + D$$

$$H + SiH_4 \rightarrow H_2 + SiH_3$$

$$H + SiD_4 \rightarrow HD + SiD_3$$

$$D + SiH_4 \rightarrow HD + SiH_3$$

$$D + SiD_4 \rightarrow D_2 + SiD_3$$

In the flow-pyrolysis experiments, however, the disilanes formed were mostly those containing even numbers of the hydrogen isotopes $Si_2H_{2\underline{n}}D_{6-2\underline{n}}$ ($\underline{n}$ = 0,1,2,3), with only small amounts of Si_2HD_5, $Si_2H_3D_3$, and Si_2H_5D. This isotopic distribution of the disilane products was rationalized in terms of disilane formation via attack of a silyl radical on silane in such a manner that the attacking radical loses a hydrogen atom while it inserts into a Si-H bond.

$$SiD_3 + SiH_4 \rightarrow \left[\begin{array}{c} .H. \\ D_2\dot{S}i\cdots\dot{S}iH_3 \\ D^{\cdot} \end{array}\right]^{\neq} \rightarrow D + HD_2SiSiH_3$$

This unusual process was preferred over silyl radical recombination as the source of disilane in order to explain the long chain length ($10^{11.5}$) estimated for the reaction. A chain reaction is required to bring the activation energy for silane pyrolysis (55.9 kcal/mol)[43] into accord with the higher estimates (77-94 kcal/mol) of the Si-H bond-dissociation energy, if cleavage into SiH_3 + H is postulated as the primary step in silane pyrolysis. However, the isotopic distribution of disilanes obtained from silane pyrolysis appears equally in accordance with a mechanism involving silylene formation, which would give only product molecules with even numbers of deuteriums and hydrogens:

$$SiH_4 \rightarrow SiH_2 + H_2$$

$$SiD_4 \rightarrow SiD_2 + D_2$$

$$SiH_2 + SiH_4 \rightarrow Si_2H_6$$

$$SiH_2 + SiD_4 \rightarrow Si_2H_2D_4$$

$$SiD_2 + SiH_4 \rightarrow Si_2H_4D_2$$

$$SiD_2 + SiD_4 \rightarrow Si_2D_6$$

John and Purnell have calculated absolute entropies and free-energy functions for singlet SiH_2 and for SiH_3.[46] They deduced an equilibrium constant for the reaction $SiH_3 + H \rightleftharpoons SiH_2 + H_2$ of $10^{15.9}$ at 600°C and $10^{11.3}$ for reaction $SiH_3 + SiH_3 \rightleftharpoons SiH_2 + SiH_4$ at the same temperature. Since disilane dissociates into $SiH_2 + SiH_4$ rather than into 2 SiH_3, and the former equilibrium favors SiH_2 even more than the latter, it was argued that dissociation of silane into $SiH_2 + H_2$ was supported. Although it appears inevitable that SiH_3 is thermodynamically unstable with respect to SiH_2 at the pyrolysis temperatures usually employed, the kinetic question concerning the primary pyrolysis mechanism cannot be decided by pointing to the relative thermodynamic stability of the pyrolysis products. The thermodynamic considerations do, however, seem hospitable to the belief that silylene should be a prominent intermediate in silane pyrolysis, even if the initial step produces silyl radicals, since disproportionation of silyl radicals could be a favorable and efficient process. Very recently, the ratio of disproportionation to recombination for silyl radicals at room temperature has been found to be 0.7.[47]

Recent chemical evidence strongly supports the view that silane pyrolysis produces, at least initially, silyl radicals, but not silylene.[48] Pyrolysis of disilane in the presence of acetylene leads to the formation of ethynylsilane $SiH_3C{\equiv}CH$ (the mechanism for this reaction is discussed in Section III.B) as the major silylene addition product. When silane was pyrolyzed in the presence of acetylene, vinylsilane $SiH_3CH{=}CH_2$ was the major product, ethynylsilane only being produced in a secondary reaction. Since vinylsilane is the expected product from addition of silyl radicals to acetylene in the presence of a good hydrogen donor such as silane, these results support the formation of silyl radicals in the initial step of silane pyrolysis. Haas and Ring also pointed out that an orbital symmetry restriction may exist for concerted elimination of H_2 from SiH_4 via a least-motion C_{2v} path.[48]

$$SiH_4 \rightarrow SiH_3 + H$$

$$SiH_3 + HC{\equiv}CH \rightarrow SiH_3CH{=}\dot{C}H \xrightarrow{SiH_4} SiH_3CH{=}CH_2 + SiH_3$$

It should be pointed out, however, that formation of vinylsilane in the preceding sequence is a chain process, and thus a very small number of silyl radicals could lead to a large amount of vinylsilane product. Most of the results on silane pyrolysis can be accommodated by a mechanism in which initial cleavage produces silyl radicals that, if not immediately trapped, can undergo disproportionation to silylene.

F. Silirane Pyrolysis

A recently discovered class of organosilicon compounds, namely, the silacyclopropanes,[2,49,50] has provided what may prove to be a widely applicable, very gentle method for the generation of dimethylsilylene.

Recently it has been found that the thermolysis of hexamethylsilacyclopropane[51] leads to the loss of the dimethylsilylene unit under very mild conditions.[52] Tetramethylethylene is obtained in high yield, and when the decomposition is carried out in the presence of suitable substrates, products expected from dimethylsilylene are obtained, as follows:

$$\text{hexamethylsilacyclopropane} \xrightarrow[73^\circ\ 15\ h]{Me_2Si(OMe)_2} Me_2Si(OMe){-}Si(OMe)Me_2\ (71\%) + Me_2C{=}CMe_2\ (95\%)$$

$$\xrightarrow[108^\circ\ 14\ h]{HSiEt_3} HMe_2SiSiEt_3\ (59\%) + Me_2C{=}CMe_2\ (71\%)$$

$$\xrightarrow[71^\circ\ 15\ hr]{HSiMe_2\text{i-Pr}} HMe_2SiSiMe_2\text{iPr}\ (40\%) + Me_2C{=}CMe_2\ (97\%)$$

$$\xrightarrow[73^\circ\ 17\ h]{H_2SiEt_2} HMe_2SiSiEt_2H\ (48\%) + Me_2C{=}CMe_2\ (82\%)$$

$$\xrightarrow[66^\circ\ 17\ h]{\text{2,3-dimethylbutadiene, THF}} Me_2Si\text{(3,4-dimethylsilacyclopent-3-ene)}\ (34\%) + Me_2C{=}CMe_2\ (37\%)$$

The mechanism of dimethylsilylene loss from hexamethylsilacyclopropane has not yet been determined. When the bisspiro compound shown in the structure that follows was subjected to pyrolysis, no dimethylsilylene was intercepted;[52] instead, a dimer of the starting material was obtained.

$$\text{bisspiro silacyclopropane} \xrightarrow[HSiEt_2]{\Delta} \text{dimer}$$

When hexamethylsilacyclopropane is pyrolyzed in the presence of styrene and α-methylstyrene, silacyclopentanes are formed[53,54] that may result from the trapping of a 1,3-diradical $Me_2\dot{C}-CMe_2-\dot{S}iMe_2$.

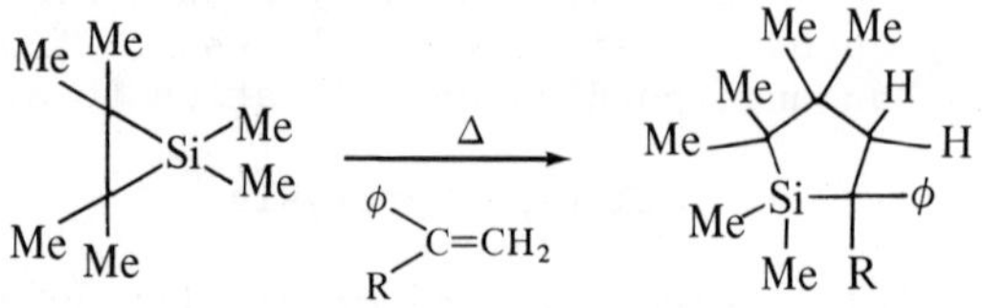

(R = H, CH_3)

Significantly, the yield of silylene addition product dropped to zero when hexamethylsilacyclopropane was heated in 2,3-dimethylbutadiene, without diluent.[52] This may be due to competition between silylene loss from a 1,3-diradical intermediate and trapping of the diradical prior to such loss. Thus until the mechanism of silylene extrustion from silacyclopropanes is elucidated, the possibility remains that intermediates other than free silylenes are transformed into the products observed.

III. THE REACTIONS OF SILYLENES

A. Silylene Insertion Reactions

In remarkable contrast with the chemistry of carbenes, almost all known thermal reactions that generate silylenes are reversible. Thus for the extrusion reactions from disilanes that involve the migration of a hydrogen, halogen, or alkoxy group, there exist corresponding insertions of silylenes into silicon-hydrogen, silicon-halogen, and silicon-oxygen bonds.[1] Since the principle of microscopic reversibility demands that the corresponding insertion and extrusion reactions pass through the same transition state if the reactants and products lie on the same potential surface for both reactions, there has been considerable recent progress toward understanding the insertion reaction via kinetic studies of silylene extrusion.

We briefly review the evidence for the occurence of direct insertion reactions for silylenes and the relative reactivity of various Si-H bonds toward silylene insertion before describing the kinetic studies of silylene insertion and extrusion reactions that shed light on the same transition state. Recent results on the relative insertion rates of various silylenes toward Si-H bonds conclude this section.

The insertion of silylenes into silicon-hydrogen bonds has been studied extensively. That direct insertion occurs:

$$R_2Si: + H\text{-}SiXYZ \rightarrow HR_2Si\text{-}SiXYZ$$

rather than hydrogen abstraction followed by radical recombination, is indicated by the absence of geminate recombination products expected for such a process:

$$R_2Si: + H\text{-}SiXYZ \rightarrow R_2SiH\cdot + \cdot SiXYZ$$

$$R_2SiH\cdot + \cdot SiXYZ \rightarrow HR_2Si\text{-}SiXYZ$$

$$R_2SiH\cdot + \cdot HSiR_2 \rightarrow HR_2Si\text{-}SiR_2H$$

$$XYZSi\cdot + \cdot SiXYZ \rightarrow XYZSi\text{-}SiXYZ$$

The occurence of direct insertion has been taken as evidence that it is the singlet state of silylenes that takes part in insertion reactions.[42] Further evidence for direct insertion by singlet silylene is the stereospecific insertion with retention of configuration found when conformationally stable 4-*tert*-butyl-1-silacyclohexanes were used as reaction substrates for photochemically generated dimethylsilylene.[55] Interestingly, photogenic dimethylsilylene has been reported to react as a nucleophile with $XC_6H_4SiMe_2H$, exhibiting a Hammett ρ constant +0.84.[56]

Two groups have studied the rate of insertion of SiH_2 from the pyrolysis of disilane into the Si-H bonds of alkylsilanes. Sefcik and Ring employed a circulating flow system at 350°C for their experiments,[57] and Cox and Purnell carried out a static pyrolysis at 320°C.[12] As Table 1 indicates, the results from the two series of experiments differ somewhat. Ring has suggested that in the static pyrolyses the insertion products may suffer secondary pyrolysis. In the circulating flow-system experiments, reaction products are condensed immediately and thus avoid secondary decomposition.

Both groups have found that substitution of methyl groups for the hydrogens of silane enhances the reactivity of the remaining Si-H bonds toward insertion by silylene. Sefcik and Ring suggested that silylene reacts as an electrophile and that the reactivity differences reflect differences in the hydridic character of the Si-H bonds. Cox and Purnell intered these reactivity differences in terms of differences in the Si-H bond-dissociation energies, but this does not conflict with the interpretation of Sefcik and Ring, since Hollandsworth and Ring had previously deduced from infrared stretching frequencies that the Si-H bond strength increases in the order $Me_3SiH < Me_2SiH_2 < MeSiH_3$.[58] Sefcik and Ring have suggested that the steric effect of a methyl group opposes the electronic effect and tends to slow the rate of silylene insertion into the Si-H bonds of methylsilanes.[59]

The transition state usually invoked for insertion of silylene into an Si-H bond involves a three-center interaction,

Table 1. Relative Insertion Rates of SiH_2 (per Si-H Bond)

Trapping Agent	Relative Insertion Rate	
	Sefcik and Ring[57]	Cox and Purnell[12]
Si_2H_6	1.0	1.0
Me_3SiH	1.6 ± 0.2	3.06
Me_2SiH_2	0.7 ± 0.1	1.59
$MeSiH_3$	0.4 ± 0.05	1.80
SiH_4	0.3 ± 0.1[a]	0.47
Si_3H_8 (primary Si-H)		3.31
Si_3H_8 (secondary Si-H)		2.46
$ClSiH_3$	<0.01[b]	

[a]From recoiling silicon atoms: P. P. Gaspar and P. Markusch, *Chem. Commun.*, 1331 (1970).
[b]Insertion product not observed.

the hydrogen bridging the two silicon atoms between which a new bond is being formed.[12]

$$SiH_2 + H\ \ SiR_3 \rightarrow \left[H_2Si \overset{H}{\cdots} SiR_3 \right]^{\neq} \rightarrow H_3Si\ \ SiR_3$$

Such a transition state has also been proposed for silylene extrusion from a disilane, the inverse of silylene insertion.[57] Vanderwielen, Ring, et al. account for the low energy of the transition state for silylene extrusion from a disilane and hence for silylene insertion into an Si-H of a silane, by invoking pentacovalent bonding to silicon in the transition state.[13] This pentacovalent bonding is believed by Ring to effect a weakening of the bonds to that center, thus leading to positive entropies of activation and a large preexponential A factor for silylene extrusion. When a hydrogen atom shifts to a silicon atom carrying a heavy group during silylene extrusion, abnormally high A factors are found, as indicated in Table 2. The activation energy for primary hydrogen migration is higher than that for secondary hydrogen migration, but the A factor is also higher for primary hydrogen migration.

$$RSiH + SiH_4 \leftarrow \left[RSiH \underset{H}{\cdots} SiH_3 \right]^{\neq} \rightleftharpoons RSiH_2SiH_3 \rightleftharpoons \left[RSiH_2 \underset{H}{\cdots} SiH_2 \right]^{\neq}$$

$$\rightarrow RSiH_3 + SiH_2$$

Table 2. Arrhenius Parameters for Polysilane Decomposition Reactions

Reaction	log A (s^{-1})	E_a (kcal/mol)	Reference
$Si_2H_6 \rightleftharpoons SiH_2 + SiH_4$	14.5	49.3	10
$MeSi_2H_5 \rightleftharpoons SiH_2 + MeSiH_3$	15.3	50.8	13
$MeSi_2H_5 \rightleftharpoons MeSiH + SiH_4$	14.1	49.9	13
$Si_3H_8 \rightleftharpoons SiH_2 + Si_2H_6$	15.7	53.0	13
$Si_3H_8 \rightleftharpoons SiH_3SiH + SiH_4$	14.7	49.2	13
$Me_3SiSiH_3 \rightleftharpoons SiH_2 + Me_3SiH$	14.48 ± 0.3	48.00 ± 0.5	60
$Me_3SiSiMe_3 \rightleftharpoons 2\ Me_3Si$	17.5	80.5	61

Paquin and Ring have proposed a model, based on the entropies of activation, for silylene extrusion from substituted disilanes where the silicon atom to which a hydrogen atom migrates undergoes rehybridization from sp^3 to dsp^3 in going to the transition state and thus becomes a trigonal bipyramid.[60] In this model, reaction is favored by a heavy group in the axial position of the pentacovalent silicon in the transition state.

The remarkable feature of silylene insertion reactions, namely, that they occur with very strong bonds such as silicon-oxygen and silicon-halogen, is easily rationalized when the relationship of the three-centered transition state to that in a very large number of molecular rearrangements of organosilicon compounds is considered.[62]

$$R_3Si\text{-}\underset{\underset{:N^-}{|}}{G} \rightarrow R_3Si\text{-}N\text{-}G:^-$$

In these rearrangments G is often a carbon atom and N an oxygen atom. In the case of silylene extrusion the stability of the silylene formed offers a driving force for migration of neutral N and cleavage of the N-G bond. The predominance of silylene formation over simple homolysis in the thermal decomposition of disilanes containing Si-H, Si-halogen, and Si-OR bonds has been related to the activation energy for silylene insertion,[63] and the simple homolysis of peralkylated disilanes has been explained in these terms. The contrast between the activation parameters for decomposition of hexamethyldisilane and the other polysilanes listed in Table 2 is dramatic.

Kinetic studies of disilane decomposition have permitted estimates of the preexponential A factors for silylene insertion reactions. The A factors for SiH_2 insertion into $MeSiH_3$ and Si_2H_6 were found to be nearly equal to the collision frequencies, and insertions into SiH_4 of MeSiH and SiH_3SiH have steric factors of at least 0.1.[13] Thus silylene insertion reactions are quite efficient, given enough energy to overcome the activation barrier. Activation energies for SiH_2 insertion into Si-H bonds are apparently nonzero. For insertion of SiH_2 into SiH_4, an activation energy of 1.3 ± 1.1 kcal/mol has been estimated.[11]

From a comparison of the A factor for silylene extrusion from $MeSiH_2SiH_3$ and from $SiH_3SiH_2SiH_3$ it was deduced that the thermal extrusion of the central SiH_2 unit of trisilane is unlikely.[13] Hence insertion of SiH_2 into silicon-silicon bonds, previously proposed, must also be considered improbable. This reaction was originally proposed to account for the high ratio of normal- to iso-tetrasilane products from the reactions of thermally generated silylene with trisilane,[14] and for the high reactivity of disilane versus silane toward the intermediates (believed to include $^{31}SiH_2$) in the reactions of recoiling silicon atoms.[20] The observation that the ratio of normal- to iso-pentasilane from the insertion of SiH_3SiH into trisilane is lower than the ratio of normal- to iso-tetrasilane from insertion of SiH_2 into trisilane was also interpreted as being indicative of insertion into silicon-silicon bonds.[57]

An interesting finding in the pyrolysis of Me_2SiSiH_3 was that at 260°C only about 70% of the silylene generated was trapped in insertion reactions.[60] From the formation of a considerable amount of polymer in the reactions of thermally evaporated silicon atoms with silanes, Skell and Owen concluded that silylenes were reasonably stable, at least at liquid nitrogen temperature.[19]

From the temperature dependence of the relative rates of insertion of SiH_2 into SiH_4 and $MeSiH_3$, a difference in the energies of activation $\underline{E}_a(MeSiH_3) - \underline{E}_a(SiH_4) = 4.2 \pm 1$ kcal/mol was determined, with $\log \underline{A}(MeSiH_3)/\underline{A}(SiH_4) - 1.7 \pm 0.3$.[60] Likewise, for insertion of SiH_2 into Si_2H_6 and Me_3SiH, $\underline{E}_a(Si_2H_6) - \underline{E}_a(Me_3SiH) = 2.9 \pm 1.5$ kcal/mol and $\log \underline{A}(Me_3SiH)/\underline{A}(Si_2H_6) = 1.4 \pm 0.5$ were determined. There was good agreement between the A factor differences deduced from insertion product ratios and those calculated from the A factors determined for the reverse reaction, silylene extrustion, combined with estimates of the overall entropy change in the reaction. The weakness in this approach to the deduction of activation parameters for the insertion reaction, based on kinetic studies of silylene extrusion, lies in the calculation of the absolute entropy of SiH_2,[46] which depends on structural parameters inferred from various spectroscopic measurements. These results do, however, support the existence of significant activation energy differences for silylene insertion reactions. Purnell and coworkers had deduced activation energies of 1.2 and 0.4 kcal/mol for insertion of SiH_2 into the Si-H bonds of silane and disilane, respectively.[11] Activation-energy differences for insertion of

SiH_2 into $MeSiH_3$, Me_2SiH_2, and Me_3SiH versus insertion into Si_2H_6 were estimated to be 0.8, 0.5 and 1.2 kcal/mol, respectively.[12] Apparently, agreement has not been reached regarding the relative activation energies of the silanes studied.

Many other insertion reactions of silylenes are known.[1] Recently discovered insertions of silylenes into strained silicon-carbon bonds are described in Section III.B.3.

Recently, the relative reactivity of silylenes toward silicon-hydrogen bonds has been studied, and the following order for silylene insertion reactions has been deduced: $SiH_2 > ClSiH > FSiH >> Cl_2Si, F_2Si$.[4] Indeed, no insertion by $SiCl_2$ of SiF_2 was detected. Since $SiCl_2$ is reported to be very reactive toward Si-Cl bonds,[1,64] it appears that Si-Cl bonds, at least in polychlorosilanes, are more reactive than Si-H bonds toward silylene insertion. However, $ClSiH_3$ is less than 0.01 times as reactive as SiH_4 toward SiH_2,[57] suggesting that SiH_2 and $SiCl_2$ show different preferences in insertion reactions. The relatively unreactive dimethoxysilylene is trapped by insertion into methanol[65] but is not trapped by trimethylsilane.[66]

$$(MeO)_2Si + MeOH \longrightarrow (MeO)_3SiH$$

$$(MeO)_2Si + HSiMe_3 \not\longrightarrow (MeO)_2HSiSiMe_3$$

Atwell has shown that a number of methylalkoxysilylenes (RO)MeSi: as well as dimethylsilylene undergo insertion into the O-H bonds of various alcohols.[67]

B. Silylene Addition Reactions

The reactions of silylenes with unsaturated hydrocarbons is an area in which spectacular progress has been made in both the finding of novel organosilicon reaction products and the understanding of reaction mechanisms. Most of the work on silylene additions is quite recent, but we take the brief time required to review the earlier work in each of the three subsections on addition reactions with olefins, dienes, and acetylenes. The addition reactions of silylenes form primary products very similar to those from the reactions of carbenes with the same substrates. This conclusion has emerged only within the past year, having been obscured by the facile rearrangements of the silylene primary adducts.

a. Addition to Olefins. While several reactions of thermally extruded dimethylsilylene with olefins have been claimed,[68] the earliest fully documented example is the gas-phase dehalogenation of dichlorodimethylsilane in the presence of ethylene reported by Skell and Goldstein in 1964.[69] The mechanism proposed for the formation of vinyldimethylsilane invoked the rearrangement of the silacyclopropane intermediate.

$$Me_2SiCl_2 + K/Na(g) \rightarrow Me_2Si$$

$$Me_2Si + CH_2{=}CH_2 \rightarrow \left[Me_2Si\langle{}^{CH_2}_{CH_2}| \right] \rightarrow Me_2SiHCH{=}CH_2$$

That dimethylsilylene was formed (this was its first report!) was supported by the formation of pentamethyldisilane when dehalogenation was performed in the presence of trimethylsilane (vide supra). Evidence for the intermediacy of a silacyclopropane intermediate was the formation of the same product from several other reaction systems for which a reaction channel leading to the silacyclopropane seems plausible.[69]

$$Me_3SiCHCl_2 + K/Na(g) \rightarrow Me_2Si\langle{}^{CH:}_{CH_3}$$

$$\downarrow \text{C–H insertion}$$

$$\left[Me_2Si\langle{}^{CH_2}_{CH_2}| \right] \rightarrow Me_2SiHCH{=}CH_2$$

$$\uparrow$$

$$Me_2Si(CH_2Cl)_2 + K/Na(g) \rightarrow Me_2Si\langle{}^{CH_2\cdot}_{CH_2\cdot}$$

In view of the possibility that a diradical intermediate intervenes in the extrusion of silylenes from silacyclopropanes (vide supra) and the formation of analogous products from the addition of silylenes to acetylenes (vide infra), the gas-phase rearrangement of 1-dimethylsilacyclopropane to vinyldimethylsilane seems quite reasonable, via $Me_2\dot{S}iCH_2\dot{C}H_2$.

Cocondensation of SiF_2 and ethylene leads to products containing two SiF_2 units, suggesting that silylene dimerization (vide infra) preceded addition.[70]

$$SiF_2 + CH_2{=}CH_2 \longrightarrow \text{(four-membered ring: } CH_2CH_2SiF_2SiF_2) + \text{(six-membered ring: } (CH_2)_4SiF_2SiF_2)$$

When SiF_2 was cocondensed with trifluoroethylene, products were obtained that were attributed to the rearrangement of a silacyclopropane intermediate formed by addition of monomeric SiF_2.[70,71]

$$SiF_2 + CF_2{=}CHF \longrightarrow [\text{1,1-difluoro-2,2,3-trifluoro-3-H-silacyclopropane}] \longrightarrow CF_2{=}CHSiF_3 + (E/Z)\ HFC{=}CFSiF_3 + HFC{=}CFSiF_3$$

It has not been established, however, that thermally generated SiF_2 undergoes reactions prior to dimerization.[1]

In 1968 cold water was thrown on the investigation of silylene-olefin reactions by the report of Atwell and Weyenberg that tetramethylethylene could not compete successfully for thermally generated dimethylsilylene with the silylene precursor sym-dimethoxytetramethyldisilane.[8]

Five years went by before the next report of an addition of a silylene to an olefin. In 1973 Ishikawa, Ishiguro, et al., reported the formation of a product believed to arise by rearrangement of a silacyclopropane intermediate, when methylphenylsilylene was photochemically generated in solution in the presence of cyclohexene.[33]

$$Me_3Si(Me\phi Si)_2SiMe_3 \xrightarrow{h\nu} Me\phi Si + Me_3SiMe\phi SiSiMe_3$$

$$Me\phi Si \xrightarrow{HSiMeEt_2} Me\phi SiHSiMeEt_2$$

$$Me\phi Si \xrightarrow{\text{cyclohexene}} [Me\phi Si\text{-norcarane}] \longrightarrow Me\phi SiH\text{-cyclohexenyl}\quad 10\%$$

An analogous product was obtained from the photochemical generation of dimethylsilylene in the presence of cyclohexene.

The occurence of the silicon analog of the classical carbene cyclopropanation of cyclohexene was supported by the observation that when methanol was added to the reaction mixture, either directly or 24 h after irradiation, a 20% yield of the methanolysis product expected from the silanorcarane was obtained.[72] Seyferth, Haas, et al. had reported previously that

$$Me\phi SiH\text{-cyclohexenyl} \xleftarrow{h\nu} Me(\phi)Si\text{-norcarane} + MeOH \longrightarrow Me\phi Si(OMe)\text{-cyclohexyl}$$

silacyclopropanes undergo facile methanolysis.[73] When solutions of the air-sensitive intermediate identified as the silanorcarane from its methanolysis product were subjected to irradiation with a high-pressure mercury lamp, cyclohexenylmethylphenylsilane was obtained in high yield. Similarly, methanolysis of the addition product of dimethylsilylene and cyclohexene gave a 6% yield of the product expected from the 7-dimethylsilanorcarane intermediate.

Several other examples have been reported recently of products from the photolysis of polysilanes in the presence of olefins. These products (see formula that follows) are rationalized in terms of silacyclopropane intermediates (formed by silylene addition) that rearrange to the observed products via 1,3-hydrogen shifts, as were invoked in the rearrangement of the silanorcaranes.[74] No 1,2-hydrogen migration products were observed.

$$Me_3Si\text{-}Si(\phi)(Me)\text{-}SiMe_3 \xrightarrow[CH_2=CH\text{-}CH_2SiMe_3]{h\nu} \left[\begin{array}{c} CH_2\text{-}CH(CH_2SiMe_3)\text{-}Si(Me)(\phi) \\ \text{(silacyclopropane)} \end{array} \right] \longrightarrow \phi MeSiHCH_2CH{=}CHSiMe_3 \quad 19\%$$

$$Me_3Si\text{-}Si(\phi)(Me)\text{-}SiMe_3 \xrightarrow[CH{=}C(CH_3)\text{-}SiMe_3]{h\nu} \left[\begin{array}{c} CH_2\text{-}C(SiMe_3)(CH_3)\text{-}Si(Me)(\phi) \\ \text{(silacyclopropane)} \end{array} \right] \longrightarrow \phi MeSiHCH_2\text{-}C(SiMe_3){=}CH_2 \quad 10\%$$

When vinyltrimethylsilane $Me_3SiCH{=}CH_2$, whose silylene adduct would be incapable of 1,3-hydrogen migration (see the addition of Me_2Si to ethylene earlier for another example), was employed as the reaction substrate, quite a different product was obtained.[74] This product of regiospecific double silylation

$$Me_3Si\text{-}Si(\phi)(Me)\text{-}SiMe_3 + CH_2{=}CHSiMe_3 \xrightarrow{h\nu} Me_3SiCH_2\text{-}CH(SiMe\phi SiMe_3)SiMe_3 \quad 14\%$$

does not seem to arise from a silylene intermediate and raises serious questions about the formation of free methylphenylsilylene in the other photochemical reactions of 2-phenylheptamethyltrisilane.

Recently, Seyferth, Annarelli, et al. have reported the transfer of dimethylsilylene from hexamethylsilacyclopropane to olefins.[54,75]

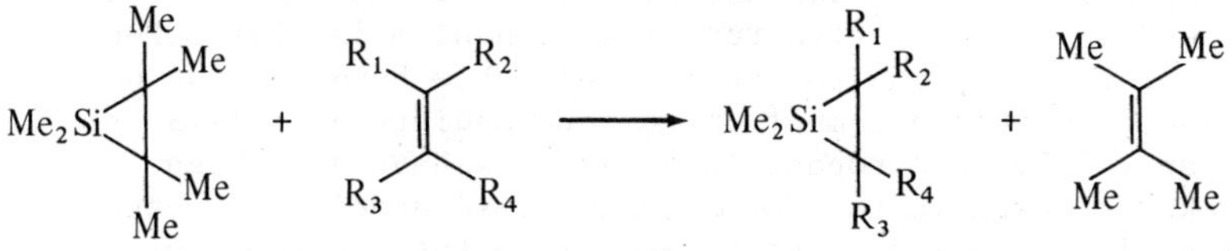

R_1	R_2	R_3	R_4
H	Me	H	Me
H	Me_3Si	H	Me_3Si
Me	Me	Me	Et
H	$(CH_2)_6$	H	$(CH_2)_6$

b. Addition to Dienes. The addition of a silylene to a 1,3-diene was first reported by Atwell and Weyenberg in 1968 and was formulated as the rearrangement product of a vinylsilacyclopropane.[8]

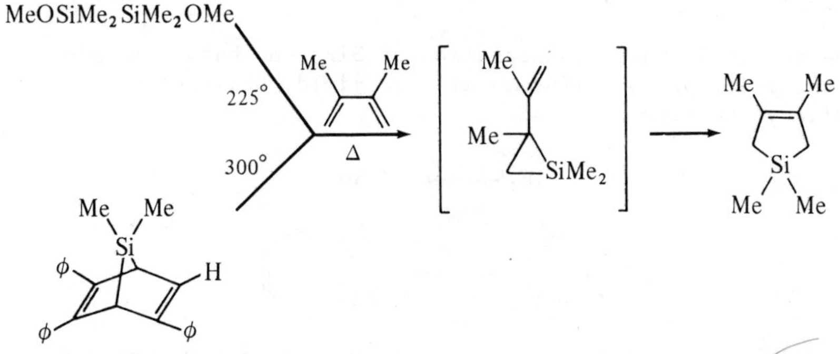

$MeOSiMe_2SiMe_2OMe$

In 1972 Chernyshev, Komalenkova, reported the reactions of $SiCl_2$ with a number of dienes.[76] The products from substituted butadienes were 1,1-dichloro-1-silacyclopent-3-enes, cyclopentadiene gave an apparent C-C insertion product, and furan yielded an adduct incorporating several silylene units:

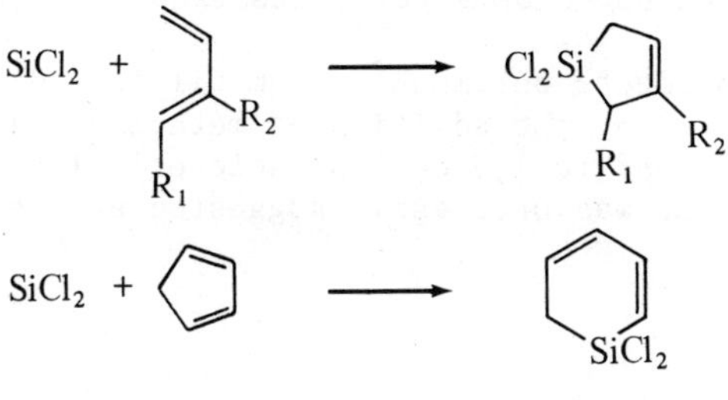

R_1	R_2
H	H
H	Me
Me	H
H	Cl

$$SiCl_2 + \text{furan} \longrightarrow \text{product } (SiCl_2)_3O$$

The structure of this last product must be regarded as having not been conclusively established. These workers also invoked as the operative reaction mechanism initial 1,2-addition of the silylene to one of the double bonds of the diene, followed by rearrangement of an intermediate vinylsilacyclopropane. Silacyclopropane intermediates have also been invoked to explain the formation of 1-silacyclopent-3-ene and 1,1-difluor-1-silacyclopent-3-ene from the reactions of recoiling silicon atoms with butadiene in the presence of PH_3,[77,78] and PF_3.[79,80]

The report by Tang and coworkers that nucleogenic $^{31}SiF_2$ undergoes addition to butadiene[79] stands in marked contrast to the behavior of thermally generated SiF_2. Thompson and Margrave

$$^{31}Si + PF_3 \longrightarrow \longrightarrow {}^{31}SiF_2$$

$$^{31}SiF_2 + \text{butadiene} \longrightarrow \longrightarrow F_2{}^{31}Si(\text{cyclopentene})$$

have reported that cocondensation of SiF_2 and butadiene gives as the major product (formed in < 2% yield, however) the disilacyclohexene:[81]

$$Si + SiF_4 \longrightarrow 2\ SiF_2$$

$$2\ SiF_2 + \text{butadiene} \longrightarrow F_2Si\text{–}F_2Si(\text{cyclohexene})$$

Virtually all reactions of thermally generated SiF_2 proceed through the dimer $(SiF_2)_2$ formed on condensation.[1] The result obtained from hot-atom experiments may reflect an electronic state of difluorosilylene different from that formed in thermal reactions, and excess kinetic energy may also play a role in the recoil experiments. However, the involvement of difluorosilylene in the hot-atom experiments requires careful scrutiny.

The variety of products obtainable from silylene-diene reactions is illustrated by the addition of methylmethoxysilylene to 2,5-dimethylfuran[82] and to 1,3-cyclooctadiene.[83] In the latter case 1,2-addition was once again suggested as the primary

MeOSiMe + Me–(furan, O)–Me ⟶ Me, OMe, Si, Me, O, Me (ring product)

MeOSiMe + (cyclooctadiene) ⟶ Me, Si, MeO (ring product)

step, and we see later that these reactions are mechanistically related, despite the fact that a conjugated diene product is obtained in the former reaction and an unconjugated diene in the latter.

Ring and coworkers have obtained 1-silacyclopent-3-enes from the addition of SiH_2, ClSiH, and MeSiH, all generated by pyrolysis of disilanes, to butadiene.[84] When trisilane was pyrolyzed in the presence of butadiene, both 1-silacyclopent-3-ene and 1,2-disilacyclohex-4-ene were obtained. Since the

$$Si_3H_8 \xrightarrow{\Delta} SiH_2 + Si_2H_6 \text{ and } SiH_3SiH_3 + SiH_4$$

$$SiH_2 + \text{(butadiene)} \longrightarrow H_2Si\text{(ring)}$$

$$SiH_3SiH \longrightarrow \underset{?}{[SiH_2{=}SiH_2]} \xrightarrow{\text{(butadiene)}} H_2Si{-}H_2Si\text{(ring)}$$

latter product is reasonably formulated as a Diels-Alder adduct, the possibility exists that silylsilylene rearranged to disilene.

Experiments designed to elucidate the mechanism of addition of silylenes to 1,3-dienes have included studies of trapping and of the stereochemistry of addition.

Concerted 1,4-cycloaddition was precluded by the finding of both <u>cis</u>- and <u>trans</u>-2,5-dimethyl-1-silacyclopent-3-ene from the addition of thermally generated silylene to <u>trans,trans</u>-2,4-hexadiene.[85]

$$Si_2H_6 \xrightarrow{\Delta} SiH_2 + SiH_4$$

$$SiH_2 + \text{(hexadiene)} \longrightarrow \text{Si, H, H (ring)} \quad \text{Si, H, H (ring)}$$

Although this result excludes concerted cycloaddition, several other paths seemed possible:

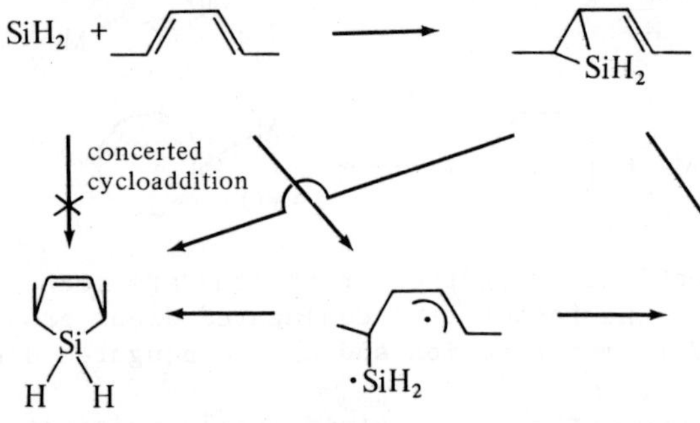

Involvement of a vinylsilacyclopropane intermediate was established by the methanol trapping experiments of Ishikawa, Ohi, et al.[86]

$$Me_3Si{-}SiMe\phi{-}SiMe_3 \xrightarrow{h\nu} Me_3SiSiMe_3 + SiMe\phi$$

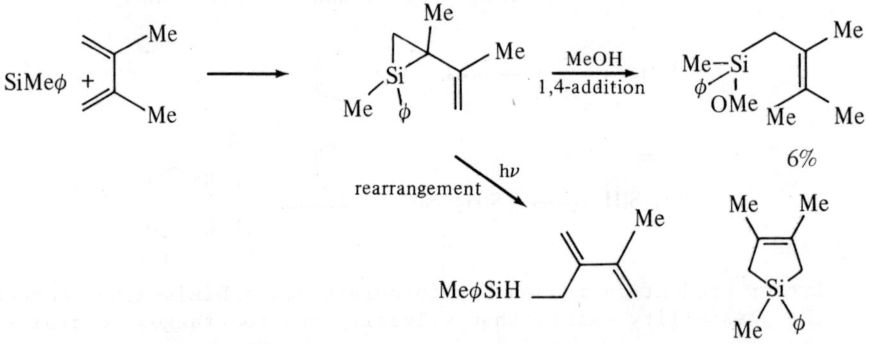

The major product from the irradiation of 2-phenylheptamethyltrisilane in 2,3-dimethylbutadiene was an adduct of the trisilane to the butadiene, once again sounding a warning as to the complexity of polysilane photolyses.

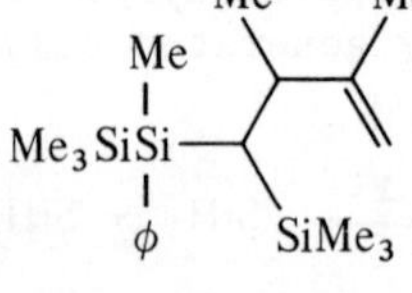

Further evidence for 1,2-addition as the primary step in the reactions of silylenes with 1,3-dienes is seen in the structure of one of the two major products from the addition of dimethylsilylene to 1,3-cyclohexadiene.[87] Rearrangement of a 7,7-dimethyl-7-silanorcarene intermediate seems clearly indicated as the source for the 3,3-dimethyl-3-sila-1,4,6-heptatriene product. For the 7,7-dimethyl-7-silanorbornene there is a possible radical pathway (path D) that circumvents the bicyclic intermediate. The exceptionally strained carbon-carbon bridge in the silanorcarene intermediate may permit C-C homolysis (path A) or concerted rearrangement (path B) to compete with Si-C homolysis (path C).

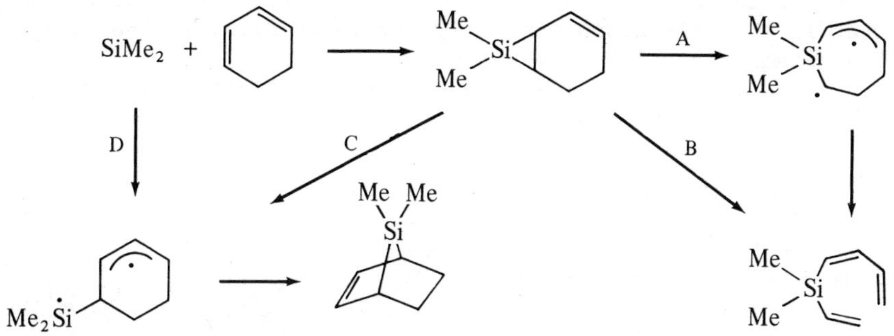

In the addition to cyclopentadiene, SiH_2, like $SiCl_2$, gives rise to only a conjugated silacyclohexadiene:[87]

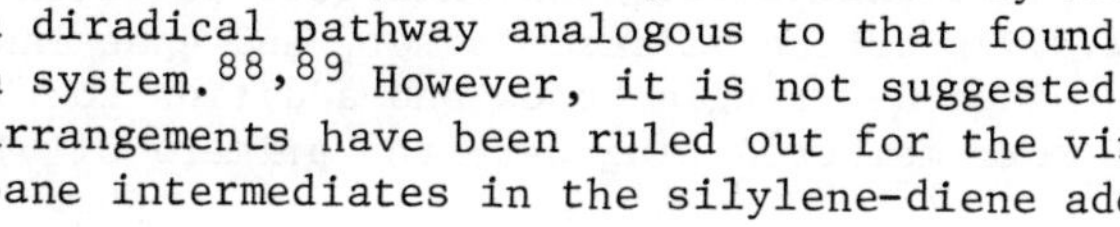

$$SiX_2 + \text{cyclopentadiene} \longrightarrow \text{silacyclohexadiene} \quad (X = H,\ Cl)$$

The reaction of cyclopentadiene with dimethylsilylene, however, produces both the 2,4- and 2,5-silacyclohexadiene. Again, initial 1,2-addition is postulated,[87] followed by rearrangement via a diradical pathway analogous to that found for the all-carbon system.[88,89] However, it is not suggested that concerted rearrangements have been ruled out for the vinylsilacyclopropane intermediates in the silylene-diene addition reactions.

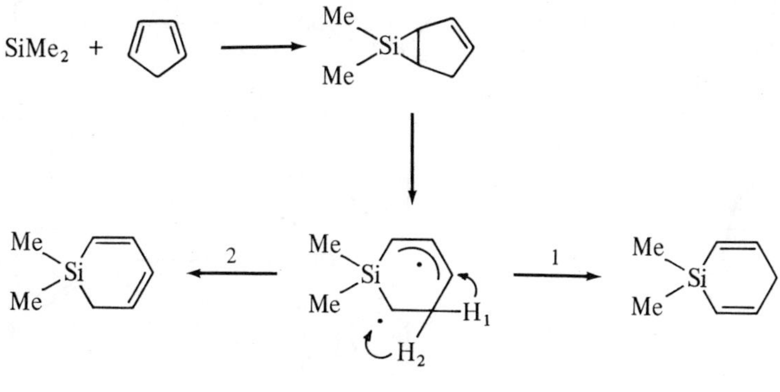

The isolation of only a single product from the additions of SiH_2 and $SiCl_2$ to cyclopentadiene and of MeOSiMe to 1,3-cyclooctadiene may be due to either conformational differences in the diradical, the operation of an entirely different addition mechanism, or the preferential loss by decomposition of the unconjugated diene product. It is provocative to note that in the all-carbon system 1,4-cyclohexadiene is thermally less stable than 1,3-cyclohexadiene.[88]

Recently two reports have appeared of the addition of silylenes to cyclooctatetraene.[90,91]

$$Cl_3SiSiCl_3 \xrightarrow[\text{flow system}]{550^\circ} SiCl_2 + SiCl_4 \qquad \text{(ref. 91)}$$

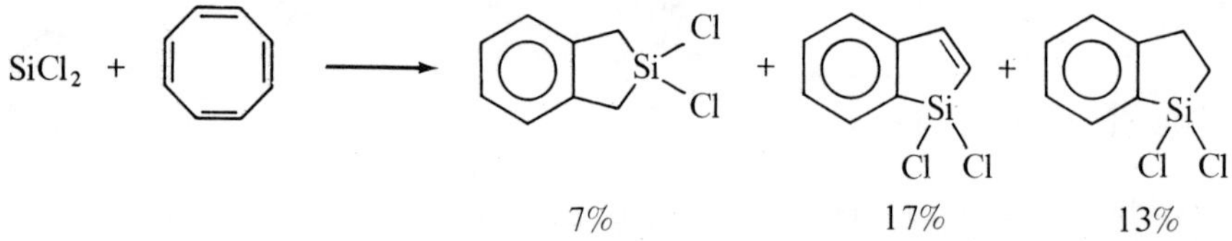

$$Me_2Si(OMe)\text{—}Si(OMe)Me_2 \xrightarrow[\text{flow system}]{550^\circ} SiMe_2 + Me_2Si(OMe)_2 \qquad \text{(ref. 90)}$$

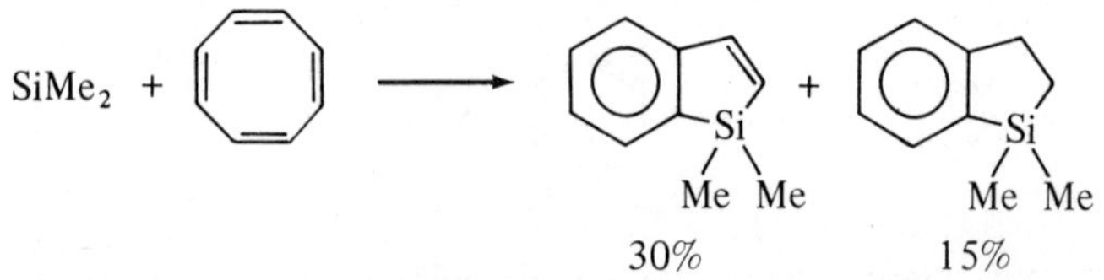

Chernyshev and coworkers interpreted their product ratios as indicating that cyclooctatetraene undergoes reaction as its bicyclo[4.2.0]octatriene valence isomer, and that insertion into the strained C-C single bonds and addition to the double bond of the four-membered ring were the primary silylene reactions.[91]

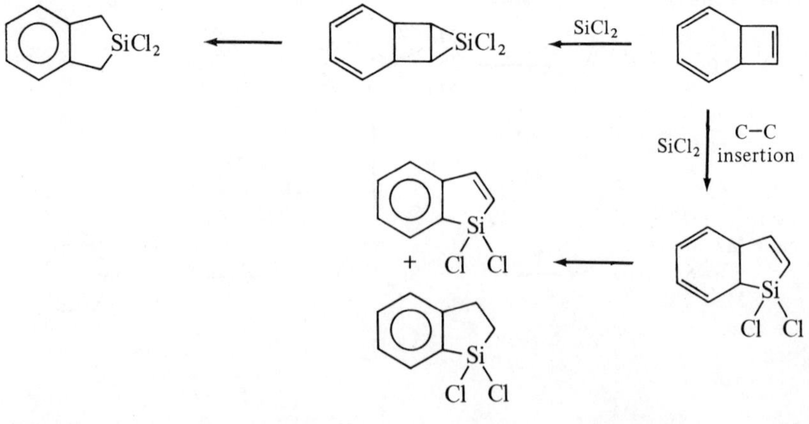

Barton and Juvet proposed a 1,2-addition to cyclooctatetraene as the initial step, followed by competing rearrangements:[90]

SiMe2 + [cyclooctatetraene] → [Me2Si bicyclic adduct] → [SiMe2 vinylsilacyclopropane]

rearrangement ← –2H →

Me Me

Although many details of the reactions of silylenes with dienes remain obscure, the general pattern seems to be emerging. As in the reactions with monoolefins, 1,2-addition leading to the formation of a silacyclopropane derivative is the primary step. The vinylsilacyclopropane intermediates can rearrange by various paths with such facility that they are rarely detected directly. In carbene reactions with 1,3-dienes, vinylcyclopropanes are obtained, since they require much more vigorous conditions for rearrangement than their silicon analogs. Thus while the products isolated from silylene-diene reactions differ among themselves and also from the products of carbene-diene reactions, the initial reactions are quite similar, at least on the superficial level of our present knowledge of silylene reaction mechanisms.

c. Addition to Acetylenes. The 15-year history of this reaction has a number of twists and turns. Now that we are beginning to understand it we can see that the end of the story resembles the beginning.

In 1962 Vol'pin and coworkers reported that they had isolated 1,1-dimethyl-2,3-diphenyl-1-silacycloprop-2-ene from the pyrolysis of a polydimethylsilylene in the presence of diphenylacetylene.[92]

$$(SiMe_2)_{55} \xrightarrow{\Delta} \underline{n}\text{-}SiMe_2 \xrightarrow{\phi C\equiv C\phi} \text{Me}_2\text{Si}(\phi C{=}C\phi)$$

Soon thereafter the product was shown to be a 1,3-disilacyclohexa-2,4-diene,[1] which Gilman, Cottis, et al. in their pioneering study of 7-silanorbornadiene pyrolysis attributed to the dimerization of a silacyclopropene intermediate.[25]

$$\xrightarrow[300^\circ]{\Delta} + SiMe_2$$

$$SiMe_2 + \phi\text{—}C\equiv C\text{—}\phi \longrightarrow \xrightarrow{\pi\text{-dimerization}}$$

In 1968 Atwell and Weyenberg carried out an elegant experiment employing a mixture of acetylenes, demonstrating that the origin of the disilacyclohexadiene could not lie in the π-dimerization of a silacyclopropene. The only mixed product isolated from a dimethylacetylene-diphenylacetylene substrate mixture could only result from dimerization of silacyclopropenes if coupling occured across Si-C bonds.[8]

$$2Me_2Si + MeC\equiv CMe + \phi C\equiv C\phi$$

The question that remains is how a product arises from the reaction of a silylene with an acetylene in which two silylene and two acetylene units have been incorporated. Nevertheless, the reaction with acetylene, despite its puzzling mechanism, quickly became a popular chemical means for the detection of silylenes. Thus the formation of MeSiOMe was deduced from the formation of the corresponding disilacyclohexadiene.[8]

$$(MeO)_2Si(Me)\text{—}Si(Me)(OMe_2)_2 \xrightarrow{\Delta} MeSiOMe + (MeO)_3SiMe$$

$$2\ MeSiOMe + 2\ \phi C\equiv C\phi \longrightarrow \text{1,4-Me}_2\text{-1,4-(OMe)}_2\text{-2,3,5,6-}\phi_4\text{-1,4-disilacyclohexa-2,5-diene}$$

Of course, insertion of the silylene into Si-O bonds of its own precursor is also observed, and in the generation of dimethoxysilylene this is the only observed reaction, even in the presence of diphenylacetylene.[8]

The chemical detection of diphenylsilylene by reaction with diphenylacetylene was mentioned earlier.[26] Dichlorosilylene[93] and methylmethoxylsilylene[94] have been trapped with unsubstituted acetylene in the gas phase, and dichlorosilylene has also been allowed to react with diphenylacetylene.[76] All of these reactions produce 1,4-disilacyclohexa-2,5-dienes.

This pattern was broken and a significant mechanistic clue was provided when in 1973 Atwell and Uhlmann reported a product containing one acetylene unit and two silylene units from the generation of dimethylsilylene in the presence of 2-butyne in the gas phase.[95]

$$MeOSiMe_2SiMe_2OMe + MeC\equiv CMe \xrightarrow[-Me_2Si(OMe_2)_2]{400^\circ,\ N_2} \text{3,4-Me}_2\text{-1,1,2,2-}(SiMe_2)_2\text{-disilacyclobutene}$$

Atwell and Uhlmann suggested that the disilacyclobutene is formed by insertion of dimethylsilylene into an Si-C bond of a silacyclopropene intermediate produced by primary 1,2-addition of dimethylsilylene to the butyne. Barton and Kilgour considered an alternative mechanism, the dimerization of dimethylsilylene, followed by cycloaddition of the resulting disilene to the acetylene.[96] Both mechanisms are depicted in the following formula.

R=Me 31% R=Et 39%

Barton and Kilgour demonstrated the involvement of the disilacyclobutene along the reaction coordinate, which leads to the disilacyclohexadiene by subjecting the disilacyclobutene to pyrolysis in the presence of acetylenes. Disilacyclohexadienes were obtained and their formation was attributed to a 2 + 4 cycloaddition of an acetylene to a disilabutadiene, itself a ring-opening product of the disilacyclobutene.

Although compounds containing unsaturated bonds to silicon such as the tetramethyldisilene and hexamethyldisilabutadiene (see the preceding formula) were not believed capable of even transient existence until just a few years ago, evidence for their formation is now rather convincing.[97,98]

Barton and Kilgour also demonstrated that silylene dimerization followed by cycloaddition can give rise to the silacyclohexadienes by allowing preformed silylene dimer to react with butyne in the liquid phase.[95] Tetramethyldisilene extrustion from disilabicyclöctadienes had previously been established by Roark and Peddle.[99]

Barton thus clearly demonstrated that the disilacyclohexadienes, which are common products of the reactions of silylenes and acetylenes, can under certain conditions be formed by a reaction sequence consisting of: (a) dimerization of silylene to disilene, (b) addition of disilene to acetylene forming disilacyclobutene, and (c) reaction of the disilacyclobutene

$$Me_2Si{-}SiMe_2 \text{ (7-silanorbornadiene, }\phi\text{)} \xrightarrow[\text{sealed tube}]{260°} Me_2Si{=}SiMe_2 + \phi{-}\phi$$

$$Me_2Si{=}SiMe_2 \xrightarrow{MeC{\equiv}CMe} \text{1,2-disilacyclobutene (Me, Me; }SiMe_2, SiMe_2) \xrightarrow{MeC{\equiv}CMe} \text{hexamethyl... 1,4-disilacyclohexadiene (Si}Me_2\text{, Me}_4)$$

with a second acetylene molecule forming the stable disilacyclohexadiene product.

In the gas-phase, however, Barton and Kilgour found evidence for the addition of tetramethyldisilene to 2-butyne, but <u>no</u> disilacyclohexadiene was detected.[100,101] From these results they concluded that different mechanisms may operate in the gas and liquid phases for the formation of disilacyclohexadienes following the reactions of silylenes and acetylenes.[100]

Whereas Barton and Kilgour showed that a silylene dimer, once formed, can undergo cycloaddition to an acetylene, it was Conlin who recently demonstrated that in the gas phase, dimethylsilylene really can dimerize, even in the presence of a considerable excess of acetylene.[102]

Low-pressure flash pyrolysis of dimethoxytetramethyldisilane in the absence of other trapping agents gives, in addition to the product of insertion of dimethylsilylene into its own precursor and other minor products, the products obtained by Roark and Peddle[99] from the rearrangement of tetramethyldisilene.[102]

$$Me_2Si(OMe){-}SiMe_2(OMe) \xrightarrow{\Delta} \text{1,3-disilacyclobutane (}Si(Me)(Me), CH_2, CH_2, Si(H)(H)\text{)} + \text{1,3-disilacyclobutane (}Si(Me)(H), CH_2, CH_2, Si(H)(Me)\text{)} \longleftarrow Me_2Si{=}SiMe_2 \longleftarrow Me_2Si{-}SiMe_2 \text{ (7-silanorbornadiene, }\phi\text{)}$$

Formation of the same major products in both reactions is consistent with a common intermediate, tetramethyldisilene. Since generation of dimethylsilylene from pyrolysis of dimethoxytetramethyldisilane is well established (<u>vide</u>

supra), the dimerization of dimethylsilylene is strongly implied.

$$2\ Me_2Si \rightarrow Me_2Si{=}SiMe_2$$

Disilacyclobutenes are also formed when SiF_2 is cocondensed with acetylenes, and dimerization of SiF_2 before addition seems very likely.[71,105]

$$SiF_2 + RC{\equiv}CR' \longrightarrow \text{(R)C=C(R')–}SiF_2\text{–}SiF_2\text{ (ring)}$$

R	R'
H	H
CF_3	H
Me	H
Et	H
Me	Me
t-Bu	H

A variety of products containing two acetylene units and two SiF_2 units is also formed,[103,104] but no 1,4-disilacyclohexa-2,5-dienes have been reported. Thus the reactions of SiF_2 with acetylenes seem to differ from those of SiH_2, alkyl-, aryl-, and alkoxysilylenes.

The fact that silylenes can dimerize suggests they are so unreactive that their concentrations can rise to the point where bimolecular reactions become feasible. It has been suggested by M. Jones, Jr. that dimerization and cross-coupling of silylenes may provide the source of new compounds unsaturated at silicon.

Even in the presence of a 10-fold excess of propyne, the disilacyclobutane rearrangement products of tetramethyldisilene are formed in combined yields of 18% compared to a 25% yield of the adduct of the silylene to propyne.[102] Thus it may be inferred that a silylene dimerizes in the presence of a large amount of a trapping reagent capable of capturing that silylene.

$$Me_2Si + HC{\equiv}CMe \rightarrow Me_2HSiC{\equiv}CMe$$

This apparent C-H insertion reaction was only recently discovered, when in 1975 Hass and Ring reported the first 1:1 adduct of a silylene and an acetylene,[48] obtained from the

$$SiH_2 + HC{\equiv}CH \rightarrow SiH_3C{\equiv}CH$$

gas-phase pyrolysis of disilane in the presence of acetylene. The suggested mechanism was addition of silylene forming silacyclopropene, followed by rearrangement:

$$SiH_2 + HC{\equiv}CH \rightarrow HC{=}CH\ (\text{ring with }SiH_2) \rightarrow H_2\dot{Si}{-}CH{=}\dot{C}H \rightarrow SiH_3C{\equiv}CH$$

Silacyclopropene formation had previously been suggested by Atwell and Weyenberg to account for the formation of vinylalkoxysilanes from the generation of silylenes in the presence of a mixture of an acetylene and an alcohol.[3,106]

$$\text{RMeSi} + \text{R'C}\equiv\text{CR'} \longrightarrow \text{(R)(Me)Si}\langle\text{CR'=CR'} \xrightarrow{\text{R''OH}} \text{R'(H)C=C(R')–Si(R)(Me)(OR'')}$$

Confirmation that a silacyclopropene can be formed by addition of a silylene to an acetylene and that alcoholysis of a silacyclopropene yields a vinylalkoxysilane has come from the work of Conlin, who obtained an air- and moisture-sensitive but otherwise rather stable compound from the flash pyrolysis of dimethoxytetramethyldisilane in the presence of 2-butyne.[107] This product, even after several weeks of storage, reacts very rapidly with methanol to give about 80% yield of cis-1,2-dimethylvinyldimethylmethoxysilane. The proton NMR spectrum and the mass spectrum of the labile product point to its structure as tetramethylsilacyclopropene, the first 1,2-adduct of a silylene to a π-bond sufficiently stable for spectroscopic detection and for separation by gas chromatography.

$$\text{Me}_2\text{Si(OMe)–Si(OMe)Me}_2 \longrightarrow \text{Me}_2\text{Si} + \text{Me}_2\text{Si(OMe)}_2$$

$$\text{Me}_2\text{Si} + \text{MeC}\equiv\text{CMe} \longrightarrow \text{Me}_2\text{Si}\langle\text{C(Me)=C(Me)} \xrightarrow{\text{air}} \textit{poof}$$

$$\text{Me}_2\text{Si}\langle\text{C(Me)=C(Me)} + \text{MeOH} \longrightarrow \text{Me(H)C=C(Me)–SiMe}_2\text{(OMe)}$$

Heating a solution of tetramethylsilacyclopropene to 105°C for an hour led to the formation of a polymer without any indication that a disilacyclohexadiene was formed.[107] Thus silacyclopropene dimerization is probably not a route to disilacyclohexadines.

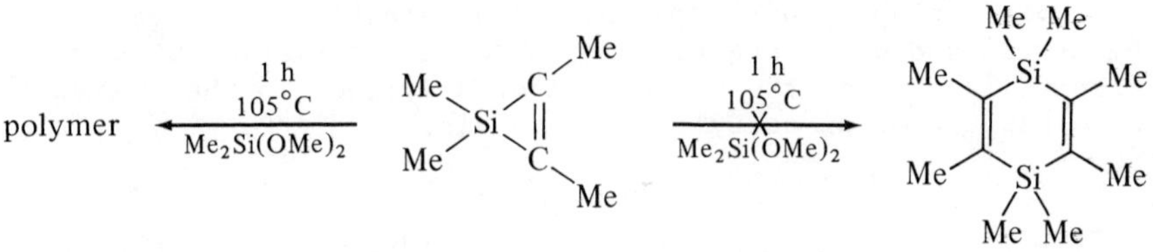

The thermal stability of tetramethylsilacyclopropene is indicated by an experiment in which heating to 70°C for 3 h led to no diminution in concentration.[107] In contrast, the half-life of hexamethylsilacyclopropane in THF solution is only 5 h at 63°C.[51]

Recently, Seyferth, Annarelli, and Vick have transferred a dimethylsilylene unit from hexamethylsilacyclopropane to bis-trimethylsilylacetylene, forming a silacyclopropene that could be isolated and completely characterized by ^1H, ^{13}C, and ^{29}Si NMR spectroscopy.[108]

Seyferth and coworkers found that the half-life of 2,3-bis-trimethylsilyl-1,1-dimethylsilacycloprop-2-ene is approximately 60 h at 75°C in benzene solution, indicating considerable stability.

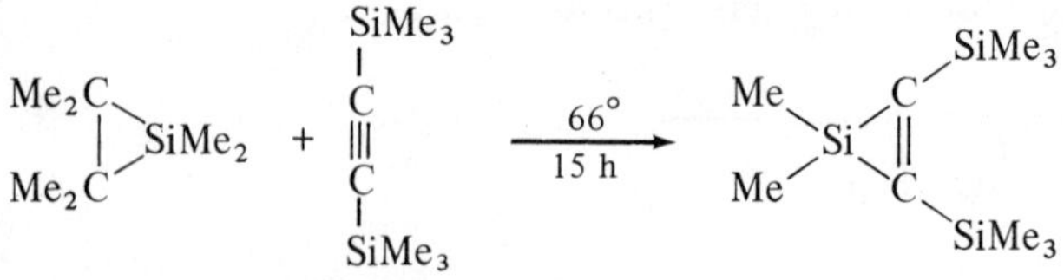

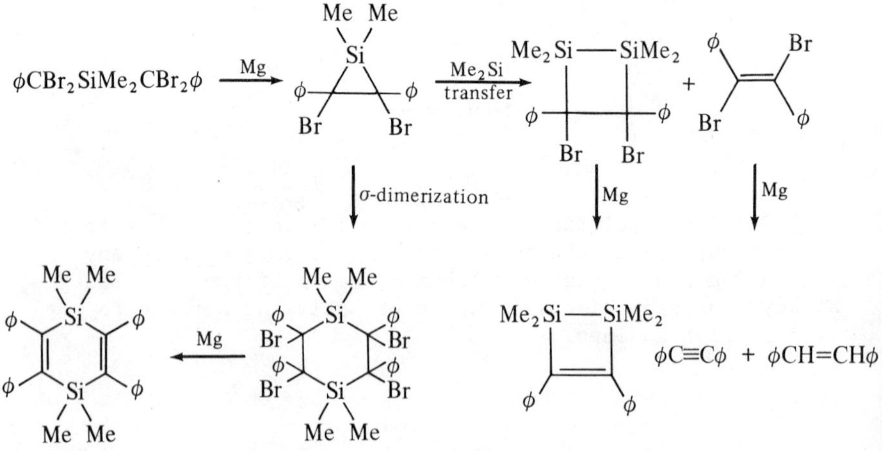

Dimethylsilylene transfer from hexamethylsilacyclopropane to bisdimethylsilylacetylene, trimethylsilyl-t-butylacetylene, 1-trimethylsilyl-1-propyne, and 1-*t*-butyl-1-propyne has also been reported to yield the corresponding silacyclopropenes.[109]

In an organometallic reaction designed to generate a silacyclopropene, Barton and coworkers found evidence for ring expansion of one silacyclopropane molecule by transfer of dimethylsilylene from another.[110] Seyferth has found dimethylsilylene transfer from hexamethylsilacyclopropane to itself, and to 2,3-bis-trimethylsilyl-1,1-dimethylsilacycloprop-2-ene.[109]

$$\text{Me}_2\text{C}\text{–}\text{CMe}_2\text{(–SiMe}_2\text{–)} + \text{Me}_2\text{Si}(\text{C(SiMe}_3\text{)=C(SiMe}_3\text{)}) \longrightarrow \text{Me}_2\text{C=CMe}_2 + \text{Me}_2\text{C–CMe}_2\text{(–SiMe}_2\text{–SiMe}_2\text{–)} + \text{Me}_3\text{SiC=CSiMe}_3\text{(–SiMe}_2\text{–SiMe}_2\text{–)}$$

Thus in solution the formation of disilacyclohexadienes from the reaction of silylenes and acetylenes may be initiated by addition of a silylene to an acetylene forming a silacyclopropene that undergoes ring expansion to a disilacyclobutene by insertion of a second silylene. Certainly there is no evidence for dimerization of silylene in solution, which would be required for the alternative route to disilacyclobutenes discussed here. As we have seen, Barton has demonstrated that in the liquid phase a disilacyclobutene can react with an acetylene molecule to give the disilacyclohexadiene.

We see that the variety of competing reactions leads to many possible results of generating a silylene in the presence of an acetylene. The silylene may dimerize, or it may add to the acetylene forming a silacyclopropene. The silylene dimer may rearrange to stable 1,3-disilacyclobutanes, but diradicals that are intermediates in this rearrangement may themselves be trapped by addition to acetylenes. The unrearranged silylene dimer, namely, a disilene, may undergo cycloaddition to the acetylene forming a 1,2-disilacyclobut-3-ene that can be isolated, but can also react with an acetylene molecule to form a disilacyclohexadiene. Silacyclopropenes formed by silylene addition to an acetylene may react with a second silylene to form a disilacyclohexadiene. The multiplicity of reaction paths not only leads to many products, but renders the product distribution extraordinarily sensitive to the reaction conditions.

Our account of the silylene-acetylene reaction began with the report by Vol'pin in 1962 that a silacyclopropene had been obtained as a stable reaction product.[92] Although this compound turned out to be a disilacyclohexadiene, Vol'pin's prediction that silacyclopropenes would be quite stable was confirmed nearly 15 years later.[107,108] Vol'pin ascribed the expected stability of silacyclopropenes to a quasiaromatic delocalization of the π-electrons into d-orbitals on the silicon atom, and this d-π/p-π overlap has been invoked in recent discussions.[2,107,108] It also appears that a silacyclopropene may appear on the reaction coordinate leading to the product isolated by Vol'pin.

In one sense the silylene-acetylene reaction is an unsolved problem, since the full sequence of elementary reactions has not been worked out for any set of reaction conditions under which disilacyclohexadienes are obtained. In another sense the problem is nearly exhausted, since a number of elementary processes has been confirmed that, if properly combined and permuted, seem capable of rationalizing most of the observed facts.

Certainly the silylene-acetylene reaction has turned out to be a complicated story, and many workers have contributed to its understanding. The richness of this reaction may be appreciated from the large number of interesting molecules that it produces: silacyclopropenes, disilaolefins, 1,3-disilacyclobutanes, 1,2-disilacyclobut-3-enes, and 1,4-disilacyclohexa-2,5-dienes.

C. Abstraction by Silylenes

No abstraction reactions of silylenes have been confirmed. This is understandable in view of the present estimates of the heats of formation of silylene and silyl radicals. Heats of formation for silylene deduced from kinetic data are: (a) SiH_2, $\Delta\underline{H}_f^\circ$ = 59.3[10], 57.9[11], 58.6 ± 3.5[13] kcal/mol, (b) MeSiH, $\Delta\underline{H}_f^\circ$ = 50.9 ± 3.5 kcal/mol,[13] and (c) SiH_3SiH, $\Delta\underline{H}_f^\circ$ = 64.5 ± 3.5 kcal/mol.[13] For SiH_3, $\Delta\underline{H}_f^\circ$ = 49.7 kcal/mol has been estimated[45] (from electron-impact appearance potentials). Thus the reaction $SiH_2 + SiH_4 \rightarrow 2\ SiH_3$ has been estimated to be <u>endothermic</u> by 32.8 kcal/mol.[46] In contrast, the insertion reaction $SiH_2 + SiH_4 \rightarrow Si_2H_6$ is exothermic by approximately 50 kcal/mol. Thus on thermochemical grounds it seems that hydrogen abstraction should not be competetive with Si-H insertion.

An experiment has been carried out recently in which the silylene precursor and reaction substrate were chosen to facilitate the detection of abstraction and radical recombination products. When dimethoxytetramethyldisilane was subjected to pyrolysis in the presence of a large excess of trimethylsilane, the only product of attack of dimethylsilylene on trimethylsilane detected was the H-Si insertion product. Neither of the _geminate_ recombination products expected from hydrogen abstraction was found, nor was the product of two consecutive abstractions.[111]

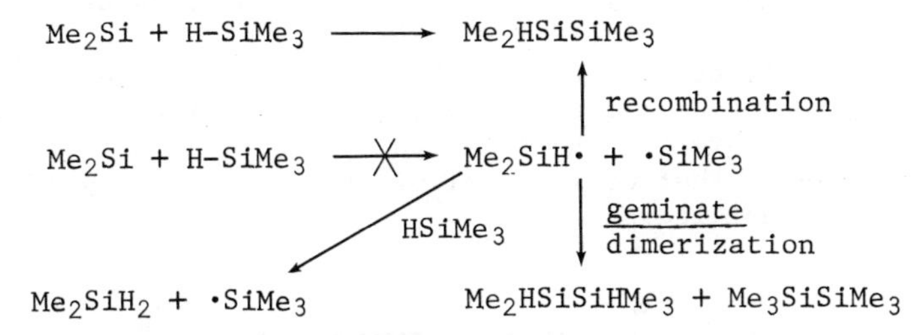

The observation that dimethylsilylene can be trapped by addition to dienes in refluxing carbon tetrachloride solution[28] strongly contraindicates the occurrence of silylene abstraction reactions. Silyl radicals are notorious for their ability to abstract halogen atoms.[112]

D. Other Reactions of Silylenes

In our discussion of silylene insertion and addition reactions, dimerization of silylene to disilenes was also mentioned. No other general classes of silylene reactions are known. There is just the hint that a silylene rearrangement may have been observed in the trapping of SiH_2SiH_2 from the generation of SiH_3SiH.[84]

$$SiH_3\text{-}SiH \rightarrow SiH_2\text{=}SiH_2$$

It has been noted that polymerization of silylenes at liquid nitrogen temperatures indicates their ability to survive many collisions with substrate molecules and thus speaks for their relative stability.[18] The same remark may be made about the dimerization of dimethylsilylene in the presence of a 10-fold excess of propyne.[102]

IV. THE STRUCTURE OF SILYLENES

The involvement of two low-lying electronic states $^1A_1$ (the first excited state) and $^3B_1$ (the ground state) in the chemistry of carbenes has been the cause of many interesting complications in the study of carbene reactions.[113] The reactions of thermally and photochemically generated silylenes thus far investigated have been discussed without invoking more than one electronic state of silylene, the singlet ground state. It has been suggested that photochemical extrusion of silylenes from polysilanes produces silylenes in excited states,[35] and the effects of scavengers on the yields of products from the reactions of recoiling silicon atoms have been interpreted in terms of the scavenging of triplet $^{31}SiH_2$[23,77] and $^{31}SiF_2$.[80]

Zeck, Tang, et al. deduced that the ground state of silylene is a singlet from the observation that in the presence of inert gas moderator, the fraction of the yield of 1-silacyclopent-3-ene that could be eliminated by nitric oxide scavenger did not increase.[23]

$$^{31}Si + \text{(butadiene)} \longrightarrow \xrightarrow{} H_2{}^{31}Si\text{(cyclopentene ring)}$$

$$PH_3$$

These workers equated the maximum reduction in the product yield due to scavenger with the amount of triplet $^{31}SiH_2$ present. They reasoned that if the ground state of silylene were the triplet, then collisions with moderator would cause singlet-triplet relaxation, thus increasing the amount of triplet present. Since no increase was observed in the amount of scavengeable intermediate in the presence of moderator, a ground singlet state was deduced for this intermediate, believed to be $^{31}SiH_2$. The argument is elegant, but as indicated earlier, the intermediacy of $^{31}SiH_2$ in the reactions of recoiling silicon atoms has been questioned and is undergoing careful investigation.[24]

That SiH_2 has a singlet ground state was recently confirmed in a rather direct manner by fixed-frequency laser photoelectron spectrometry.[114] In the experiment, photoionization of SiH_2^- produced both low-lying electronic states of silylene, as deduced from the energy spectrum of the liberated electrons. A singlet-triplet separation $\leq$ 0.6 eV ($\leq$ 14 kcal/mol) was deduced from the electron-photodetachment spectrum. The overlapping vibrational progressions in the $^2B_1$ SiH_2^- $\rightarrow$ $^3B_1$ SiH_2 + e^- transition led to a greater uncertainty in the singlet-triplet separation than in the value 19.5 ± 0.5 recently reported for CH_2 employing a similar technique.[115]

The ground state and the singlet-triplet splitting for SiH_2 had previously been assigned by two theoretical calculations. Jordan's semiempirical valence-bond calculation of 1966 predicted a $^1A_1$ singlet ground state for silylene with $\angle$HSiH = 95.4° lying 16,180 cm^{-1} (46 kcal/mol) below the first excited triplet state $^3B_1$, $\angle$HSiH = 137.8°.[116] An _ab initio_ SCF CI calculation by Wirsam in 1972 predicted a 97.46° bond angle for the singlet ground state, which was placed 0.201 eV (4.8 kcal/mol) below the excited triplet state, whose bond angle was calculated to be 123.53°.[117] Recent near-Hartree-Fock results of Meadows and Schaefer predict a 94.3° singlet ground state, lying 18.6 kcal/mol below the 117.6° $^3B_1$ triplet state.[118] The singlet-triplet separation for SiH_2 was reduced to 10.0 kcal/mol by applying a semiempirical correction[118] that was necessary to reconcile the theoretically predicted singlet-triplet splitting for CH_2 with Lineberger's experimental result.[116] Compared with the massive number of calculations on methylene,[113] theoretical activity has been quite modest in the silylene business.

The bond angle in the ground singlet state of SiH_2 has been determined by double-flash UV spectroscopy to be 92°5', with bond lengths of 1.516 Å.[119] The absorption spectrum of SiH_2 is under continued investigation.[120]

V. FUTURE DEVELOPMENTS

Various interesting and useful reactions of silylenes are now known, and a determined start has been made at unraveling their reaction mechanisms. Still, the state of our mechanistic knowledge of silylene reactions is at about the stage that carbene chemistry was in 1960. Since chemists of the 1970s have a vastly expanded theoretical and experimental repertoire, it is clear that we have a long way to go. For only one silylene reaction, namely, insertion, do we have any information about transition-state structure. Even for the insertion reaction we have only limited and somewhat contradictory information about substituent effects on the reactivity of the silylene and its reaction substrates. We are even worse off in our understanding of silylene addition reactions. It has been established that 1,2-additions occur, but we still do not know whether they are concerted, or instead involve diradical intermediates. Our knowledge of even the reactions that are believed to generate silylenes is in a primative state. Only for disilane pyrolysis is the formation of free silylenes firmly established. For polysilane photolysis and silirane thermolysis, both of great importance because of the mild conditions under which they are carried out, there is the question as to whether silylenoid excited states or diradicals are formed that can mimic free silylenes in their reactions. Our ability to *use* these reactions is limited by our lack of knowledge of both silylene generation and silylene reactivity.

Thus it is easy to predict that there will be a considerable effort expended in the next few years to expand our mechanistic knowledge of reactions that produce and consume silylenes. The first order of business is to clarify the mechanisms of the new mild silylene generation procedures, to verify the liberation of silylenes, and to characterize the initial states in which they are produced. Then these reactions can be used in kinetic and other mechanistic studies to elucidate silylene reaction mechanisms. Within a few months we should certainly expect to see the results of the application of the Skell rule to the silylene addition reaction.

A strong stimulus for this work has been the discovery of the synthetic utility of silylene addition reactions, through the production of a variety of unsaturated organosilicon compounds otherwise accessible with difficulty. It is clear that silylenes will become widely utilized as synthetic reagents in organosilicon chemistry, now that mild and convenient methods are becoming available for their generation under moderate conditions. The synthetic activity will, of course, lead to the discovery of a host of new reactions. Entirely new reactions such as silylene rearrangements will be sought.

If recent estimates of the singlet-triplet separation in silylene are correct and this splitting is smaller than that for methylene, then there is a rich chemistry of triplet silylenes waiting in the wings. This was pointed out by Margrave for SiF_2 at a time when the singlet-triplet splitting for SiH_2 was believed to be much larger than that of CH_2.[71]

Not only does our present knowledge indicate that silylenes and carbenes have a number of reactions in common, but the development of both areas as fields of research bear a marked similarity to each other. Both enjoyed a first wave of interest in which the broad features of their reactions were explored. The difficult task, the hard but rewarding work of detailed elucidation of reaction mechanism, commenced only when it became clear that carbenes were useful building blocks, and this is also the case for silylenes. It is the second phase of silylene chemistry on which we are now embarked.

VII. ADDENDIUM

Dimethylsilylene has recently been generated by pyrolysis of pentamethyldisilane.[121] Curiously, no heptamethyltrisilane, the product of insertion of dimethylsilylene into the Si-H bond of its precursor was detected. This raises the question whether this silylene reaction product is less stable than the silylene precursor. Trapping of the silylene by insertion into Si-H and Si-Cl bonds of monosilanes was observed. Pyrolysis of 1,1,2-trimethyldisilane yields methylsilylene as well as dimethylsilylene, the former in greater yield.

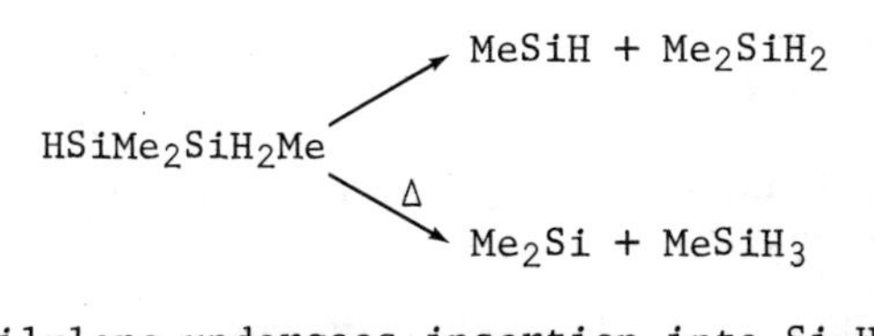

Methylsilylene undergoes insertion into Si-H and Si-Cl bonds of monosilanes. Dimethylsilylene is also formed via chlorine atom migrations in the pyrolysis of chloropentamethyldisilane.[122]

$$ClSiMe_2SiMe_3 \xrightarrow{\Delta} Me_2Si + Me_3SiCl$$

The pyrolysis of $Cl_2SiMeSiClMe_2$ and $ClSiMe_2SiClMe_2$ was also studied.

VII. ACKNOWLEDGMENT

I am grateful to Dr. Robert T. Conlin for discussions ranging widely over silylene chemistry and beyond. Dr. Donald P. Paquin read this manuscript and contributed helpful suggestions for its improvement. Discussions with Professors Maitland Jones, Jr. and Thomas J. Barton, Dr. Michael Sefcik, Dr. Rong-juh Hwang, Roy Lewis, Becky Cornett, Bruce Cohen, and Steve Lockhart have also been most helpful.

VIII. REFERENCES

1. P. P. Gaspar and B. J. Herold, "Silicon, Germanium, and Tin Structural Analogs of Carbenes," in _Carbene Chemistry_, 2nd ed., W. Kirmse (ed.), Academic, New York, 1971, p. 504.

2.* D. Seyferth, _J. Organomet. Chem._, _100_, 237 (1975).

3. W. H. Atwell and D. R. Weyenberg, _Intra-Science Chem. Rept._, _7_, 139 (1973).

4. R. L. Jenkins, A. J. Vanderwielen, S. P. Ruis, S. R. Gird, and M. A. Ring, _Inorg. Chem._, _12_, 2968 (1973).

5. E. Hedaya, _Acc. Chem. Res._, _2_, 367 (1969).

6. R. T. Conlin, "1. Mechanisms of Dimethylsilylene Addition Reactions. 2. Carbenoid Insertions into Carbon-Hydrogen bonds Activated by Group IV Metals," Ph.D. thesis, Washington University, August 1976.

7. W. H. Atwell and D. R. Weyenberg, Angew. Chem. Int. Ed. Engl., 8, 469 (1969).

8. W. H. Atwell and D. R. Weyenberg, J. Am. Chem. Soc., 90, 3438 (1968).

9. W. H. Atwell, L. G. Mahone, S. F. Hayes, and J. G. Uhlmann, J. Organomet. Chem., 18, 69 (1969).

10. M. Bowrey and J. H. Purnell, Proc. Roy. Soc. Lond., A, 321, 341 (1971).

11. P. John and J. H. Purnell, J. Chem. Soc. Faraday Trans. I, 69, 1455 (1973).

12. B. Cox and J. H. Purnell, J. Chem. Soc. Faraday Trans. I., 71, 859 (1975).

13. A. J. Vanderwielen, M. A. Ring, a nd H. E. O'Neal, J. Am. Chem. Soc., 97, 993 (1975).

14. M. Bowrey and J. H. Purnell, J. Am. Chem. Soc., 92, 2594 (1970).

15. P. Estacio, M. D. Sefcik, E. K. Chan, and M. A. Ring, Inorg. Chem., 9, 1068 (1970).

16. R. B. Baird, M. D. Sefcik, and M. A. Ring, Inorg. Chem., 10, 883 (1971).

17. P. S. Skell and P. W. Owen, J. Am. Chem. Soc., 89, 3933 (1967).

18. P. W. Owen and P. S. Skell, Tetrahedron Lett., 1807 (1972).

19. P. S. Skell and P. W. Owen, J. Am. Chem. Soc., 94, 5434 (1972).

20. P. P. Gaspar and P. Markusch, Chem. Commun., 1331 (1970).

21. P. P. Gaspar, P. Markusch, J. D. Holten, III, and J. J. Frost, J. Phys. Chem., 76, 1352 (1972).

22. J. D. Holten, III and P. P. Gaspar, unpublished results.

23. O. F. Zeck, Y. Y. Su, G. P. Gennaro and Y.-N. Tang, J. Am. Chem. Soc., 96, 5967 (1974).

24. R.-J. Hwang, "The Role of Silylene in the Reactions of Recoiling Silicon Atoms," Ph.D. thesis, Washington University, December 1976.

25. H. Gilman, S. G. Cottis, and W. H. Atwell, J. Am. Chem. Soc., 86, 1596 (1964).

26. H. Gilman, S. G. Cottis, and W. H. Atwell, J. Am. Chem. Soc., 86, 5584 (1964).

27. T. J. Barton, A. J. Nelson, and J. Clardy, J. Org. Chem., 37, 895 (1972).

28. A. Laporterie, J. Dubac, P. Mazerolles, and M. Lesbre, Tetrahedron Lett., 4653 (1971).

29. T. J. Barton, J. L. Witiak, and C. L. McIntosh, J. Am. Chem. Soc., 94, 6229 (1972).

30. M. Ishikawa and M. Kumada, Chem. Commun., 612 (1970).

31. M. Ishikawa and M. Kumada, Chem. Commun., 489 (1971).

32. H. Sakurai, Y. Kobayashi, and Y. Nakadaira, J. Am. Chem. Soc., 93, 5272 (1971).

33. M. Ishikawa, M. Ishiguro, and M. Kumada, J. Organomet. Chem., 49, C71 (1973).

34. H. Sakurai, Y. Kobayashi, and Y. Nakadaira, J. Am. Chem. Soc., 96, 2656 (1974).

35. B. G. Ramsey, J. Organomet. Chem., 67, C67 (1974).

36. M. Ishikawa and M. Kumada, J. Organomet. Chem., 42, 325 (1972).

37. M. Ishikawa, T. Takaoka, and M. Kumada, J. Organomet. Chem., 42, 333 (1972).

38. O. M. Nefedov and M. N. Manakov, Angew. Chem., Int. Ed. Engl., 5, 1021 (1966).

39. P. Mazerolles, M. Joanny, and G. Tourrou, J. Organomet. Chem., 60, C3 (1973).

40. D. R. Weyenberg, L. H. Toporcer, and A. E. Bey, J. Org. Chem., 30, 4096 (1965).

41. D. R. Weyenberg, L. H. Toporcer and L. E. Nelson, J. Org. Chem., 33, 1975 (1968).

42. P. S. Skell and E. J. Goldstein, J. Am. Chem. Soc., 86, 1442 (1964).

43. J. H. Purnell and R. Walsh, Proc. Roy. Soc. Lond., A, 293, 543 (1966).

44. P. Potzinger, A. Ritter, and J. Krause, Z. Naturforsch., 30a, 347 (1975). In this paper new bond-dissociation energies suggested are D(Si-H) = 89 ± 4, D(Si-C) = 85 ± 4, D(Si-Si) = 75 ± 8 and D(Si-Cl) = 116 kcal/mol for Me_3SiCl and 104 Cl_3SiCl.

45. M. A. Ring, M. J. Puentes, and H. E. O'Neal, J. Am. Chem. Soc., 92, 4845 (1970).

46. P. John and J. H. Purnell, J. Organomet. Chem., 29, 233 (1971).

47. P. Potzinger, private communication; see B. Reimann, A. Matten, R. Laupert, and P. Potzinger, "Zur Reaktion von Silylradikalen. Das Verhältnis Disproportionierung/ Rekombination," in press.

48.* C. H. Haas and M. A. Ring, Inorg. Chem., 14, 2253 (1975).

49. R. L. Lambert, Jr. and D. Seyferth, J. Am. Chem. Soc., 94, 9246 (1972).

50. D. Seyferth, R. L. Lambert, Jr., and D. C. Annarelli, J. Organomet. Chem., 122, 311 (1976).

51. D. Seyferth and D. C. Annarelli, J. Am. Chem. Soc., 97, 2273 (1975).

52.* D. Seyferth and D. C. Annarelli, J. Am. Chem. Soc., 97, 7162 (1975).

53. D. Seyferth, R. L. Lambert, Jr., D. C. Annarelli, C. K. Hass, S. C. Vick, and D. P. Duncan, 32nd Southwest Regional Meeting, American Chemical Society, Fort Worth, Texas, December 1-3, 1976, Abstr. SYMP-55, p. 122.

54. D. Seyferth and D. C. Annarelli, J. Organomet. Chem., 117, C51 (1976).

55. H. Sakurai and M. Murakami, J. Am. Chem. Soc., 94, 5080 (1972).

56. H. Sakurai, S. Komiya, and Y. Nakadaira, 28th Annual Meeting of the Chemical Society of Japan, April 2, 1973, Abstr. III-1226; see also reference 34, footnote 7.

57. M. D. Sefcik and M. A. Ring, J. Am. Chem. Soc., 95, 5168 (1973).

58. R. P. Hollandsworth and M. A. Ring, Inorg. Chem., 7, 1635 (1968).

59. M. D. Sefcik and M. A. Ring, J. Organomet. Chem., 59, 167 (1973).

60. M. A. Ring and D. P. Paquin, 32nd Southwest Regional Meeting, American Chemical, Society, Fort Worth, Texas, December 1-3, 1976, Abstr. SYMP-58, p. 122. See also D. P. Paquin, "The Kinetics of Polysilane Decomposition and Their Reverse Silylene Insertion Reactions," Ph.D. thesis, University of California, San Diego, and San Diego State University, 1976.

61. I. M. T. Davidson, J. Chem. Soc., Chem. Commun., 323 (1973).

62. A. G. Brook, Chem. Res., 7, 77 (1974).

63. I. M. T. Davidson, J. Organometal. Chem., 24, 97 (1970).

64. P. L. Timms, Inorg. Chem., 7, 387 (1968).

65. D. R. Weyenberg and W. H. Atwell, Pure Appl. Chem., 19, 343 (1969).

66. R.-J. Hwang and P. P. Gaspar, unpublished results.

67. W. H. Atwell, U.S. Patent 3,478,078, November,11, 1969.

68. O. M. Nefedov and M. N. Marakov, Zh. Obshch. Khim., 34, 2465 (1964); O. M. Nefedov, T. Szekely, G. Garzo, S. P. Kolesnikov, M. N. Manakov, and V. I. Shiryaev, Int. Symp. Organosilicon Chem., Sci. Commun., Prague, 1965, 65. Chem. Abstr., 65, 12298b (1966).

69. P. S. Skell and E. J. Goldstein, "Silacyclopropanes," J. Am. Chem. Soc., 86, 1442 (1964).

70. J. C. Thompson, J. L. Margrave, and P. L. Timms, J. Chem. Soc., Chem. Commun., 566 (1966).

71. J. L. Margrave and P. W. Wilson, Acc. Chem. Res., 4, 145 (1971).

72. M. Ishikawa and M. Kumada, J. Organomet. Chem., 81, C3 (1974).

73. D. Seyferth, C. K. Haas, and D. C. Annarelli, J. Organomet. Chem., 56, C7 (1973).

74.* M. Ishikawa, F. Ohio, and M. Kumada, Tetrahedron Lett., 645 (1975).

75. D. Seyferth, 172nd National Meeting, American Chemical Society, San Francisco, Calif., August 29-September 3, 1976, Abstr. ORGN-73.

76. E. A. Chernyshev, N. G. Komalenkova, and S. A. Bashkirova, Dokl. Akad. Nauk SSSR, 205, 868 (1972).

77. G. P. Gennaro, Y. Y. Su, O. F. Zeck, S. H. Daniel, and Y.-N. Tang, J. Chem. Soc., Chem. Commun., 637 (1973).

78. P. P. Gaspar, R.-J. Hwang, and W. C. Eckelman, J. Chem. Soc., Chem. Commun., 242 (1974).

79. Y.-N. Tang, G. P. Gennaro, and Y. Y. Su, J. Am. Chem. Soc., 94, 4355 (1972).

80. O. F. Zeck, Y. Y. Su, and Y.-N. Tang, J. Chem. Soc., Chem. Commun., 156 (1975).

81. J. C. Thompson and J. L. Margrave, Inorg. Chem., 11, 931 (1972).

82. M. E. Childs and W. P. Weber, J. Org. Chem., 41, 1799 (1976).

83. M. E. Childs and W. P. Weber, Tetrahedron Lett., 4033 (1974).

84. R. L. Jenkins, R. A. Kedrowski, L. E. Elliot, D. C. Tappen, D. C. Schlyer, and M. A. Ring, J. Organomet. Chem., 86, 347 (1975).

85.* P. P. Gaspar and R.-J. Hwang, J. Am. Chem. Soc., 96, 6198 (1974). The reported 1:1 product ratio was in error. Trans, trans-2,4-hexadiene gives a 6:1 cis/trans product ratio, while cis, trans-2,4-hexadiene gives a 2:1 cis/trans product ratio (ref. 24).

86.* M. Ishikawa, F. Ohi, and M. Kumada, J. Organomet. Chem., 86, C23 (1975).

87.* R.-J. Hwang, R. T. Conlin, and P. P. Gaspar, J. Organomet. Chem., 94, C38 (1975).

88. R. J. Ellis and H. M. Frey, J. Chem. Soc., A, 553 (1966).

89. R. S. Cooke and U. H. Andrews, J. Org. Chem., 38, 2725 (1973).

90. T. J. Barton and M. Juvet, Tetrahedron Lett., 3893 (1975).

91. E. A. Chemyshev, N. G. Komalenkova, S. A. Bashkirova, A. V. Kisin, and V. I. Pchelintsev, Zh. Obshch. Khim., 45, 2221 (1975).

92. M. E. Vol'pin, Yu. D. Koreshkov, V. G. Dulova, and D. N. Kursanov, Tetrahedron, 18, 107 (1962).

93. E. A. Chernyshev, N. G. Komalenkova, and S. A. Bashkirova, Zh. Obshch. Khim., 41, 1175 (1971).

94. E. G. Janze, J. B. Pickett, and W. H. Atwell, J. Am. Chem. Soc., 90, 2719 (1968).

95. W. H. Atwell and J. G. Uhlmann, J. Organomet. Chem., 52, C21 (1973): Octamethyl-1,4-disilacyclohexa-2,5-diene is also formed under these conditions (T. Barton, private communication).

96. T. J. Barton and J. A. Kilgour, J. Am. Chem. Soc., 96, 7150 (1974).

97. L. E. Gusel'nikov, N. S. Nametkin, and V. M. Vdovin, Uspekhi Khimi, 43, 1317 (1974).

98. L. E. Gusel'nikov, N. S. Nametkin, and V. M. Vdovin, Acc. Chem. Res., 8, 18 (1975).

99. D. N. Roark and G. J. D. Peddle, J. Am. Chem. Soc., 94, 5837 (1972).

100.* T. J. Barton and J. A. Kilgour, J. Am. Chem. Soc., 98, 7746 (1976).

101. In the gas-phase flow-system experiments (refs. 96, 100) an oxidation product of the disilacyclobutene

Me_2Si $SiMe_2$ (ring: O; =C(Me)–C(Me)=) was obtained along with a product of addition to 2-butyne of a diradical $MeH\dot{S}iCH_2\dot{S}iMe_2$ from the rearrangement of tetramethyldisilene (ref. 99).

102. R. T. Conlin and P. P. Gaspar, J. Am. Chem. Soc., 98, 868 (1976).

103. C. S. Liu and J. C. Thompson, Inorg. Chem., 10, 1100 (1971).

104. C. S. Liu, J. L. Margrave, J. C. Thompson, and P. L. Timms, Can. J. Chem., 50, 459 (1972).

105. C. S. Liu, J. L. Margrave, and J. C. Thompson, Can. J. Chem., 50, 465 (1972).

106. W. H. Atwell, U.S. Patent 3,576,024, April 20, 1971.

107.* R. T. Conlin and P. P. Gaspar, J. Am. Chem. Soc., 98, 3715 (1976).

108.* D. Seyferth, D. C. Annarelli, and S. C. Vick, J. Am. Chem. Soc., 98, 6382 (1976).

109.* D. Seyferth and S. C. Vick, J. Organomet. Chem., 125 C11 (1977).

110. Private communication from T. J. Barton, August 1976.

111. R.-J. Hwang and P. P. Gaspar, unpublished experiments.

112. D. Cooper, J. Organomet. Chem., 7, P26 (1967); R. A. Jackson, Adv. Free Radical Chem., 3, 231 (1969).

113. P. P. Gaspar and G. S. Hammond, in Carbenes, Vol. II, R. A. Moss and M. Jones, Jr. (eds.), Wiley, New York, 1975, p. 207.

114.* A. Kasdan, E. Herbst, and W. C. Lineberger, J. Chem. Phys., 62, 541 (1975).

115. P. F. Zittel, G. B. Ellison, S. V. ONeil, E. Herbst, W. C. Lineberger, and W. P. Reinhardt, J. Am. Chem. Soc., 98, 3731 (1976).

116. P. C. Jordan, J. Chem. Phys., 44, 3400 (1966).

117. B. Wirsam, Chem. Phys. Lett., 14, 214 (1972).

118.* J. H. Meadows and H. F. Schaefer, III, J. Am. Chem. Soc., 98, 4383 (1976).

119. I. Dubois, Can. J. Phys., 46, 2485 (1968).

120. I. Dubois, G. Duxbury, and R. N. Dixon, J. Chem. Soc. Faraday Trans., 2, 71, 799 (1975).

121. I. M. T. Davidson and J. I. Matthews, J. Chem. Soc. Faraday Trans. I, 72, 1403 (1976).

122. I. M. T. Davidson, and M. E. Delf, J. Chem. Soc. Faraday Trans. I, 72, 1912 (1976).

8

THEORY OF REACTIVE INTERMEDIATES AND REACTION MECHANISMS

K. N. HOUK

Louisiana State University, Baton Rouge, Louisiana 70803

I. INTRODUCTION

Theory is an increasingly helpful tool for the study of reactive intermediates. In fact, if an intermediate is sufficiently reactive, then theory is the only technique that can provide information about the geometry, energy, and intimate behavior of the species. Quantum mechanics (or some of its practitioners) sometimes promises more than it delivers, but this situation is rapidly changing. This change results from theoretical advances rather than from any newly discovered modesty on the part of theoreticians.

The years 1975-1976 were marked by rapid assimulation of theoretical principles and techniques by organic chemists. Indeed, organic theory now often leads instead of follows experiment. I have chosen to subdivide this chapter into three parts. The first part discusses selected papers that provide direct insight into the structures and reactivities of intermediates, such as carbocations, carbanions, and carbenes. The second part is a critical evaluation of the theoretical calculations of surfaces of important organic reactions as well as of certain general principles about surfaces of reactions, particularly photochemical ones, published in the last 2 years. Third, several papers that reveal new theoretical principles and insights as well as new techniques are discussed. In each case I am writing for the organic chemist, not the theoretician, so that some points are elaborated in more detail than would be appropriate for an article on recent advances in theory *per se*.

II. CARBOCATIONS

Wolf, Harch, Taft and Hehre reported *ab initio* (STO-3G) calculations and accompanying gas-phase measurements on the stabilities of cyclopropylcarbinyl and benzyl cations.[1] The calculated differences in stabilities of primary, secondary, and tertiary cations (relative to the corresponding neutral hydrocarbons) are shown in Figure 8.1. The calculations predict a small preference for the benzyl cation for the tertiary system and a large difference for the primary. Experimental ion cyclotron resonance (ICR) studies show that the cyclopropylcarbinyl is actually slightly preferred for the tertiary system (ΔG^0 = -0.8 kcal/mol), but the secondary system favors the benzyl system (ΔG^0 = +4.8 kcal/mol). Thus phenyl is better able to stabilize a primary or secondary cation, but cyclopropyl is better able to stabilize a tertiary cation. Perhaps of greatest significance is the fact that calculated π charges and ^{13}C chemical shifts do correlate with each other but do not correlate *with stability*. The simple orbital model proposed by Taft and co-workers to explain the better stabilization of primary cations by phenyl, but tertiary by cyclopropyl, does not, in my opinion, fully rationalize this phenomenon. In effect, they suggest that phenyl has such a large stabilizing effect on the primary system that additional methyls do not help much, whereas the smaller effect of cyclopropyl can be aided by methyl. However, this model does not provide insight into why the tertiary cyclopropylcarbinyl system is more stable than the cumyl cation. Perhaps the geometry of the cumyl system (nonplanar) does not allow full interaction of the phenyl group with the tertiary cationic center.

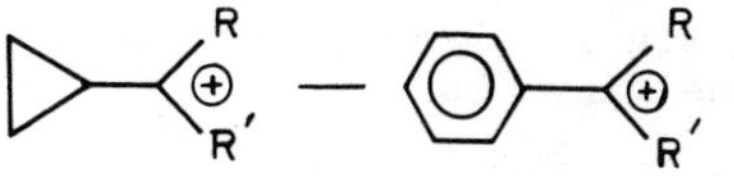

R	R′	ΔE (kcal/mole)
H	H	+20.2
Me	H	+9.8
Me	Me	+2.4

FIGURE 8.1. ΔE for cyclopropylcarbinyl and benzyl cations

An extraordinarily extensive calculation has been reported on the vinyl cation by Weber, Yoshimine and McLean.[2] Although performed on an extremely small cation, the results of this work show what can be done with MO calculations today (at least with the computational power of IBM-San Jose available) and provide some insight into the validity, or lack thereof, of approximate calculations on cations. Figure 8.2 shows the geometries calculated by *ab initio* SCF methods with an extended basis set, assuming (on the left) a linear CCH and (on the right) a bridged cation with C_{2v} symmetry. At the SCF level the bridged cation is 5.6 kcal/mol less stable than the classical cation. However, even at the "Hartree-Fock limit," *ab initio* calculations neglect correlation energy. That is, in Hartree-Fock calculations each electron feels the average repulsion of the other electrons in the molecule. In fact, the electron repulsion in a molecule is much less (by an amount called the "correlation energy") than is calculated by the Hartree-Fock method. Configuration interaction (CI) calculations calculate some or, in principle, all of this correlation energy. In the case of the vinyl cation, extensive CI reduced the energy of the bridged ion more than the "classical" structure so that both structures shown in Figure 8.2 have the same energy, and a barrier less than or equal to 1.6 kcal/mol separates these species. The authors estimate that these two species differ by no more than 1-2 kcal/mol.[2] Is the vinyl cation classical or nonclassical? It is both! Furthermore, if generalizations about correlation energies are possible, one might conclude that good SCF calculations will underestimate the stability of nonclassical cations relative to classical.

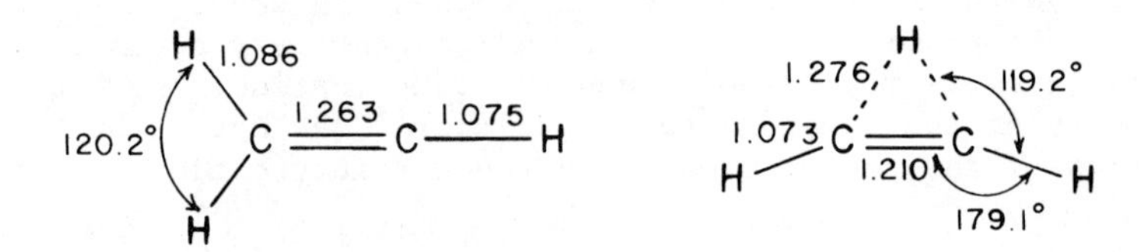

FIGURE 8.2. Most stable planar and bridged C_{2v} geometries of the vinyl cation

Less "benchmark" in nature but perhaps of more immediate practical significance, it has been shown that _ab initio_ minimal basis set (STO-3G) calculations can reproduce both aromatic substituent constants (σ^+)[3] and aliphatic substituent constants (σ_I).[4] Streitweiser, Hehre, et al. calculated the energy of the conversion of substituted benzenium ions to the corresponding benzene derivatives. The difference between the energy of this reaction in a substituted case and that for benzene itself (ΔE) correlates reasonably well with the σ^+ substituent constant

$$\Delta E = (-22.83 \pm 1.63)\sigma^+ + (0.40 \pm 0.85)$$

Some deviations are well understood. For example, along the series Me, Et, _i_-Pr, _t_-Bu, the experimental solution-phase electron-donor abilities decrease ($\sigma_p^+ = -0.311, -0.295, -0.280, -0.256$), whereas the donor abilities calculated from the relationship shown above increase in magnitude ($\sigma_p^+(\text{calc}) = -0.354, -0.407, -0.456, -0.491$, for methyl through _t_-butyl, respectively). As had been shown earlier by calculations and by measurements of gas-phase basicities of substituted benzenes,[5] the Baker-Nathan order of electron release by alkyl groups in solution (Me > Et > _i_-Pr > _t_-Bu) is an artifact of solvation. Calculated or gas-phase measurements indicate that the intrinsic order is the so-called inductive order (_t_-Bu > _i_-Pr > Et > Me).[5]

For aliphatic substituent constants (σ_I), a correlation between differences in heats of the reaction, $XCH_2CH_2NH_2 + H^+ \rightarrow XCH_2CH_2NH_3^+$, and the substituent constant, σ_I, of the substituent was found.[4]

These results are of particular significance since various semiempirical techniques have previously failed to account quantitatively for substituent effects. A relatively economical and generally available ab initio calculational technique[6] can now be used to determine theoretically the role of substituents on molecule stabilities and reaction rates. Theoretical assessments of field, inductive, and resonance effects are now possible. General assessments of the use of these techniques for cations[7] and more general problems[8] have now appeared.

III. CARBANIONS

Many fewer calculations have been reported for carbanions than for carbocations. The reason is undoubtedly due in part to the fact that calculations with small basis sets indicate that the HOMO's of these molecules have positive energies; that is, the calculations predict spontaneous dissociation to a radical plus an electron. Nevertheless, calculations do give reasonable information about the relative energies of various anions, and a small avalanche of papers has recently addressed the question of why carbanions adjacent to sulfur are much more stable than carbanions adjacent to oxygen. The answer to this question, of interest in itself, is perhaps of general significance to the seemingly never-ending debate over the role of d orbitals in a particular phenomenon.

Streitweiser and Williams,[9] Csizmadia, Wolfe, et al.,[10] Lehn and Wipff,[11] and Epiotis, Yates, et al.,[12] carried out ab initio calculations with and without d orbitals on equilibria such as: $CH_3XH \rightleftarrows HXCH_2^- + H^+$, when X is S or O.[9-12] Although inclusion of d orbitals on S greatly stabilizes the carbanion, the neutral compound is similarly stabilized. The conclusion is that d orbitals are unimportant in enhancing the acidities of hydrogens α- to sulfur in sulfides.[9-12] Instead S stabilizes carbanions by polarization[9-11] and by the longer C-S than C-O bond.[10] More specifically, Bernardi, Csizmadia, et al.,[10] Lehn and Wipff,[11] and Epiotis, Yates, et al.,[12] found from an investigation of conformational energies that the interaction of the carbanion lone pair with the σ^*_{SH} orbital has an appreciable stabilizing effect on the carbanion. Thus low-lying vacant orbitals associated with sulfur, but not vacant d orbitals, help stabilize carbanions. Epiotis and co-workers discussed this model in greatest detail and concluded that the anion lone pair interacts more strongly with a σ^*_{SH} orbital than a σ^*_{OH} because the lone pair and the σ^*_{SH} are closer in energy than the lone pair and the σ^*_{OH} and also because the overlap of the lone pair with the σ^*_{SH} orbital is greater than that with the σ^*_{OH}.[12] The difference in overlap arises from the greater distance of the lone pair from the substituent on sulfur. The substituent (e.g., H) atomic orbitals in the σ^*_{SH} or σ^*_{OH} bond overlap in an antibonding fashion with the lone-pair orbital.[12]

I would be remiss if I failed to mention a 1972 communication that established the principles and philosophy invariably used to explain conformational phenomena in MO terms, including the thiocarbanion case described earlier. This paper,[13] in which the authors' names and addresses constituted a full 10% of the total length, described the stabilizing influence of two-electron interactions (e.g., between an occupied lone pair and a low-lying vacant σ^* orbital) and the destabilizing influence of four-electron interactions (e.g., between an occupied lone-pair orbital and an occupied σ orbital). The balance between these determines the preferred conformation of molecules.[13]

A careful exploration of the methyl radical, anion, and cation surfaces using *ab initio* 4-31G calculations provided an explanation based on Walsh-Mulliken correlation diagrams for the variations in degrees of pyramidalization and barriers to inversion as one of the HCH angles contracted below 120^0.[14] The methyl anion has a moderate preference for the pyramidal structure in the 4-31G calculations (each C-H bond is 23^0 out of a plane perpendicular to the C_3 axis). The barrier to inversion is 8 kcal/mol in these calculations. When one of the HCH angles (α) is fixed and the remaining bond lengths and angles are varied, the degree of planarity and the barrier to inversions change markedly. Table 1 shows the equilibrium angles (β in degrees) by which one C-H bond deviates from the HCH plane and the barriers to inversion (ΔE_{inv} in kcal/mol) for both the methyl anion and radical as a function of α.

As α is constricted from an equilibrium value near tetrahedral to smaller values, the equilibrium value of β increases (the molecule becomes more highly pyramidal), and the barrier to inversion increases. The methyl radical, which is normally planar, becomes nonplanar with a small barrier to inversion when the angle α is constricted. Inspection of the Walsh-Mulliken correlation diagrams, where orbital energies are plotted as a function of angle β for fixed α, indicates that this phenomenon is due mainly to differences in the behavior of the HOMO of the methyl anion or the singly occupied MO (SHOMO) of the methyl radical. The three highest occupied and the lowest unoccupied molecular orbitals of CH_3 are shown schematically in Figure 8.3. The shapes of the MO's correspond roughly to those for a planar molecule with $\alpha = \angle HCH = 100^0$.

For a trigonal species, the 3a' and 1a" orbitals would be degenerate, but at $\alpha < 120^0$ the 3a' drops below the 1a" in energy. For a fixed value of α the energy of 1a" is approximately constant for all values of β, and this orbital has

TABLE 1. Equilibrium-out-of-plane Angles, β, and Barriers to Inversion

	Anion		Radical	
α	β	ΔE_{inv}	β	ΔE_{inv}
60°	77°	33[a]	-	-
90°	72°	20.3	32.4°	1.1
109.47°	63.4°	9.4	5°	0.1
120°	56.6°	4.6	0°	0
135°	32.8°	0.4	0°	0

[a] Extrapolated values

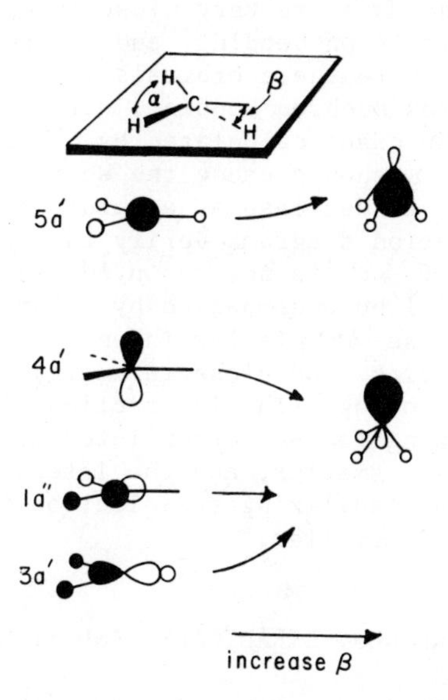

FIGURE 8.3. HOMO's and LUMO's for the methyl radical or anion

little to do with the shape of the molecule. The 3a' orbital rises rapidly as β is increased, and this energy change causes the methyl cation to prefer planarity for all angles of α. However, 4a' decreases in energy with an increase in β (i.e., increase in pyramidalization). It is this decrease in energy that causes the methyl anion to prefer a nonplanar structure. In the methyl radical, where the 4a' orbital is singly occupied, the decrease in one 4a' electron energy on bending is slightly less than the increase in the two 3a' electron energies. Thus the methyl radical is planar, but out-of-plane deformations are energetically easy. This type of reasoning is thoroughly familiar.[15] However, it was found that the decrease in energy of the 4a' orbital with increase in β is more rapid as the angle α contracts.[14] Thus the greater tendency for nonplanarity in anions or radicals where one angle is small (e.g., in the cyclopropyl anions or radical) results from the greater decrease in the HOMO or SHOMO (4a') energy on bending than in systems without a constricted angle, α.[14] The origin of this difference was not speculated on by Schleyer et al. but can be interpreted in terms of Levin's perturbation treatment of inversion barriers.[16] Levin noted that the 4a' orbital drops in energy upon bending as a result of mixing of the 4a' orbital with the LUMO (5a'). The large barrier to inversion in phosphines, where 4a' and 5a' are very close in energy and, as a result, mix substantially on bending, and the small barrier to inversion in amines, where these orbitals are much further apart and do not mix as much on bending, were discussed in these terms.[16] In the cases calculated by Schleyer, Allen, et al., the decrease in α should cause the energies of the 3a' and the 5a' orbitals to decrease as a result of increased H-H bonding. The correlation diagrams verify that this occurs in 3a', but the energy of 5a' is not shown.[14] When α is small, the increase in β will be accompanied by a large decrease in the 4a' energy, because 4a' is far in energy from 3a', which pushes 4a' up on bending, and close in energy to 5a', which pushes 4a' down on bending. The interaction of 4a' and 3a', which is a destabilizing four-electron interaction, is decreased as α is made smaller, and the interaction of 4a' with 5a', which is a stabilizing two-electron interaction, is increased as α is made smaller.

IV. CARBENES, DIRADICALS, AND YLIDES

Several important papers have been published in which various types of *ab initio* calculations have been used to determine the geometries and electronic properties of nonisolable species. Davis, Goddard and Bergman reported *ab initio* GVB calculations on various points on the C_3H_4 surface and deduced not only some interesting details about the transformation of cyclopropenes and vinyldiazomethanes to vinylmethylenes but also some novel and general features of the electronic structures of diradicals.[17] The relative energies of various states and crude representations of the electron distributions in these states are given in Figure 8.4. The ground state of the vinylmethylene

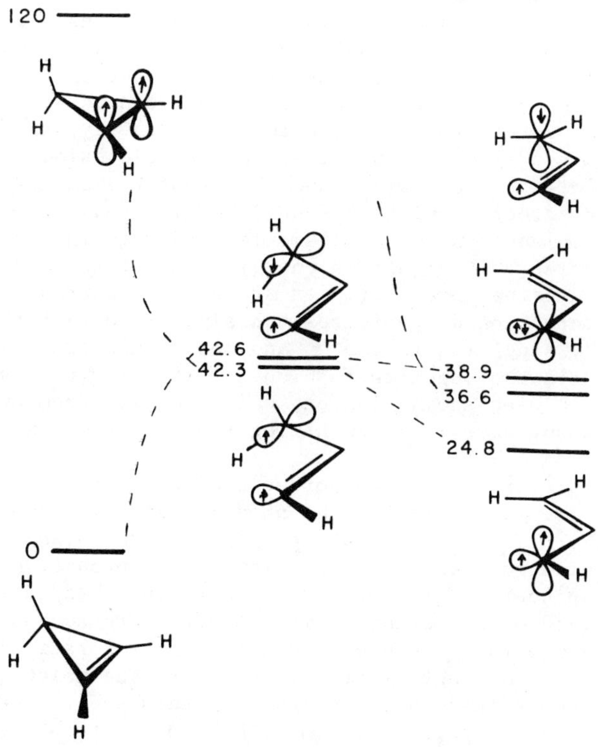

FIGURE 8.4. The GVB calculated energies of various C_3H_4 species

species is essentially a triplet methylene with an appended vinyl group. That is, perhaps surprisingly, the electron in the p orbital is not delocalized into the adjacent vinyl π system. Similarly, the lowest singlet state, which is calculated to lie 12 kcal/mol above the ground-state triplet, is like the lowest singlet state of methylene, with a lone pair in the σ hybrid orbital and a vacant π p orbital. Once again there is essentially no delocalization of the vinyl π electrons into this vacant p orbital. An excited singlet, at 14 kcal/mol above the ground-state triplet, has essentially a $\sigma\pi$ diradical structure, with the double bond localized between C-1 and C-2.[17]

The localization calculated in these cases is a very interesting phenomenon that results from the differences in exchange interactions in various localized (so-called resonance) structures. That is, the extent of mixing of valence-bond structures such as $\dot{C}$-C=C and C=C-$\dot{C}$ depends on the difference in energy between these two structures. For the allyl radical these two structures are equal in energy and extensive mixing occurs. However, in the vinylmethylene case with one σ unpaired electron and one π, these structures differ in energy mainly by the difference in exchange interactions between the two unpaired electrons. For the triplet state the lower-energy structure has the two unpaired electrons as close as possible, since the most stable triplet structure will have the greater exchange interaction. The structure with the unpaired electrons at opposite ends of the molecule has a much smaller exchange interaction and is hence considerably higher in energy. The latter mixes insignificantly into the lowest triplet state. In contrast, the exchange interaction destabilizes the singlet so that the lowest energy singlet with one σ and one π unpaired electron is localized in the opposite fashion. Just below this singlet there is the closed-shell singlet, which suffers from exchange interactions but benefits from having two electrons in a hybrid orbital.[17]

The result is similar to Borden's discussion of the singlet and triplet states of the planar trimethylenemethane "diradical."[18] The ground-state triplet has the planar D_{3h} symmetry, which puts the unpaired electrons, in part, on common atoms. The planar singlet electronic structure has these electrons localized on different atoms. A structure such as a localized methyl radical on one CH_2 and an allyl radical on the other CH_2's and the central atom results.[18] The molecule will distort in such a fashion as to maximize this separation (minimize the exchange repulsion). Thus the ground-state singlet has one CH rotated by 90^0.[18]

The results of both the Davis and Borden groups can be generalized by stating that the electronic structures of open-shell singlets will frequently have localization of the unpaired electrons on different parts of the molecules; this is true even if symmetrical molecular geometries are assumed, but the molecular geometries of these species may be expected to distort in a direction that will enhance this separation. On the other hand, triplet states will have the unpaired electrons localized in the same part of the molecules (but of course in different orbitals).

Returning briefly to the insights about cyclopropene pyrolysis that can be gleaned from the Davis group calculations (Figure 8.4), the ground state apparently begins to open to the $\sigma\sigma$ diradical shown in the center of the figure, but rotation of the methylene leads to the excited $\sigma\pi$ singlet. This can reclose or decay to the carbene-like states. On the other hand, diazopropene photolysis should give only the carbene-like states.[17]

Calculations by <u>ab initio</u> SCF methods on the acyclic C_3H_2 species have been reported by Hehre, Pople, et al.[19] One of these species, the propargylene molecule, is found to have a localized singlet carbene-like structure (Figure 8.5) with an essentially localized triple bond. The lowest triplet state of this species is much different in geometry, having essentially the allene-like geometry shown in Figure 8.5, but with some CCC bending.[19] This structure does not seem to correspond to the generalization made earlier from the Davis group and Borden work, since the unpaired electrons have maximum separation. It may well turn out that a calculation including CI would produce a geometry of the triplet having both unpaired electrons more localized on the same carbon and with a more localized triple bond at the opposite end of the molecule.

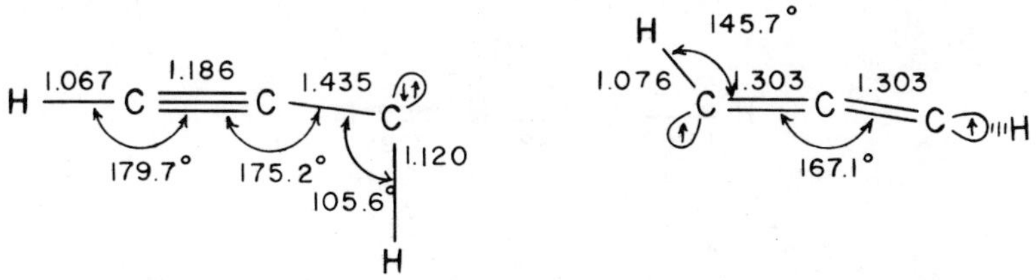

FIGURE 8.5. Singlet and triplet propargylenes[19]

The geometries of a number of nitrile ylides have been calculated in our laboratories by <u>ab initio</u> SCF methods by Pierluigi Caramella.[20] These calculations, like those of Davis and colleagues, teach a general lesson, namely, that delocalization may be less important than s-p mixing (hybridization) in determining the most favored geometry of a molecule. The species considered, the nitrilium ylide, imine, and oxide, are shown in their most favored geometries (by 4-31G optimizations) in Figure 8.6.[20] The ylide is highly bent at C-1 and resembles a carbene in that the HOMO is more or less a hybrid lone-pair orbital at C-1, whereas the LUMO is more or less a vacant p orbital at C-1. The more sophisticated calculations by Davis and co-workers on the related vinylmethylene species might indicate that the nitrile ylide will be even more localized than is indicated in Figure 8.6. The nitrile imine is planar but lies on a very flat surface, with bending using very little energy. The oxide is linear but requires only 1.3 kcal/mol to distort the HCN angle to 165^{0}. The trends here are certainly correct although the exact structures may be somewhat inaccurate. In the simplest terms the trend can be explained in terms of the electronegativity of atom 3. That is, the preferred structure for the nitrile oxide places the lone pair at the electronegative oxygen and gives a molecule with essentially a CN triple bond and CO single bond, whereas the bent nitrile ylide structure places the lone pair in an orbital on C-1 with appreciable s character and gives a molecule with two CN double bonds.

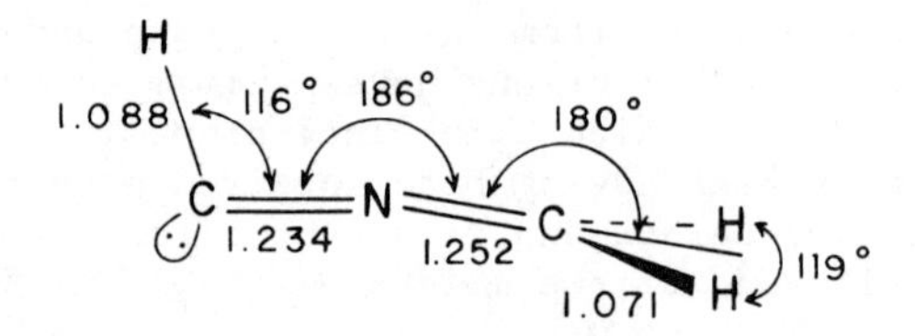

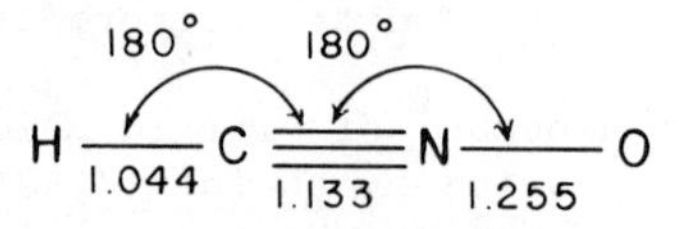

FIGURE 8.6. 4-31G Optimized geometries of nitrile ylide, nitrile imine and nitrile oxide

The results are of general importance as far as the understanding of molecular geometries is concerned, but of greater immediate importance was the insight these calculations gave into the regioselectivities of reactions of these species. To summarize in the simplest terms, the oxygen terminus of the nitrile oxide is the more nucleophilic end, combining with electrophiles and the most electrophilic site of alkenes. On the other hand, the bending of the HCN angle is not too difficult, so that the oxygen terminus will not be a great deal more nucleophilic than the carbon for cycloadditions with late transition states.[20] This bending in the transition state will be even more important for the imine. For the nitrile ylide, C-1 is the more nucleophilic terminus. The regioselectivities observed in cycloadditions of these species could be explained on this basis.[20] In contrast to other 1,3-dipoles, C-1 of the nitrile ylide is both the more nucleophilic and the more electrophilic terminus.

Arduengo and Burgess have provided dramatic computational evidence for the effect that substituents have on the geometries of 1,3-dipolar species.[21] Thiocarbonyl ylide (thione methylide, CH_2SCH_2) is a planar molecule with an energy minimum separated by a 13 kcal/mol conrotatory barrier from the more stable thiirane. However, the unsymmetrically substituted species, $(Me_2N)_2CSC(CHO)_2$, is calculated by MINDO/3 to be most stable in a geometry with the electron-deficient terminus $[C(CHO)_2]$ rotated by 90^0 with respect to the CSC plane. The strong donor-acceptor unsymmetrical substitution renders this zwitterionic structure more stable than either the planar or the cyclic form. An X-ray crystallographic determination of the structure of a highly asymmetrically substituted thiocarbonyl ylide verified this prediction.[21]

V. CYCLOBUTADIENE

Theoreticians were apparently thrown into a tizzy by several independent reports that low-temperature matrix-isolated cyclobutadiene was square, as indicated by its IR spectrum. No ESR spectrum, indicative of a triplet species, could be detected.[22] Virtually all theoretical calculations[23] predicted that the lowest singlet state was rectangular and the lowest triplet state square, but most calculations, as well as the experimental data of Pettit,[24] indicated that the rectangular singlet was lower in energy than the square triplet. An exception was the generalized valence bond (GVB) calculation performed by Newton,[22c] which predicted that the triplet was 7.7 kcal/mol above the singlet.

Dewar and Kollmar suggested on the basis of MINDO/3 calculations that the rectangular singlet was lower in energy than the square triplet, but that cyclobutadiene was formed as the square triplet by the photolytic methods of generation.[25] The calculations indicated that the triplet lies in a 3.5 kcal/mol well, which at 4 K would prevent conversion of the square triplet to the ground-state singlet.

On the other hand, Borden has discussed how inclusion of electron repulsion in SCF calculations can render the square singlet more stable than the square triplet.[26] He showed that configuration interaction (CI), a mixing of higher-energy electronic configurations (orbital occupancies) with the lowest-energy electronic configuration, can stabilize the lowest singlet state more than the lowest triplet.[26] Nevertheless, various SCF calculations and X-ray structures of highly substituted isolable cyclobutadienes suggest that the rectangular singlet is the most stable species.[23]

The apparent paradox has not yet been fully resolved, but the work of Maier, Hartan, et al. has shown that the species whose IR spectrum had been attributed to free cyclobutadiene was actually a charge-transfer complex of cyclobutadiene and CO_2. Only two bands in the IR spectrum can be attributed to free cyclobutadiene.[27] Thus the argument that cyclobutadiene is square because the four IR bands predicted for the D_{4h} molecule are the only ones observed is weakened. Fewer bands than are predicted for either the D_{4h} or the D_{2h} geometries are seen, indicating that a number of bands (perhaps the missing bands required for D_{2h} symmetry) are too weak to be observed in the spectrum. However, Maier concluded that the IR spectrum was consistent with the square singlet nature of the ground state.

Theory has fared rather poorly in recent applications to the cyclobutadiene problem. That is, the squareness of the lowest singlet state was not "predicted" by calculations until after experimental data suggested that it should be. Meanwhile, although some uncertainty has been raised as to the meaning of the experimental results, theory has not definitively predicted the geometry or multiplicity of the ground state of the molecule.

VI. SURFACES OF ORGANIC REACTIONS

Certain qualitative generalizations about surfaces discovered in the last few years are discussed in this section. In addition, a few detailed investigations of especially important reactions are covered. Perhaps the most general theoretical development in the last 2 years has been in the understanding of excited state surfaces. The number of papers in this area has been considerable, but those of Salem, with Turro, Dauben, and Devaquet, and those of Michl stand out.

Early in 1975 Dauben, Salem, et al. reviewed the general concept that the shapes of excited state surfaces could be deduced from knowledge (experimental, calculated, or divined) about the relative energies and symmetries of ground and excited states of reactants and products.[28] Since an excited-state reaction frequently involves the breaking of a single or a double bond (the latter is really a rotation), the understanding of the various states available to a species with two more-or-less isolated orbitals that are similar in energy is crucial to the understanding of these surfaces. Such "diradical" states have been analyzed in detail by Salem and Rowland,[29] and Michl.[30] We give here a brief discussion of these ideas to lay the groundwork for the discussion of the hottest new papers.

The central difficulty in dealing with diradical states is the absolute necessity of including configuration interaction in a MO description, or alternatively, to use a valence-bond description that, unfortunately, is somewhat less familiar to organic chemists. To illustrate the connection between these methods, suppose that we start with a strongly bonding orbital, represented in a MO picture as $\psi_1 = \varphi_a + \varphi_b$, where φ_a and φ_b are atomic orbitals on, for example, two hydrogen atoms, so that $\varphi_a + \varphi_b$ is the bonding combination of these orbitals (throughout this discussion we neglect normalization). Putting two electrons into ψ_1 gives the following electronic configuration $\psi_1{}^2 = [\varphi_a + \varphi_b](1) \times [\varphi_a + \varphi_b](2)$, where the numbers in parentheses are indices of electrons. Multiplying out this expression gives $\psi = \varphi_a(1)\varphi_a(2) + \varphi_b(1)\varphi_b(2) + \varphi_a(1)\varphi_b(2) + \varphi_a(2)\varphi_a(1)$. Written this way, it is clear that ψ_1 is equally covalent and ionic in character. That is, $\varphi_a(1)\varphi_a(2)$ and $\varphi_b(1)\varphi_b(2)$ are ionic contributions having both electrons on one atom, whereas $\varphi_a(1)\varphi_b(2)$ and $\varphi_a(2)\varphi_a(1)$ are covalent contributions, having the electrons equally shared by both atoms. Now, as the atoms are pulled apart, a doubly occupied ψ_1 ceases to be a good representation for the electronic structure. In other words, a simple valence-bond picture, with one electron in φ_a and the other one φ_b, that is, $\varphi_a(1)\varphi_b(2)$, would be a better representation. The MO picture forces the lower-energy orbital to be doubly occupied unless configuration interaction is taken into account. That is, at bonding distance there is a much higher energy configuration, with two electrons in an antibonding orbital, $\psi_2 = \varphi_a - \varphi_b$. The doubly excited configuration $\psi_2{}^2 = [\varphi_a - \varphi_b](1) \times [\varphi_a - \varphi_b](2)$ can be multiplied out to give $\psi_2{}^2 = \varphi_a(1)\varphi_a(2) + \varphi_b(1)\varphi_b(2) - \varphi_a(1)\varphi_b(2) - \varphi_a(2)\varphi_b(1)$, which is equally covalent and ionic, but these contributions are combined with a negative sign. As the bond is stretched, $\psi_2{}^2$ drops rapidly in energy, and as it does so, begins to mix with $\psi_1{}^2$. At diradical geometries, where the a-b bond is stretched sufficiently so that there is essentially no interaction between φ_a and φ_b, the singlet ground state of the radical can best be represented as $\psi_1^2 - \psi_2^2 = \varphi_a(1)\varphi_b(2) + \varphi_a(2)\varphi_b(1)$, which has only covalent, and no ionic, contributions and is essentially the same as the single valance-bond representation of the diradical, $\varphi_a(1)\varphi_b(2)$. The mixing of configurations, ψ_1^2 and ψ_2^2, is a simple type of configuration interaction.[29,30]

In general, a diradical will have four states, which in the valence-bond representation are the diradical (or covalent) singlet and triplet, ^1D and ^3D, respectively, $\varphi_a(1)\varphi_b(2)$, depending on the spins of electrons 1 and 2, and higher energy zwitterionic states, $\varphi_a(1)\varphi_a(2) + \varphi_b(1)\varphi_b(2)$, called Z_2, and $\varphi_a(1)\varphi_a(2) - \varphi_b(1)\varphi_b(2)$, called Z_1.[28] Figure 8.7 shows the behavior of these states schematically as the energies of the component atom orbitals, φ_a and φ_b, are made increasingly different, with φ_a lower in energy than φ_b.

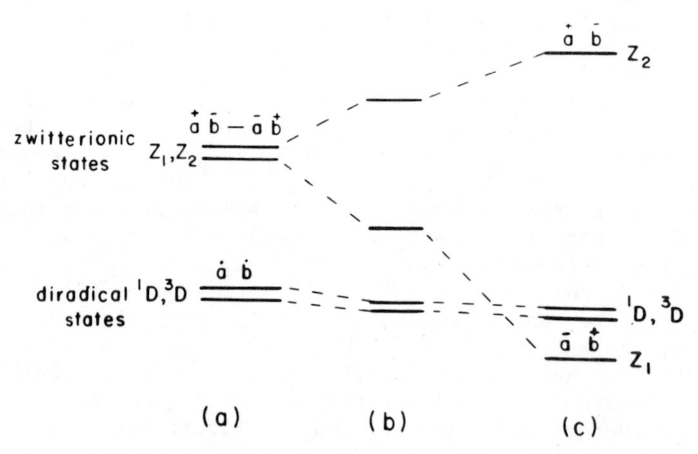

FIGURE 8.7. Diradical states: (a) φ_a, φ_b degenerate, (b) φ_a slightly lower than φ_b, (c) φ_a much lower than φ_b[28]

When φ_a is much lower in energy than φ_b, the zwitterion Z_1 may drop lower in energy than ^1D and ^3D because the lower electron repulsion in ^1D and ^3D is more than compensated for by putting both electrons in the lower-energy orbitals, φ_a.[28-30]

A startling application of these ideas in the area of photochemistry was made by Salem and co-workers in 1975.[31] As is well known, the lowest excited singlet of ethylene has an energy minimum at a geometry where one end of the molecule has been rotated by 90° with respect to the other. With such a geometry the ordering of states is like that shown near the left of Figure 8.7. If the perpendicular ethylene is rendered unsymmetrical by substitution of a donor at one end or an acceptor at the other end, or both, then the ordering of states will become more like that shown nearer the right of Figure 8.7. The extraordinary conclusion obtained from calculations by Salem and co-workers was that even a slight "unsymmetricalization" (e.g., pyramidalization of one terminus), is sufficient to cause a dramatic "sudden polarization" of the lowest excited singlet state, Z_1.[31] The chemical consequences of this phenomenon are profound, but the consequences of this phenomenon with regard to excited states are perhaps even more so. To quote Salem and colleagues, "Under no circumstances should one rely on ordinary chemical intuition to estimate the extent of charge separation in excited Z_1 states."[31]

The application of these ideas made in the first paper was to the photocyclization of 1,3,5-hexatrienes.[31] Calculations on the Z_1 (first excited singlet) state of this molecule indicated that as the molecule is rotated about the C_3-C_4 bond, there is a sudden polarization when the dihedral angle between the two allyl fragments is near 90^0. As shown in Figure 8.8, a cyclization mechanism involving the Z_1 excited state can be written where the planar vertical singlet relaxes to the orthogonal charge-separated species. At this point the allyl anion portion snaps shut to a cyclopropyl carbanion, which then combines with the allyl cation portion to give the final bicyclo-[3.1.0]hexene.[31] Such a mechanism for diene and triene photocyclizations had been proposed earlier by Dauben based on experimental results.

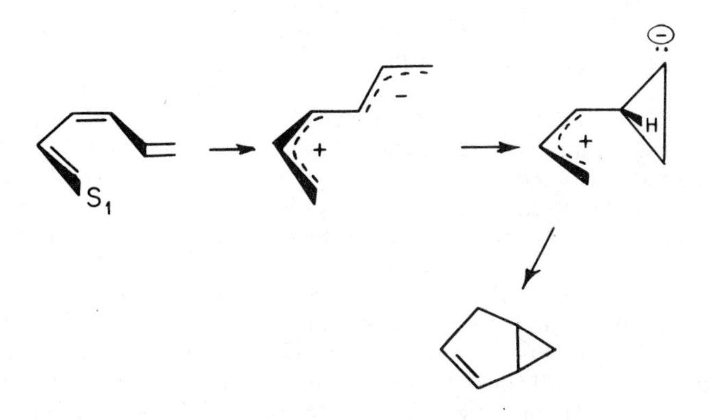

FIGURE 8.8. The "sudden polarization" mechanism of the hexatriene photocyclization

More extensive calculations were carried out on the butadiene photocyclization to bicyclobutane.[32] Although neither group of authors did so, it is interesting here to compare the Salem group's results to *ab initio* calculations on the butadiene electrocyclization to cyclobutene reported by Grimbert, Segal, et al.[33] Bruckmann and Salem performed calculations on ground (^1D) and excited (Z_1 and Z_2) states of butadiene in a conformation where one terminus was rotated by 90^0 with respect to the remainder of the molecule. The ground state of this species is essentially a localized methylene radical joined to an allyl radical. The two zwitterionic statss can be represented in the extreme as $^+CH_2-CH \overset{---}{} {}^-CH \overset{---}{} CH_2$ and $^-CH_2-CH \overset{---}{} {}^+C \overset{---}{} CH_2$. Bruckmann and Salem concluded that either of these could be the first excited state of the methylene-allyl geometry of butadiene, depending on the extent of pyramidalization of the methylene terminus. As shown on the left-hand side of Figure 8.9, either the methylene-anion, allyl-cation, II, or the methylene-cation, allyl-anion, I, can be the lowest singlet excited state, depending on the exact geometry of the species. The allyl portions are drawn as a localized anion or cation since the geometrical optimizations showed this to be the case.[32] In the optimized geometries II is 9.2 kcal/mol more stable than I and conversion of I to II involves a 6.6 kcal/mol barrier. Both II and I are simply different points on the Z_1 surface, corresponding to isomeric forms of zwitterion Z_1 represented earlier in Figure 8.8. Tracing the photochemical course of the butadiene to bicyclobutane conversion in greater detail, the lowest singlet of *transoid* butadiene is stabilized by methylene rotation. For the parent butadiene the species II is more stable, and Bruckmann and Salem assume that this can cyclize at 2 and 4 in a π^2 disrotatory fashion. The resulting cyclopropyl-cation, methylene-anion can cyclize to bicyclobutane subsequently or in concert with the first bond formation. The methylene-cation, allyl-anion species, if formed, also cyclizes, but in a conrotatory fashion since the 4π electron allyl system is involved.[32] In fact, using a ground-state analogy, the allyl anion is far more likely to cyclize than is the allyl cation, but as internal conversion to the D state may accompany cyclization, this ground-state analogy may be inappropriate.

Bruckmann and Salem carried out some calculations on substituted species and considered possible effects of substitution in a qualitative fashion.[32] They noted that terminal methyl substitution stabilizes I more than II so that both species are about the same in energy. A second methyl (4-methyl-1,3-pentadiene) should render I more stable than II, so that only conrotatory cyclization should be observed. Dauben's experimental results involved relatively unsymmetrical donor-substituted cases and followed the latter course. Bruckmann and Salem predicted the opposite (disrotatory) course for symmetrical, or

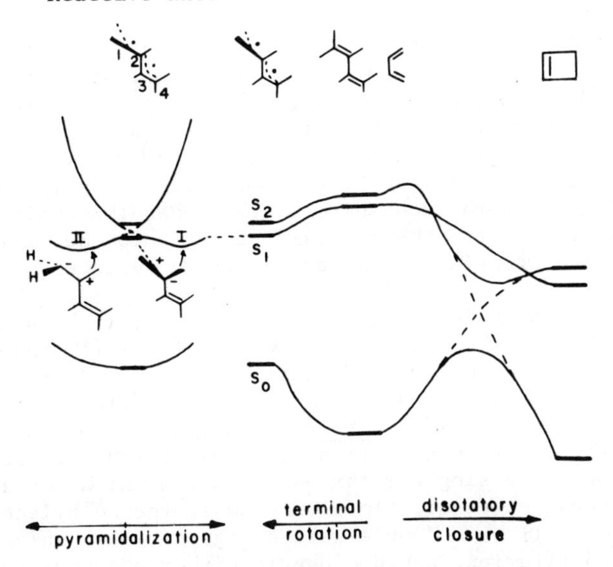

FIGURE 8.9. Ground and excited singlet states of butadiene (center), methylene-allyl (left), and cyclobutene (right). The left summarizes the Bruckmann-Salem calculations;[32] the right the Grimbert-Segal-Devaquet calculations.[33]

acceptor-substituted butadienes. This is a remarkable prediction, and one of the few clearcut predictions that has been made for photochemical reactions since the Woodward-Hoffmann rules were created. That is, qualitative surface arguments developed in the last few years have been extraordinary at creating insight and rationalizing experimental results, but here we have a risky prediction that will undoubtedly be tested soon.

Turning to the Devaquet et al. calculations on the photoelectrocyclization of cisoid butadiene to cyclobutene,[33] the right-hand side of Figure 8.9 shows the results of these calculations. For butadiene itself the lowest excited singlet state (S_1) may be represented as mainly a $\pi_2 \rightarrow \pi_3$ (HOMO-LUMO) excitation, whereas S_2 is not well represented by a single configuration. Instead, S_2 is an antisymmetric combination of a doubly excited configuration $(\pi_2 \rightarrow \pi_3)^2$ and a singly excited configuration $(\pi_1 \rightarrow \pi_3)^1 - (\pi_2 \rightarrow \pi_4)^1$. As rotation occurs, S_2 first rises, but then correlates with the ground state of cyclobutene. [In the simple Woodward-Hoffmann orbital correlation diagram the $(\pi_2 \rightarrow \pi_3)^2$ configuration correlates with the ground configuration of cyclobutene.] The ground configuration of butadiene correlates with S_2 of cyclobutene. [In the Woodward-Hoffmann diagram the correlation is with $(\pi \rightarrow \pi^*)^2$.] The important point of these calculations is that near the halfway point an intended crossing is avoided, leading to the large barrier in the ground state and a "Michl funnel" in S_2 (which, however, at this point is the lowest excited singlet). This behavior had been predicted previously by a host of earlier workers but is given additional credence by these relatively sophisticated calculations. The final conclusion about the mechanism is that when the excited state S_1 begins to rotate, the molecule passes into the excited funnel, from which crossing to the ground state surface and activationless closure occurs.[33]

It is particularly interesting to compare the right-hand side of Figure 8.9 to the left-hand side. On the left, Salem has considered a single methylene rotation in transoid butadiene, while on the right Devaquet et al. have considered the double methylene rotation of a cisoid butadiene. It is conceivable that the electrocyclization of cisoid butadiene might also be initiated by a single methylene rotation which was not considered by Devaquet et al. In such a case, the methylene-allyl pair might well be a double-well energy minimum, which, depending on substitution, could close in either a conrotatory or disrotatory fashion, or could cyclize to bicyclobutane. The story of butadiene photochemistry undoubtedly has many chapters still unwritten.

There are further ramifications of the zwitterionic nature of biradicaloid excited singlets that are exciting to contemplate. As discussed by Salem and Stohrer,[34] if the zwitterion (e.g., Z_1 in Fig. 8.7) is sufficiently stabilized by asymmetric substitution or by polar solvents, there will be an intended crossing of the ^1D and Z_1 states, resulting in a minimum in the ground surface, and a double minimum in the excited surface. The behavior is shown in Figure 8.10.[34] Salem and Stohrer discussed the twisting of ethylene, where ^1D and Z_1 are the ground and excited singlet surfaces. In case (b) the ground-state surface can show a minimum of highly polar nature. The behavior is a result of the avoided crossing of potential surfaces, which has been discussed more generally by Salem, Leforestier, et al.[35]

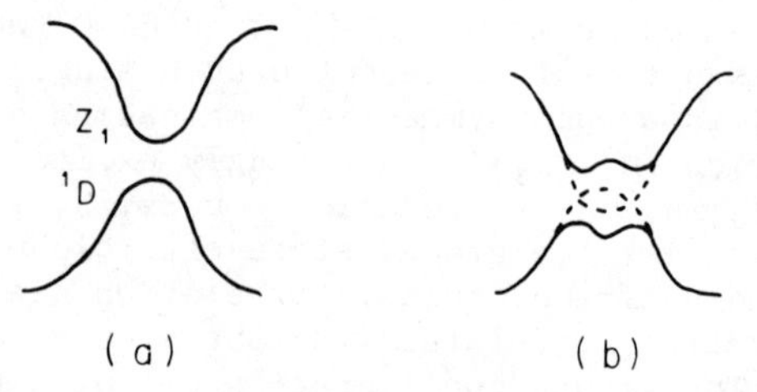

FIGURE 8.10. The ^1D and Z_1 states for a polar case in (a) nonpolar solvent, or (b) polar solvent

As an illustration of the further use of state-correlation diagrams and considerations of state crossings, we cite the works of Grimbert and Salem[36] and Turro, Farneth, et al.[37] on the α-cleavage of ketones. What follows is an amalgamation of these papers and earlier work.[28] Figure 8.11 shows a qualitative correlation diagram for the α-cleavage of an acyclic (or unstrained cyclic) ketone and of a cyclobutanone. For an acyclic ketone, the ground state of the molecule correlates with a diradical, called a σ,σ diradical. The lowest excited triplet and singlet states, which are of the n,π^* type, do <u>not</u> correlate with the ground state of the diradical species, because the last is of the σ,σ type. That is, both unshared electrons in the diradical are in σ-orbitals, so that the n,π^* state, which has one odd n (or σ-type orbital) and one odd π-electron, is of different symmetry. Thus the conversion of the n,π^* states to the lowest-energy diradical states is symmetry-forbidden if the system has a plane of symmetry. The conversion of the $^3\pi,\pi^*$ state to the diradical triplet is symmetry-allowed, whereas the π,π^* singlet (not shown) correlates with a higher-energy zwitterionic state (ion pair) of the cleaved molecule. Although the cleavage of either n,π^* state is symmetry-forbidden, the $^3n,\pi^*$ and $^3\pi,\pi^*$ states can mix in the region of the crossing if the system deviates from planar symmetry. On the other hand, the singlet must intersystem cross to the $^3\pi,\pi^*$ surface for α-cleavage to proceed. Even though this is a relatively favorable $n,\pi^* \rightarrow \pi,\pi^*$ process, the spin-orbit coupling that enables the intersystem crossing will still leave this process less facile than the adiabatic process resulting from mixing of the two triplet surfaces. Thus the cleavage of $^3n,\pi^*$ states of alkanones is more facile than that of $^1n,\pi^*$ states.[28,36]

In the case of cyclobutanones, Turro, Farneth and Devaquet showed that essentially the same process can occur, but that another, photochemically allowed process becomes thermochemically feasible. The process is that shown on the right-hand side of Figure 8.11. Because of the strain of the cyclobutanone, which is relieved on cleavage, the σ,σ and σ,π forms of the linear diradical both become energetically accessible intermediates. In the limit of no interaction between the radical centers of the two fragments, these states would be degenerate. The formation of the σ,π diradicals from the n,π^* states is symmetry-allowed and should occur readily. Thus both the singlet and triplet states of cyclobutanones undergo α-cleavage. The decarbonylation of cyclobutanones, which is not usually observed with unstrained ketones, was also explained by an elaboration of these diagrams. Even though the decarbonylation of an equilibrated bent acyl radical is appreciably

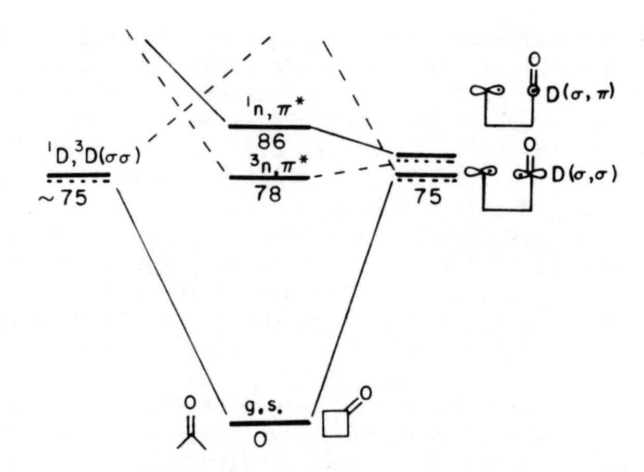

FIGURE 8.11. Correlation diagrams for α-cleavage of an acyclic ketone and cyclobutanone

endothermic, the linear diradical that is initially formed in the α-cleavage can spontaneously lose CO, since it is formed in a state from which formation of CO and the trimethylene diradical is exothermic.[37]

VII. SINGLET OXYGEN REACTIONS

M. J. S. Dewar has described detailed calculations on the surfaces for singlet oxygen reactions with alkenes.[38] In his inimitable style, Dewar provokes thought and generates ideas despite ardent devotion to his semiempirical (MINDO/3) method. The MINDO/3 is an internally consistent model of chemistry. That this model deviates in significant ways from reality is becoming increasingly apparent (see later). Nevertheless, an exhaustive exploration of this model for the important, perplexing, and remarkably complex reactions of singlet oxygen with alkenes, provides a number of important insights that might not have been obtained from pure ratiocination.

The MINDO/3 calculations by Dewar and Thiel,[38] and perturbation arguments by Inagaki and Fukui,[39] have led to theoretical respectability for the formation of perepoxides (peroxiranes) in the cycloadditions of singlet oxygen to alkenes. For the addition of singlet oxygen to ethylene, the MINDO/3 calculated heats of reaction of various species along the minimum energy reaction path are shown in Figure 8.12. No direct pathway to dioxetane could be found. The calculated activation energy for the overall reaction is quite large, and it would appear from the activation energies for decomposition of the peroxirane intermediate (28 and 34 kcal/mol) that this species should be isolable. However, small rings are too stable in MINDO/3.

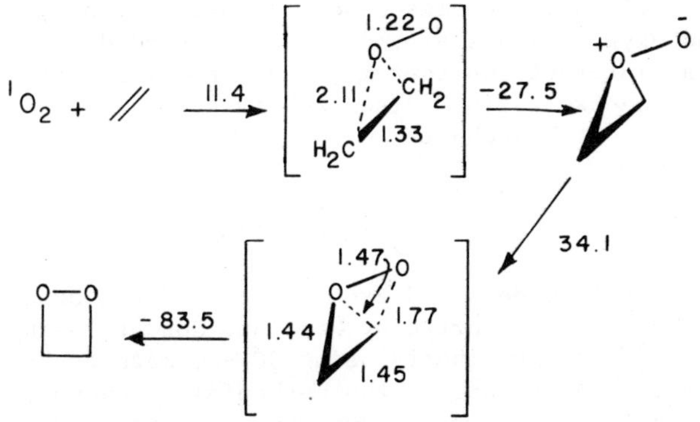

FIGURE 8.12. The MINDO/3 mechanism of singlet oxygenation of ethylene. Numbers are the calculated energies of each step.

In the reaction with propene, peroxiranes are calculated to be intermediates in the formation of "ene" reaction products, CH_2=CH-CH_2OOH. The activation energy for formation of dioxetane is 12.7 kcal/mol higher than that for formation of "ene" product, and this is consistent with experimental observation of mainly ene products with alkylethylenes.

For the reaction with vinylamine, a different mechanism was found, involving the zwitterion, H_2N^+=CH-CH_2OO^- (E_{act} = 41.1 kcal/mol), which collapsed to ene product, HN=CH-CH_2OOH, peroxirane, and dioxetane with nearly equal activation energies (23.6, 23.9, and 26.8 kcal/mol, respectively). However, the likelihood that the peroxirane stability is overestimated is mentioned here. One might also wonder how much peroxirane stabilities are overestimated in the surfaces referred to earlier.

Dewar and Thiel comment briefly on the important role of the alkene HOMO and the singlet oxygen LUMO (π^*_{OO}) in determining the favored transition state for formation of the peroxirane, and Inagaki and Fukui have analyzed this and other interactions in detail. The CNDO/2 partial surfaces suggest that the peroxirane is not a true intermediate, but lies in a flat sloping valley on the energy surface and collapses effortlessly into dioxetane. This species is called a "quasi-intermediate" and is related to Hoffmann's 1,4-biradical "twixtyl" in that no barriers exist for conversion to other species, but a peroxirane may wander about aimlessly on a plateau for a time before collapsing to dioxetane. The formation of dioxetane is thus considered by Inagaki and Fukui to be concerted in the sense that no true intermediate intervenes between reactants and products, but stepwise in the sense that one oxygen initially becomes bonded to both carbons, and then the second oxygen swings around to become bonded to carbon also. The orbital interactions that make this possible represent a case of "pseudoexcitation" discussed in greater detail in a later section of this chapter.

VIII. OZONE REACTIONS

Two moderately detailed examinations of the ozonolysis mechanism have been reported by Wadt and Goddard using GVB calculations[40] and by Hiberty using SCF-MO methods.[41] Both of these investigations support the mechanism of ozonolysis suggested by Criegee 28 years ago and since refined by elaborate experimental investigations by Bailey, Murray, Kuczkowski, and others.

Wadt and Goddard studied the electronic structures of various states of methylene peroxide (CH_2OO), commonly known as carbonyl oxide. As for ozone studied earlier, a singlet biradical structure is indicated for methylene peroxide by the GVB calculations rather than the more familiar resonance hybrid of zwitterionic species. Goddard and Wadt emphasize the fact

singlet biradical | resonance hybrid

ozone

carbonyl ylide (methylene peroxide)

that the singlet biradical representation is superior to the zwitterionic resonance hybrid representation. The latter is, of course, a throwback to the Lewis electron-pair representation of electronic structure. It has been recognized for some time that species such as ozone and other so-called "1,3-dipoles" are not very polar. It is also worth mentioning here that the singlet biradical character of species such as ozone, methylene peroxide, and diazomethane, which is emphasized by GVB calculations, is much different from one common definition of biradicals, which defines these as species with a small singlet-triplet split resulting from near degeneracy of, and very small interaction between, singly occupied orbitals. Thus the calculated splittings between the lowest singlet (ground-state) and lowest triplet states of ozone, methylene peroxide, and diazomethane are 1.4 eV, 0.84 eV, and 3.7 eV, respectively.[40,42] In comparison, ethylene and formaldehyde have splittings of 4.2 eV and 3.5 eV,[42] while "common diradicals," such as perpendicular ethylene, oxygen, methylene, and, perhaps, trimethylene and tetramethylene, have triplet ground states. By this criterion, methylene peroxide is the most diradical-like of the 1,3-dipoles listed here. While in the midst of this digression on diradicals and the applicability of this terminology to "1,3-dipolar" species, the calculations of Hayes and Siu on some of these species should be noted. Hayes and Siu defined the percent diradical character in "1,3-diradicals" as the percent of $\psi_A{}^2$ that mixes with a $\psi_S{}^2$ configuration using SCF orbitals for ψ_S and ψ_A.[43] As discussed earlier, for two slightly interacting atomic orbitals, φ_A and φ_B, the MO calculations force two electrons into whichever combination, $\psi_S = \varphi_A + \varphi_B$ or $\psi_A = \varphi_A - \varphi_B$, is lower in energy. For a "pure" diradical, configurations with two electrons in either one of these orbitals, ψ_A or ψ_B, have appreciable ionic character and are thus poor representations of the diradical. If ψ_A and ψ_B are degenerate, then $\psi_S{}^2 - \psi_A{}^2$ is the proper representation, since this combination eliminates ionic terms. As φ_A and φ_B interact more strongly, then $\psi_S{}^2$ becomes a better representation of the ground state of the "diradical," which has become more closed-shell in nature. Hayes and Siu defined the diradical character in terms of the percent of ψ_A mixed into the ground state in a CI calculation. For ozone, carbonyl ylide, and allyl anion, these percents are 30%, 38%, and 8%, respectively. The first two numbers agree qualitatively with Goddard's contention that ozone, and, to a greater extent, carbonyl oxide (methylene peroxide) should be represented as singlet diradicals.

How do these considerations relate to our extensive SCF investigations of the MO's of 1,3-dipoles?[44,45] Figure 8.13

summarizes the generalizations we have made about 1,3-dipoles based on semiempirical and ab initio calculations[20,44] and on a perturbation treatment where XY and Z fragments are united to form the π system of the generalized 1,3-dipole, X-Y-Z.[45] The HOMO of these systems is related to the ψ_A orbital of a 1,3-diradical, while the LUMO is related to ψ_S. The order is inverted from that expected for through-space interaction of the terminal AO's (atomic orbitals) because of strong mixing of the terminal AO's with the AO on the central atom. We found that the HOMO of these molecules is polarized toward the more electronegative terminus, and the LUMO is polarized oppositely.

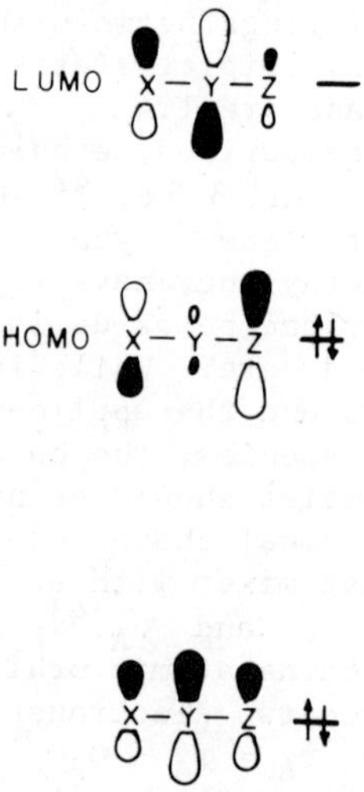

LUMO X—Y—Z

HOMO X—Y—Z

X—Y—Z

FIGURE 8.13. MO's of 1,3-dipoles.[44,45] For the parent molecules, XY = HCN, NN, H_2CNH, H_2CO, OO, and Z = CH_2, NH, O

The Hayes and Goddard works indicate that this one-configuration (single-determinant) representation may not be adequate for these species in certain cases. However, we can make certain generalizations about when CI, or mixing of the $LUMO^2$ configuration into the $HOMO^2$ configuration, is important. The importance of this CI will be greatest when the HOMO and LUMO are closest in energy and are most nearly distributed over the same regions in space. For the allyl anion this will not be of much importance because the HOMO-LUMO gap is very large and the HOMO is located only on the terminal atoms, with the LUMO concentrated mostly on the central atom. The latter arises from the extraordinarily good donor properties of the Y moiety, which is CH^- in this system. As the donor properties of the Y fragment decrease ($\approx CH^- > S > NH$ or $N > O > CH_2$), the HOMO-LUMO gap will decrease, and the $HOMO^2$-$LUMO^2$ interaction will increase because the LUMO becomes more concentrated at the terminal atoms. Second, as the electronegativity of the termini increases, the HOMO-LUMO gap will increase. This can be understood in either of two ways. We may note the larger coefficients at the termini in the HOMO than in the LUMO; electronegative atoms lower the HOMO energy more than that of the LUMO. Alternatively, one may note that electronegative termini will place ψ_A and ψ_S at low energy, where the central donor atom will more effectively mix with, and raise, the ψ_S energy. Thus along the series CH_2OCH_2 (carbonyl ylide), CH_2OO (carbonyl oxide), OOO (ozone), the HOMO-LUMO gap increases, $LUMO^2$ mixes less into $HOMO^2$, and the diradical character decreases. Similarly, along the series CH_2NHCH_2 (azomethine ylide), CH_2NHNH (azomethine imine), CH_2NHO (nitrone), NHNHO (azoxy), and ONHO (nitro), the diradical character decreases. Interestingly, relatively highly polarized systems (e.g., N_2O) should have less diradical character (less CI) than less polarized systems (e.g., CH_2N_2). Thus for highly polarized systems the generalizations about nucleophilicity and electrophilicity made on the basis of HOMO and LUMO coefficients are strong predictions, unlikely to be reversed by CI, whereas for less polarized systems CI is more important and possible reversals (or diminution) of predictions could occur by inclusion of CI. Nevertheless, the trends we have noted based on MO reasoning should be preserved in more elaborate treatments with CI.

This long digression has been included here to show how the recent valence-bond work of Goddard does not abolish insights obtained from MO considerations and to make explicit for these cases the oft-repeated statement that valence-bond and MO treatments are convergent if carried to sufficient levels of accuracy.

Returning to the Goddard insights into ozonolysis, these can be conveniently summarized as shown in Figure 8.14. The energies of the various species shown here were obtained by thermochemical estimates, but agree for the most part with GVB calculations including CI.[40] Several features of this diagram are of particular interest. First, the biradical designation for ozone does not imply that it reacts by radical mechanisms, just as the 1,3-dipolar designation should not imply that the species reacts via polar (ionic) mechanisms. The energies of the diradical intermediates are considerably above the transition-state heat of formation, which can be estimated from the 4-5 kcal/mol activation energy for the reaction of ozone with simple alkenes.[40] It was suggested that such a biradical could be formed in hindered systems and lead to epoxide formation. It is energetically possible for a diradical to intervene between the primary ozonide (1,2,3-trioxolane) and the methylene peroxide plus formaldehyde, and Goddard and Wadt suggest that it does so.[40]

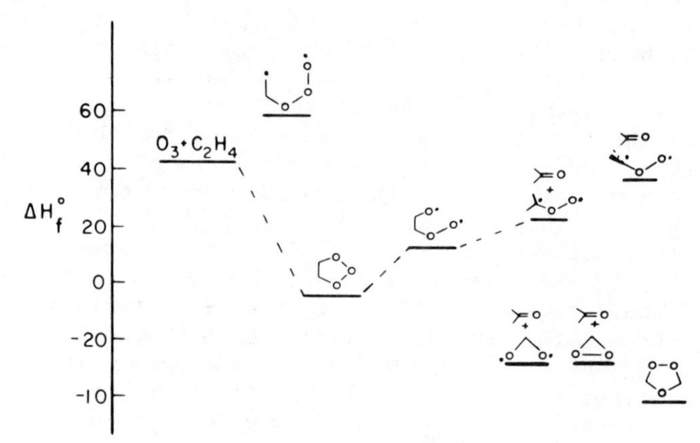

FIGURE 8.14. Wadt and Goddard's thermochemically calculated heats of formation of various species on the ozonolysis surface[40]

Aside from extensive considerations of the structure of methylene peroxide, the new insight contained in the Wadt-Goddard work concerns the stabilities of the cyclic form of methylene peroxide, dioxirane, and the acyclic isomer, dioxymethylene. Although these species have not been detected or trapped, Wadt and Goddard suggest that experiments that have been carried out are equivocal. Wadt and Goddard discuss a number of gas-phase products as being the result of dioxymethylene intermediates.[40]

Hiberty also focused attention on the carbonyl oxide (methylene peroxide) intermediate and its formation from 1,2,3-trioxolane. Hiberty used <u>ab initio</u> MO calculations, but included CI, which showed that a doubly excited configuration, with two electrons transferred from the in-plane orbital on the terminal oxygen to the π-LUMO, contributes importantly to the ground state.

The 1,2,3-trioxolane is calculated to be by far the most stable primary adduct that can be formed from ozone and ethylene. Other species occasionally invoked as intermediates, such as dioxetane oxide, peroxyoxirane, or acyclic adduct, are calculated to be over 100 kcal/mol less stable than the 1,2,3-trioxolane. The most stable conformation of this trioxolane is an envelope conformation that is calculated to decompose in a concerted fashion to planar carbonyl oxide and formaldehyde with an activation energy of about 11 kcal/mol and an exothermicity of 7.8 kcal/mol.[41]

The carbonyl oxide, which must be able to maintain its conformation (configuration) about the CO bond in the Bailey mechanism and various elaborations of this mechanism, has a barrier to rotation about this bond of 12.4 kcal/mol in Hiberty's work[41] and 36.5 kcal/mol in Wadt and Goddard's GVB calculations with CI.[40] The latter authors also thermochemically estimated a barrier of 15.5 kcal/mol.[40] Given the low activation energies expected for cycloaddition of carbonyl oxides to ketones, this would appear to be a sufficient barrier to preserve the configuration about the CO bond.

IX. DIELS-ALDER AND 1,3-DIPOLAR CYCLOADDITION MECHANISMS

The mechanism of the Diels-Alder reaction has been of great interest for at least 50 years. Recent mechanistic investigations seemed to converge on a concerted mechanism,[46] but some proponents of the diradical intermediate mechanism were undaunted. Without doubt, no long-lived intermediates are involved in most Diels-Alder reactions, but perhaps a very short-lived intermediate or a highly unsymmetrical (or asynchronous) transition state could be involved. Theory would seem to be the ideal means to prove the nature of the rate-determining transition state. Three calculations reported recently agree that no intermediates are involved in the reaction, but the calculated transition states are very different. Figure 8.15 shows the butadiene-ethylene Diels-Alder transition states calculated by Salem and co-workers,[47] and by Dewar, Griffin et al.[48] Burke, Leroy, et al. have also carried out calculations which give a transition state much like Salem's.[49] Figure 8.15 also shows the transition states of the 1,3-dipolar cycloaddition of fulminic acid to ethylene calculated by Poppinger[50] and Dewar.[51] A curiously consistent pattern is revealed here. The semiempirical calculations by Dewar (MINDO/3 with CI for the Diels-Alder reaction, MINDO with CI for the 1,3-dipolar cycloaddition), indicate extremely unsymmetrical transition states for both cycloadditions. *Ab initio* calculations by Salem for the Diels-Alder reaction (STO-3G + 3 x 3 CI) and by Poppinger (STO-3G) for the 1,3-dipolar cycloaddition predict highly synchronous transition states. We have investigated this problem in greater detail by comparing various symmetrical and unsymmetrical geometries of approach of addends to one another using various semiempirical and *ab initio* techniques.[52] The results are rather startling; *ab initio* (STO-3G) and EHT calculations favor geometries where both newly forming bonds are of equal length, whereas MINDO/3, MINDO/2, and CNDO/2 calculations favor geometries where one of these bonds is much shorter than the other.[52] That is, there seems to be a bias in the latter three calculations for biradicaloid transition states for these two allowed [4+2] cycloadditions, whereas ab initio and EHT calculations favor symmetrical transition states. One difference between the two groups of calculations that we have noted is that ab initio and EHT calculations include overlap, while the MINDO's and CNDO neglect all interatomic overlap. Furthermore, as discussed in the third part of this chapter, a synchronous allowed transition state has less π closed-shell repulsion than an asynchronous one. That is, the four-electron repulsion arising from the interaction of filled orbitals is smaller in the synchronous transition state because the HOMO's of the two addends are of different symmetry and do not overlap appreciably unless the transition state becomes highly unsymmetrical. A second fact of importance is the observation by Gordon and coworkers that MINDO methods underestimate long-range bonding interactions relative to short-range interactions.[53] Thus there may be an additional unnatural bias in MINDO calculations in favor of one short and one long bond as opposed to the medium-length bonds in cycloaddition transition states.

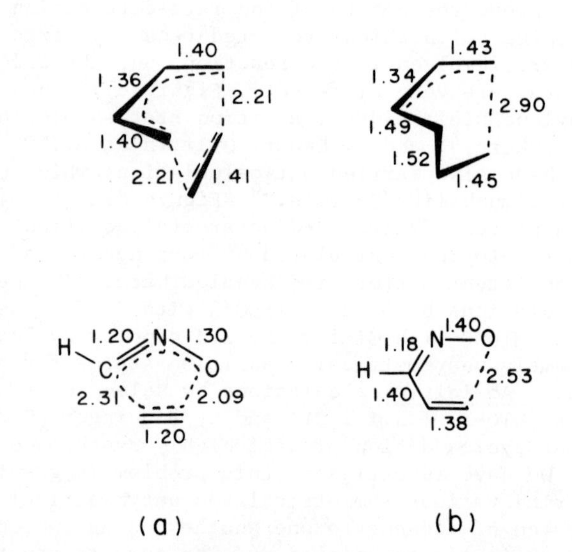

FIGURE 8.15. Diels-Alder and 1,3-dipolar transition states calculated by: (a) ab initio and (b) semiempirical techniques.

Salem has commented on the fact that his 3 × 3 CI may overcorrelate the diradical (asynchronous) geometries relative to the concerted (synchronous). This and the preceding considerations imply that synchronous allowed cycloaddition transition states are, in fact, of lower energy than asynchronous ones.

Salem and co-workers explored the Diels-Alder surface in some detail. Because the 42-dimension surface could not be fully explored, the procedure used was to fix certain variables (e.g., CH bond lengths) and to vary others to create a coarse grid of points along certain reasonable reaction pathways. These calculations were performed by STO-3G + 3 × 3 CI, and important points were recalculated by 4-31G + 3 × 3 CI.[47] In other words, assumptions were made, but an impressive amount of detail was obtained, undoubtedly at no mean cost. Several important conclusions were made. First, the symmetrical transition state shown in Figure 8.15 lies on the minimum-energy pathway from reactants to products. However, at an energy of only a few kcal/mol higher there is a transition state leading to an extended diradical (the radical termini are maximally separated). The barrier to closure of this diradical to cyclohexene consists essentially of the energy required to rotate about the single bond that converts the extended diradical to a *cisoid* diradical having the two radical centers in proximity. The latter is essentially a stretched cyclohexene, with a long 3-4 single bond. It is also possible to proceed over a low-energy barrier from the symmetrical transition state to the diradical surface, but Salem et al. find it dynamically unlikely that both bonds would begin forming and then one bond would begin to unform midstream.[47]

Recent *ab initio* calculations on 1,3-dipolar cycloadditions of fulminic acid[50] and of diazomethane[54] to substituted alkenes suggest that substituents distort the bond formation in the transition states in the expected direction, but that the degree of this distortion is remarkably small. For example, the forming CC and CO bond lengths in the calculated transition state of the fulminic acid-ethylene reaction are 2.37 Å and 2.14 Å, respectively, whereas these have changed to 2.35 Å and 2.15 Å in the transition state of aminoacetylene, leading to 5-aminoisoxazole.

X. NUCLEOPHILIC SUBSTITUTIONS AND ADDITIONS

Numerous semiempirical and *ab initio* calculations have been performed on simple nucleophilic substitution reactions. Many of these are referred to by Keil and Ahlrichs[55] and Dannenburg.[56] One of the more thought-provoking investigations yet reported is that of Dannenburg,[56] who carried out INDO calculations for the reactions of water with protonated alcohols (MeOH, EtOH, and _i_-PrOH), and of fluoride with alkyl fluorides (MeF, EtF, and _i_-PrF). The calculations have special relevance to the question of what is the electronic nature of the various solution species proposed by Winstein 20 years ago:

$$\mathrm{RX} \rightleftharpoons \underset{\text{intimate ion pair}}{\mathrm{R^+X^-}} \rightleftharpoons \underset{\text{solvent-separated ion pair}}{\mathrm{R^+\|X^-}} \longrightarrow \underset{\text{dissociated ions}}{\mathrm{R^+ + X^-}}$$

$$\mathrm{R^+\|X^-} \xrightarrow{\ S\ } \mathrm{RS} \qquad \mathrm{R^+ + X^-} \xrightarrow{\ S\ } \mathrm{RS}$$

There are certain serious difficulties encountered with the quantitative treatment of reaction paths with INDO methods (e.g., see Section IX). Nevertheless, an important qualitative trend is observed: For attack of water on protonated methanol or ethanol, there was a smooth decrease in energy as water approached, but for protonated isopropanol a distinct energy minimum was found at a newly forming CO bond distance of 3.6 Å. At this distance the other CO bond is about the same as that in protonated isopropyl. In the fluoride reaction with methyl fluoride a one-step reaction is found, whereas ethyl fluoride gives two deep minima at 3.0 and 3.6 Å, and isopropyl fluoride gives a single minimum at 3.0 Å. In the case of isopropanol $-H^+$ plus water, the energy minimum corresponds to a rather close approach of the water oxygen to the isopropyl methyl groups, and Dannenburg suggests that this corresponds to a solvent-separated ion pair![56] That is, the "intimate ion-pair" is suggested to be a protonated intermediate, or "anion-stabilized intermediate," where the terminology implies that the proton (or other Lewis acid) has stabilized the latent anionic leaving group. It is suggested that the solvent-separated ion pair is a species in which the solvent has become closely associated with the back-side of the protonated substrate. It is renamed an anion-cation stabilized intermediate since the incipient anion is stabilized by protonation and the incipient cation is stabilized by nucleophilic water or fluoride. This picture is rather different from the common conception of ion pairs as cations with different arrangements of leaving groups, solvents, and nucleophiles around the cationic center.

New calculations on nucleophilic addition to carbonyls have been reported by Lipscomb and co-workers.[57] These calculations were designed to test the: (a) stringency of the requirement that nucleophiles attack carbonyl groups from a direction perpendicular to the carbonyl plane on a line through the carbonyl carbon ("orbital steering") and (b) role of nucleophile solvation on such reactions. The calculations were performed by a method called PRDDO, which neglects some small multi-center integrals and generally gives reasonable agreement with, but is much faster than, minimal basis set ab initio calculations (e.g., STO-3G). For the reactions of F^-, OH^-, and OMe^- with formyl fluoride, formaldehyde, and formamide, there were no observable barriers to reaction whatsoever, implying no gas-phase activation energy. The addition of water molecules around fluoride, formyl fluoride, and the difluoromethoxide ion produced marked stabilization of these species and a dramatic decrease in the calculated exothermicity of the reaction (from -187 kcal/mol to -33 kcal/mol), implying that solvent reorganization may produce the activation energies for these reactions. This is particularly likely since the 4-31G calculations indicate a small exothermicity (-28 kcal/mol) that might more easily be converted into endothermicity, or at least the presence of a barrier between reactants and products if solvent were included. Concerning the question of preferential direction of attack of nucleophiles on carbonyls, the calculated minimum energy pathways do, indeed, indicate perpendicular attack, but the energetic preference for this approach (as opposed to a nonorthogonal approach) is relatively small, suggesting that there is little energetic advantage to perfectly orthogonal attack. For example, in the attack of methoxide on formamide at a carbonyl carbon, with a methoxide oxygen separation of 2.38 Å, an attack 20^0 from perpendicular is about 10 kcal/mol less advantageous than perpendicular. The "force constant" for bending the O=C---O angle with d_{C---O} = 2.38 Å is 0.014 (kcal/mol)/deg^2, as compared to the value of 0.062 (kcal/mol)/deg^2 at the equilibrium value (1.55 Å) of d_{C---O}. Lipscomb et al. do suggest that solvent may restrict the approach path,[57] a conclusion that seems particularly appropriate if the entire barrier to reaction is caused by solvent.

These calculations were predated by theoretical studies by Bürgi, Lehn, and Wipff,[58] and by the crystal-structure investigations summarized by Bürgi.[59] The ab initio calculations concerned the addition of hydride to formaldehyde; although carried out on a small-model system, the Bürgi calculations provided much detail about the surface since many degrees of freedom were included. At moderate distances the hydride approaches from above the H_2CO plane, but at an angle approximately 109.5^0

(tetrahedral) from the CO bond. Even at a hydride-carbonyl carbon distance of 1.5 Å (vs. the covalent distance of 1.12 Å), a deviation of the hydride by 10° from the most favored line of approach is only 3 kcal/mol more difficult.

Bürgi has recently summarized his work relating the X-ray crystal structures of molecules to the most favored path of nucleophilic addition.[59] By comparing the structures of molecules containing tertiary amine and carbonyl moieties separated by different distances, Bürgi and co-workers were able to deduce the relationships between the distance between the nucleophilic amine and carbonyl carbon atoms, and other structural features such as the C=O distance and the extent of carbonyl pyramidalization. According to this work the preferred angle of nitrogen approach is 107° from the CO bond. There is a smooth relationship between the N-C distance and CO length and degree of pyramidalization. The last two begin to change appreciably only after the N-C distance is less than 2.5 Å.[59]

At this point the discussion of Baldwin's empirical rules[60] of ring closure seems appropriate. Although these are not really theoretical in the sense of other papers discussed here, "Baldwin's rules" have already had great impact on organic chemistry. The theoretical basis of these empirical relationships is, in part, known, and the rules will undoubtedly instigate further theoretical inquiry. The rules predict when intramolecular nucleophilic ring closures should be favored. For example, nucleophilic attack on a saturated center (S_N2) is favored when the nucleophile can attack at an angle of 180° from the CX bond vector, where X is the leaving group. Such an alignment is difficult when five- or six-membered rings are formed.[60] For intramolecular attack on trigonal (sp^2) centers (e.g., a carbonyl carbon), the nucleophilic attack occurs on a line perpendicular to the plane of the ligands attached to the trigonal center and at an angle of approximately 109.5° from a line defined by the trigonal center and the atoms attached by a multiple bond. This follows directly from the Bürgi work discussed already. For digonal (sp) centers the approaching nucleophile attacks at an angle 60° from the triple bond. The last conclusion is the most surprising and the least supported. Baldwin suggests that these rules will also hold for cationic or radical attack. As yet there is little theoretical justification for this, but the empirical evidence is compelling.[60]

Returning to the theoretical treatment of nucleophilic reactions, studies have been reported that model rather directly the active sites of several enzymes.[61-63] These papers are of importance because they provide examples of the size and complexity of systems for which meaningful calculations are possible.

Umeyama, Imamura, et al. investigated the mechanism of α-chymotrypsin action by CNDO/2 methods.[61] The active site was modeled by placing acetate anion, imidazole, and ethanol in the relative positions found for aspartate-102, histidine-57, and serine-195, respectively, in the X-ray structure of α-chymotrypsin, shown schematically in Figure 8.16. At the active site substrates such as phenylacetates are hydrolyzed. In this mechanism, the hydroxyl of serine-195 attacks the carbonyl. The question of general interest is the mode of operation of the "charge-relay" mechanism, in which protons are transferred from imidazole to aspartate, and from serine to imidazole. That is, these proton transfers could either be "coupled" (synchronous) or "uncoupled" (asynchronous), as explored theoretically in simple model systems, by Gandour, Maggiora, et al.[64] Umeyama and co-workers found that the proton transfer from imidazole to aspartate precedes deprotonation of serine. The presence of a substrate carbonyl near the serine hydroxyl facilitates proton transfer from the serine hydroxyl. Thus the serine attack on the carbonyl is coupled with serine hydroxyl deprotonation.[61]

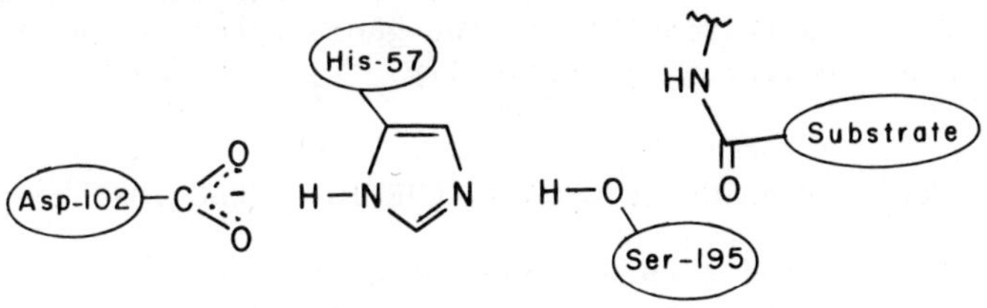

FIGURE 8.16. The important functional groups in the active site of serine proteinases

Scheiner, Kleier, and Lipscomb,[62] and Scheiner and Lipscomb[63] have studied this same charge-relay system using the PRDDO method mentioned earlier. Formate, imidazole, and methanol served as active site models, and formamide served as a protein substrate model. Proton transfer from serine to imidazole, and from imidazole to formate, is found to be an activated process, with the calculated activation energy about twice that observed experimentally. Once the serine is deprotonated, its attack on the carbonyl of formamide is an unactivated exothermic process.[62] The decomposition of the tetrahedral intermediate was found to involve transfer first of a proton from aspartic acid to imidazole, then movement of the imidazolium toward the tetrahedral intermediate nitrogen, followed by protonation of amide nitrogen by imidazolium.

A coupled transfer of both protons was calculated to require an impossibly high activation energy. Loss of ammonia gives the acyl enzyme (here, methyl formate). A water molecule was added at this stage of the calculations, and formic acid was substituted as the acyl enzyme model. Water could be situated in a position hydrogen-bonded to, and easily deprotonated by imidazole. As deprotonation proceeded, attack of the incipient hydroxide on formic acid (acyl enzyme) occurred without activation. Subsequent steps follow those for deammonolysis of the first tetrahedral intermediate. The calculations not only explain how the aspartate-imidazole dyad can activate the serine hydroxyl, rendering it a powerful nucleophile, but provide details about the ubiquitous function of imidazole as a base.

This work is most interesting since it indicates how complex a mechanism can be calculated. At the same time, the work emphasizes how far theoretical chemistry is from dealing quantitatively with real biological systems. The use of simple molecules as models for enzyme pieces can reveal important features of mechanism. On the other hand, these studies depended on the knowledge about the active site obtained from X-ray crystallographic data. No mechanistic prediction could have been made; indeed, no calculation of even qualitative significance could have been financially feasible if the geometry of the active site had to be predicted theoretically.

XI. DYNAMICS OF ORGANIC CHEMICAL REACTIONS

In 1973 Wang and Karplus reported a dynamical study of the insertion of singlet methylene into hydrogen.[65] The pictures they presented were a dramatic reminder to organic chemists that molecules do not undergo reactions by smooth distortions along the minimum-energy surface from reactants to products, but instead follow a variety of trajectories from reactants to products, the trajectory followed being a complex function of the vibrational, rotational, and translational energies of a molecule or system of molecules.

Chapuisat and Jean have provided a second example of a dynamical study of an organic reaction--the isomerization of cyclopropane.[66] The mechanism of this isomerization has been of extraordinary interest to both experimentalists and theoreticians. In 1968 Hoffmann made definite predictions about the stereochemistry of cyclopropane isomerizations and provided a theoretical outline of the potential energy surface involved in this isomerization.[67] One outstanding deduction of this work was that the edge-to-edge diradical, EE, shown in Figure 8.17, was the most stable of the "diradical species" and should close in a conrotatory fashion because the HOMO of this diradical is antisymmetric about a symmetry plane that bisects the molecule. The surface for this reaction has been explored thoroughly by Salem and his groups, and Chapuisat and Jean have now provided a dynamical study.[66] This is a particularly appropriate system for a dynamical investigation since the conrotatory, disrotatory, and monorotatory pathways of degenerate isomerization are similar in energy. In addition, the surface calculations predict no intermediate, only a transition state resembling a diradical species (EE). It has been suggested that many diradicals are not true intermediates but may behave kinetically like intermediates if they have a significant lifetime due to "aimless wandering" about a relatively flat surface and can isomerize before finding a downhill path to a stable product.

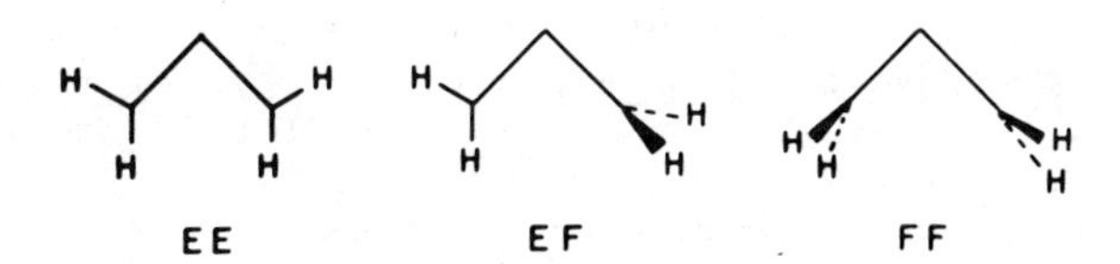

FIGURE 8.17. Important diradical species in cyclopropane isomerization: EE = edge-to-edge; EF = edge-to-face; FF = face-to-face

The dynamical study of cyclopropane isomerizations followed the procedures typically used in this area.[66] An analytical function was fitted to the ab initio surface, and this was followed by a classical mechanical study of the trajectories that a molecule would follow, given some specific rotational, vibrational, and translation energies (initial conditions). It is not possible to do justice to this study without pictures, and the interested reader should consult the original references for these. However, some general conclusions can be stated here. For cyclopropane to isomerize, at least one half of the ~ 60 kcal/mol activation energy must be placed in the asymmetric CH_2 twisting vibration that causes rotation of two CH_2 termini toward the EE diradical transition state. That is, even if the total internal energy is greater than 60 kcal/mol, one C-C bond will simply continue to open to the "FF diradical" and reclose without isomerization. If the twisting vibrational energy is 12 to 20 kcal/mol, "optical" isomerization (conrotatory opening followed by conrotatory closure) occurs, but the trajectory followed does not pass directly through the transition state; rather, the C-C bond length oscillates as the molecule bounces through the "mountain pass" separating reactant from products. An unreactive band is observed for asymmetric vibrational energies of 22 to 32 kcal/mol, and reactivity returns for energies of 32 to 35 kcal/mol. Here a more ideal trajectory is followed, more or less directly from reactant through the EE transition state to products. At vibrational energies above 30 kcal/mol, insufficient energy is left in CC opening modes to allow isomerization. The molecule undergoes oscillatory distortion without isomerizing. The conclusion from this work is that dynamics do not change the conclusions obtained from the static potential energy surface. Most importantly, there is no kinetic-intermediate behavior produced by dynamics. That is, both conrotatory "optical" isomerization (through the EE diradical) and monorotatory "geometrical" isomerization (through the EF diradical) involve a single rotation of both (EE) or one (EF) termini between reactants and products. There is no dynamic interconversion of the EE and EF diradicals and no behavior expected for equilibrating diradicals. This fits well with Berson's experimental studies of simple cyclopropanes (e.g., optically active _trans_-1,2-dideuterocyclopropane), but experimental studies of more highly substituted cyclopropanes suggest intermediate-like behavior. Chapuisat and Jean promise dynamical studies of substituted cases.[66]

Finally, although dynamical studies have not been carried out as yet, it is interesting to compare the tetramethylene surface (two ethylenes ⇄ cyclobutane) to the trimethylene (cyclopropane isomerization). Segal has reported _ab initio_ calculations on tetramethylene that indicate both the _gauche_ and _anti_ tetramethylene biradicals to be energy minima separated from each other by a barrier of 3.6 kcal/mol and from dissociation to two ethylenes by 3.6 and 2.3 kcal/mol, respectively.[68] The _gauche_ diradical can close to cyclobutane, but with a barrier of ~ 2 kcal/mol. These barriers are caused essentially by the barrier to rotation about the central CC bond in this species. A dynamical study here could tell whether these diradicals can equilibrate, or whether the small barriers are effectively smoothed out by the "momentum" of the reacting species.[68]

XII. THEORETICAL PRINCIPLES, INSIGHTS, BRAINSTORMS, AND TECHNIQUES

In this final section several important theoretical principles discussed in the recent literature are summarized. My main goal here is to emphasize how the interaction of occupied orbitals of one molecule and vacant orbitals of the second, or similar interactions between pieces of molecules, is merely one, albeit important, type of interaction that will determine the stabilities of reactive intermediates and stable molecules.

Kitawa and Morokuma have developed an elegant procedure for partitioning the energy changes occurring in bimolecular interactions.[69] They define the energy changes as E_{es}, the electrostatic (Coulombic) attraction or repulsion arising from the charge distributions of the isolated molecules when they are moved into bonding distances, E_{pl}, the polarization energy lowering that results from reorganization of the MO's of one molecule caused by the charge distribution of the other, and vice versa, E_{ex}, the exchange (or closed-shell) repulsion that arises from the mixing of filled orbitals on the two molecules, and E_{CT}, the energy lowering resulting from the mixing of filled orbitals on one molecule with the vacant orbitals of the other, and vice versa. Morokuma and co-workers have developed a program to dissect the ab initio calculated energy changes into the energy components listed here.

The Morokuma energy components encompass the energy changes that occur in a Hartree-Fock calculation or various approximations to it. In addition to these changes, there will be changes in correlation energy, the correction taking into account the fact that electron motions are correlated, rather than being just the result of each electron feeling the repulsion produced by the average position of all of the other electrons, as is calculated by the Hartree-Fock-Roothan method. The correlation energy can be calculated by configuration interaction, but some of the correlation energy might well be considered as polarization, charge transfer, and so on. Another part of the correlation energy which can be calculated by perturbation methods is the dispersion energy, which arises from the interaction of excited configurations on both molecules with the ground configuration of the complex of two molecules. In some papers the dispersion energy is treated implicitly as part of the polarization energy. In the remainder of this chapter I briefly try to show how each of these interaction energies has been treated in various ways by different authors and where the various interactions are likely to be important.

XIII. EXCHANGE OR CLOSED-SHELL REPULSION

The interaction of two filled orbitals, φ_a and φ'_a, which have energies ϵ_a and ϵ'_a, results in destabilization by an amount

$$\frac{-4\left(H'_{aa} - \epsilon_{av}S'_{aa}\right)\left(S'_{aa}\right)}{1 - S'^{2}_{aa}}$$

which is the amount of closed-shell, or exchange repulsion arising from the interaction of two closed-shell (filled) orbitals. The quantity ϵ_{av} is the average of the orbital energies of φ_a and φ_a'. This expression has been known and used for some time.[71] For example, it has been referred to earlier here in connection with the rationalization of preferred conformations. However, Epiotis and Yates suggested an important new general feature of this term in their discussion of aromaticity.[72] The qualitative consequences of the equation shown above are easily discerned. First, the extent of repulsion will increase as ϵ_{av} increases. That is, two high-lying MO's repel each other more than do two lower lying MO's. The repulsion also increases approximately as the square of the overlap between interacting orbitals increases. The latter is, of course, a symmetry-dependent term, which has important consequences for cyclic interactions. Epiotis and Yates showed that aromatic systems are not only stabilized more than non- or antiaromatic systems, but the aromatic systems are less destabilized by closed-shell repulsions. We have discussed our studies of the same phenomenon in synchronous and asynchronous Diels-Alder transition states in an earlier section of this chapter.[52] Both aromatic molecules and aromatic (Woodward-Hoffmann-allowed) transition states will suffer less closed-shell repulsion than nonaromatic.[52] The surprising part of this is that the apparently more crowded species experiences less repulsion. Furthermore, this type of repulsion is omitted in semiempirical calculations neglecting overlap. If a nonsymmetry-dependent repulsive force is substituted in the calculations, then incorrect predictions of the relative stabilities of atomatic and antiaromatic molecules or transition states may be a result.

XIV. CHARGE-TRANSFER, CONFIGURATION INTERACTION, PARADOXICAL INTERACTIONS, AND PSEUDOEXCITATION

These terms, along with polarization, have been the subjects of several important papers in the last 2 years. The principles in these different papers are often the same but have been presented in different formalisms or applied to different examples, so that it is difficult for the novice to ferret out congruences and differences.

Inagaki, Fujimoto, et al. have invented the terms "pseudoexcitation" and "paradoxical interaction" to describe cases of bimolecular interactions where HOMO-HOMO and LUMO-LUMO interactions may stabilize the transition state, in addition to the more familiar HOMO-LUMO interactions.[73] The Inagaki, Fujimoto, and Fukui (IFF) concept can best be shown graphically, and Figure 8.18 describes schematically the various electronic configurations involved. In a transition state of a reaction involving one nucleophilic species (the donor, D) and one electrophilic species (the acceptor, A), the interaction between the donor HOMO and the acceptor LUMO, the familiar frontier orbital interaction, can be described in terms of a configuration interaction formalism.[75] Thus the interaction of the ground configuration with the charge-transfer configuration (Fig. 8.18b) gives rise to stabilization, $\Delta E(CT)$, which can be calculated (neglecting overlap) as:

$$\Delta E(G\text{-}C) = \frac{H_{GC}^2}{E(G) - E(C)}$$

The extent of charge-transfer stabilization depends on the size of the interaction between the orbitals differing in occupation, namely, the donor HOMO and the acceptor LUMO. There can be other interactions involving transfers of electrons from various filled orbitals on one molecule to vacant orbitals on the other, but because of the energy difference in the denominator, the HOMO(D) → LUMO(A) configuration contributes most to stabilization of the D-A complex.

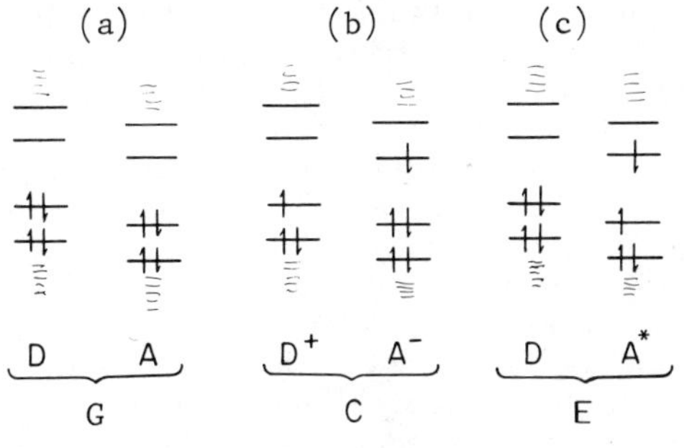

FIGURE 8.18. Ground (G), charge-transfer (C), and monoexcited (E) configurations

In addition to these second-order interactions, there can be third-order interactions, such as between a, b, and c in Figure 8.18. The stabilization energy arising from this third-order interaction is:[73]

$$\Delta E(G\text{-}C\text{-}E) = \frac{H_{G-C}\, H_{G-E}\, H_{C-E}}{[E(G) - E(C)][E(G) - E(E)]}$$

The three matrix elements in the numerator involve interaction between the: (a) ground and charge-transfer configurations, a function of HOMO(D)-LUMO(A) interaction, (b) ground and monoexcited configurations, a function of HOMO(A)-LUMO(A) interaction (related to the transition probability), and (c) charge-transfer and monoexcited configurations, a function of the HOMO(D)-HOMO(A) interaction. There is a similar interaction involving the excited donor, which will depend on LUMO(D)-LUMO(A) interactions.

Inagaki, Fujimoto, and Fukui (IFF) refer to these third-order interactions as pseudoexcitations, since the ground-state interactions take on some of the characteristics of excited-state interactions. The LUMO-LUMO and HOMO-HOMO interactions are paradoxical in the sense that they are interactions expected for excited-state reactions, whereas the treatment involves ground-state reactions. Since the denominator of $\Delta E(G\text{-}C\text{-}E)$ involves a product of two energy separations, the third-order stabilization is expected to be considerably smaller than the second-order interaction.

The IFF group described the cases in which the third-order interactions will be non-negligible.[73] If there is very strong charge transfer due to the small magnitude of $E(G) - E(C)$, then the third-order interaction may become significant. This is conceived by IFF as "excitation induced by charge transfer." We should note, however, that whereas $H_{G\text{-}C}$ is large only for Woodward-Hoffmann allowed transition states, $H_{C\text{-}E}$ will be large only for Woodward-Hoffmann forbidden transition states. The specific cases discussed by IFF are somewhere in between these extremes. For example, the cycloadditions of singlet oxygen, benzyne, electron-deficient alkenes, ketenes, and isocyanates to electron-rich alkenes are Woodward-Hoffmann forbidden if they occur through a parallel plane approach of the two addends. Those authors suggest that they begin in a fashion that maximizes donor HOMO-acceptor LUMO interaction; that is, a π-complex is formed with both terminal orbitals of the donor HOMO overlapping with one terminus of the acceptor LUMO.[76] At this stage charge transfer is very strong, stronger than expected on the basis of isolated reactants if stretching of addends has occurred.[76] "Pseudoexcitation" can now stabilize a pathway from this π-complex to the [2 + 2] adduct.[73] That is, at a point intermediate between the π-complex and the adduct, both $H_{G\text{-}C}$ and $H_{C\text{-}E}$ interactions can be large.

Epiotis came to conclusions similar, but not identical, to those of IFF. As both Epiotis and ourselves[75,76] have discussed, the energy between the ground and charge-transfer configurations drops drastically as the donor and acceptor molecules approach one another, and in the extreme, this difference could become negative.[74] In the case of an allowed reaction (pericyclic or otherwise), this formal configuration crossing matters little because these two configurations mix strongly so that the ground state goes smoothly to products. However, Epiotis suggests that a charge-transfer configuration may drop below the formal ground configuration in the case of forbidden pericyclic reactions. This configuration can be further stabilized by interactions with excited configurations, leading to an energetically favorable pathway to products, despite the formal forbiddenness of the reaction.[74] As we have pointed out,[76] alternative pathways would seem to be preferred in such cases. Furthermore, as the IFF treatment makes apparent, a maximum stabilization can be obtained for a transition-state geometry intermediate between allowed and forbidden geometries.

XV. POLARIZATION

This term and the term "polarizability" are frequently used without clear identification, but a number of important recent papers have pointed out the physical origin of these phenomena and their chemical significance. Libit and Hoffmann showed how the shapes of the π-orbitals of propene can be formally built up from intramolecular mixing of the π-orbitals of ethylene with those of a methyl group.[77] In 1975 Imamura and Hirano discussed polarization resulting from both intermolecular orbital mixing ("dynamic orbital pumping") and from the influence of a charge on a molecule ("static orbital pumping"),[78] while in 1976 the IFF group described their "orbital mixing rule"--polarization again--and applied it to several chemical phenomena.[79]

An electrostatic perturbation on a molecule, or a perturbation caused by overlap of the filled and vacant orbitals of a perturber with those of the molecule, A, causes polarization of molecule A. Polarization arises from the mixing of various orbitals on A(i,j...) induced by the orbitals of P. This "second-order" mixing takes the following form:[77-79]

$$C_{ji} = \frac{H_{ik}H_{jk}}{\left(\epsilon_i - \epsilon_j\right)\left(\epsilon_i - \epsilon_k\right)}$$

The coefficient, C_{ji}, is a measure of the amount of orbital j (on A) mixed into i (on A) under the influence of orbital k (on P). A simple application of this equation is demonstrated in Figure 8.19. In the center of this diagram are shown the σ and σ^* MO's of H_2 as an example of orbitals i and j of molecule A. On the left of H_2 is shown a low-lying filled orbital, which will exemplify the case where a donor orbital interacts with a "polarizable" bond--here a σ bond of H_2. The donor orbital (k) causes σ^* to be mixed into σ in a bonding fashion at the right-hand hydrogen atom, the site of perturbation. The polarization is that which might have been expected in simpler terms--namely, in terms of charge densities, the perturbed hydrogen molecule has a charge density polarized away from the electron-rich perturber. The result can be easily

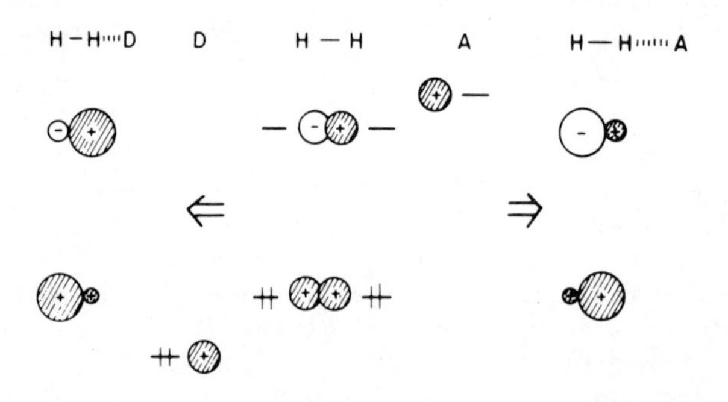

FIGURE 8.19. Polarization of H_2 induced by (left) a low-lying donor orbital, and (right) a high-lying acceptor orbital

generalized, as has been done in various ways in the references already cited,[77,79] and in our discussion of aromatic polarization.[80]

An orbital i will be polarized by a lower-lying orbital of a perturber by mixing in all higher energy orbitals in an antibonding fashion at the site of perturbation, and in all lower energy orbitals in a bonding fashion. Thus σ^* mixes in σ in a bonding fashion and is polarized toward the donor perturber. In the case of an acceptor, lying higher in energy than i, the quantity $\varepsilon_i - \varepsilon_k$ is negative, so that i mixes in all higher lying orbitals, j, in a bonding fashion, and all lower lying orbitals in an antibonding fashion.[80] Several generalizations should also be recognized: (a) Two matrix elements are important; polarization is most effective when both orbitals of A overlap considerably with the perturber orbital, (b) The perturber polarizes most the orbital to which it is closest in energy, and (c) The major polarizations of i are caused by those orbitals j that are closest in energy to i.

There is a second form of polarization, caused not by mixing of the perturber orbitals with those of A, but by the electrostatic effect of the charges of the perturber. Imamura and Hirano refer to the orbital mixing effect as a "dynamic orbital pumping" and the electrostatic effect as "static orbital pumping."[78] The latter has been described in detail by these workers. The extent of mixing of orbitals j into i can be calculated as:

$$C'_{ji} = \frac{H'_{ij}}{\epsilon_i - \epsilon_j}$$

This arises because H'_{ij}, the interaction of i with j, is negative in the presence of a positive charge perturber. This can be quantitatively demonstrated as shown in Figure 8.20 for a positive charge interacting with a hydrogen molecule. An even simpler generalization is possible here than for "dynamic orbital mixing." A positive charge will cause orbital i to mix in all higher lying orbitals in a bonding fashion and all lower lying orbitals in an antibonding fashion. Exactly the opposite is true for a negative charge. Here bonding or antibonding is determined by the site that experiences the greatest electrostatic attraction, or repulsion, for the positive or negative charge.[78]

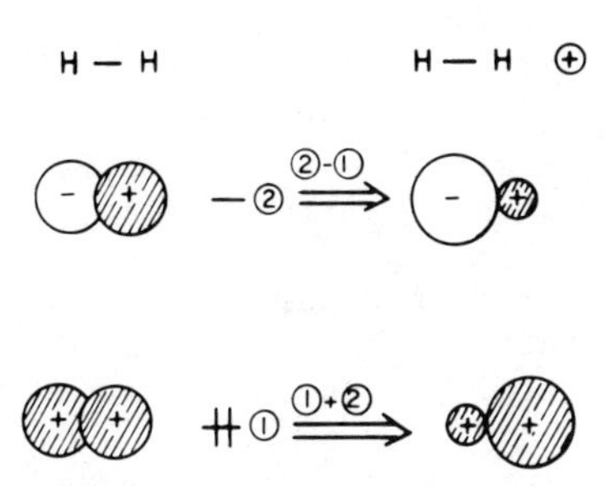

FIGURE 8.20. Polarization of H_2 induced by a positive charge

XVI. EVALUATIONS OF THE RELATIVE IMPORTANCE OF POLARIZATION RELATIVE TO CHARGE TRANSFER, ELECTROSTATIC, AND EXCHANGE INTERACTIONS

Because so many important insights have been gained from considerations of the influence on charge-transfer (frontier or HOMO-LUMO) interactions on chemical structure and reactivity, it is easy to lose sight of the fact that polarization and other factors may frequently be of pivotal importance in determining these factors.

One approach to the quantitative evaluation of different types of interactions is that developed by Kitawa, Morokuma, and co-workers.[69] Of the investigations reported by this group, one is of particular interest to the understanding of organic chemical phenomena. It has been observed experimentally that alkylation of amines increases their basicity, and this has been attributed to polarization effects. That is, as the alkyl group is increased in size, a more polarizable electron cloud is available to stabilize the positive charge. Similarly, polarization of alkyl groups has been invoked to explain the increasing acidity (increasing alkoxide stability) of alcohols as the size of the alkyl group is increased. Umeyama and Morokuma found that for the series of amines from ammonia through trimethylamine, the stabilizations due to electrostatic attraction and charge transfer are large and similar in magnitude ($\sim -92 \pm 7$ kcal/mol), and the polarization term is smaller, but varies most along the series from ammonia (-27 kcal/mol) to trimethylamine (-65 kcal/mol). Thus the increase in polarization along the series from ammonia to tertiary amine accounts, in large part, for the increased basicity.[81] Qualitatively, as the size or number of the alkyl groups increases, more vacant orbitals become available to mix with the filled orbitals and polarize electron density toward the positively charged proton. Similar treatments of hydrogen bonding[83] and "charge-transfer" complexation[70] indicate that all four types of interactions contribute comparably to stabilization, and in the latter case dispersion energy estimated separately is as important as the total SCF stabilization.[70]

These ideas of polarization have been applied to explain various chemical phenomena and will undoubtedly gain even greater importance as the ideas are assimilated by organic chemists. For example, following our qualitative explanation of Diels-Alder catalysis by Lewis acids,[84] Imamura and Hirano showed how the change in shape of an electron-deficient alkene LUMO could be explained by the ideas of "orbital pumping" or polarization.[78]

Fukui and Inagaki have considered catalysis in a more complete way.[85] In a formidable-looking series of equations,[86] they develop generalizations for the understanding of interactions between the orbitals (actually electronic configurations) of a catalyst and those of two substrate molecules. Although discussed in terms of configuration interaction, the model is basically similar to the polarization model discussed in the preceding paragraphs except that the focus is now on energy changes caused by the interaction of catalyst (C) orbitals with the orbitals of two molecules (A and B) simultaneously. In such a case there will be usual stabilizing interactions of C with A and of C with B, but, in addition, there are certain stabilizing terms due to the interaction of C with both A and B simultaneously.

Fukui and Inagaki considered two extreme cases. Suppose the catalyst is an acceptor with respect to both substrate molecules. Then the interaction of the substrate filled orbitals with the vacant orbitals of the catalyst will provide the most important stabilizing interactions. If second-order energy terms arising from charge transfer are sufficiently strong, there can also be a third-order stabilization of significant magnitude. The case is much like "pseudoexcitation" induced by strong charge transfer referred to earlier, but here we have a stabilizing interaction resulting from "pseudocharge transfer" induced by charge transfer. In configuration-interaction terms the interaction of a first configuration, A^+C^-, with a second one, B^+C^-, can lead to a third-order stabilization. The magnitude of this interaction will depend not only on the strength of the A^+C^- and B^+C^- mixing with the ground configuration, but on the interaction between the HOMO's (or other vacated orbitals) of A and B. The final result is that the complex can be stabilized by HOMO-HOMO interactions between A and B, making a formally forbidden reaction allowed. Similarly, if the catalyst is a donor relative to A and B, the LUMO-LUMO interaction between the substrates A and B can be facilitated, again rendering a forbidden reaction feasible. Fukui and Inagaki applied this reasoning to a variety of catalytic processes, all of which are formally Woodward-Hoffmann-forbidden reactions in the absence of catalyst.[85]

We have applied the concept of polarization,[80,87] combined with the use of frontier MO theory, to explain the unusual regioselectivity observed by Paquette and co-workers in the photochemical di-π-methane rearrangements of substituted benzonorbornadienes.[80,88] The experimental results were puzzling at first sight since the products do not obviously arise from the more stable diradical intermediates that might be postulated to be formed upon initial bridging. However, a rationalization of the regioselectivities observed in these reactions could be constructed through considerations of the preferred initial bonding modes dictated by the orbital coefficients of the aromatic ring polarized by polar substituents.

XVII. ORBITAL DISTORTION AND OTHER TREATMENTS OF SYN/ANTI STEREOSELECTIVITY

In 1975 Liotta published two thought-provoking communications explaining the stereoselectivity of certain reactions using the "orbital distortion" technique.[89] The method is essentially an empirical technique that attempts to relate the distortion of frontier MO's to the direction from which preferential nucleophilic and electrophilic attack occurs. Anh, Eisenstein, et al. had previously shown that alkyl groups placed unsymmetrically with respect to the plane of a carbonyl group caused an unsymmetrical distribution of charges,[90] and Klein used orbital mixing ideas to rationalize the stereochemistries of nucleophilic additions to cyclohexanones, and of electrophilic additions to methylenecyclohexanones.[91]

Liotta proposed that a π-system asymmetrically substituted (above or below the nodal plane of the π-system) would distort in such a way as to maximize bonding and minimize antibonding in the HOMO and LUMO.[89] Liotta suggested that because the donor orbital mixes with the HOMO of the alkene in an antibonding fashion, orbital distortion will occur in such a fashion as to relieve antibonding, so that the HOMO π-orbital will become more concentrated on the side of the plane away from the donor substituent orbital. Electrophilic attack should then occur away from the donor substituent, in the direction of highest HOMO density. Similar ideas were applied to LUMO's to predict the preferential stereochemistry of nucleophilic attack.[89] Liotta applied these ideas to a few examples in his papers, and he and Burgess have carried the gospel of orbital distortion far and wide in lectures and private communications. There is no doubt that the technique is very successful in accounting for the stereoselectivities of a variety of reactions. However, the theoretical basis of the technique is still somewhat obscure. That is, distortion of the kind suggested by Liotta[89] must result from σ-π mixing, and in a variety of calculations Liotta and Burgess, as well as Pierluigi Caramella,[93] in our laboratories, have found that the extent of σ-π mixing induced by asymmetrically located perturbers is miniscule, apparently far too small to account for the relatively high stereoselectivities experimentally observed. As suggested by Liotta and Burgess, it is conceivable, however, that the extent of σ-π mixing may increase as the reagent approaches the π-system.

Inagaki and Fukui have used second-order perturbation theory to explore σ-π mixing in norbornene and related systems.[79,92] They concluded that the strained $C_1C_7C_4$ bridge in this molecule induces σ-π mixing in the following way: Since the π-orbital is higher in energy than the σ-bridge donor orbital, the π-orbital will mix in all lower lying orbitals (e.g., in the C_2C_3 σ orbital) in a bonding fashion. The resulting mixing, although very small in absolute terms, causes the π-electron density near the bridge to be smaller than that away from the bridge. Inagaki and Fukui give electron-density plots that show greater electron density as required.

Finally, we have investigated this problem and concluded that secondary orbital interactions rather than any direct asymmetric distortion of the π-system may be responsible for apparent orbital distortion.[93] The HOMO of a donor-substituted π-system will consist of the π-orbital mixed in an antibonding fashion with the donor orbital. When an electrophile approaches the top of the π-system, there will be no substantial secondary orbital interactions with the substituent orbital, but on approach from the bottom, the electrophile will be confronted by substantial antibonding secondary orbital interactions. In effect, the orbital electron density will be distorted to some extent as suggested by Liotta, because the antibonding interaction diminishes electron density in the region of the node introduced by this interaction. However, the effect of this interaction is of a much shorter range than that suggested by the Liotta treatment. Application of this idea to the norbornene system studied by Inagaki and Fukui gives a different explanation of the greater electron density on the _exo_ face of norbornene; namely, the $C_1C_6C_5C_4$ bridge mixes in an antibonding sense with the norbornene π-orbital, and an electrophilic reagent approaching from the bottom will experience repulsive secondary orbital interactions.

To summarize this section, we note that the idea of orbital distortion is a second-order effect like that of polarization, which is undoubtedly of chemical significance. At the moment the theoretical origins of distortion are not fully understood, and the magnitude of this effect has not been determined.

XVIII. COMPUTATIONAL TECHNIQUES

I wish to conclude with a few brief undocumented comments on the state of computational techniques as applied to organic reactions. At this writing (January, 1977) two computer programs most generally used for the study of organic molecules and organic reactions are the Gaussian 70, written by Pople's group, and MINDO/3, written by Dewar's group. The former encompasses a spectrum of sophistication (approaching the Hartree-Fock limit) and cost, and the latter is said to be the ultimate in parameterization of semiempirical INDO method. Other excellent techniques, such as the ab initio GVB method and Lipscomb's PRDDO, are used less, probably because of lesser accessibility and, in the former case, to cost factors. Others, such as PCILO, are used extensively in certain areas of investigation, such as conformations of biomolecules. More sophisticated and vastly more expensive CI packages are used less generally despite the promising results that can be obtained, as reported herein. The problem with all financially feasible methods is that they are approximations, so that one can never rely totally on the predictions generated from these calculations. Instead, chemical judgment still needs to be applied, and one must be forever wary that the method used is deviating from reality in some unforeseen way. On the other hand, pure chemical judgment is not without its pitfalls, since many forms of chemical judgment are as yet uncalibrated. Many effects are often interrelated, and perhaps only a calculation, even an approximate one, is capable of determining the relative importance of various effects, all of which could contribute, in principle, to a given chemical phenomenon.

There is a definite division of opinion on the relative merit of ab initio and semiempirical schemes. Nowadays, theoreticians use ab initio (STO-3G) calculations much in the same fashion that Hückel calculations were used several decades ago. Energy differences obtained by these calculations are believed to be qualitatively accurate, but firm belief is only attached to calculations carried out with extended basis sets and considerable CI. In the other camp, which consists of Dewar and several theoretical organic chemists interested not so much in elegance of theory but in utility of results, MINDO/3 is believed to be a chemically reliable technique whose lesser cost more than compensates for its shortcomings.

However, as we have noted in this chapter, MINDO/3 may be misleading in certain predictions, but such cases are still poorly defined. Although it might be hoped that various computational techniques would converge on the same prediction (preferably corresponding to reality), this does not seem to be the case as far as ab initio techniques and MINDO/3 are concerned.

An interesting case in point has been given by Strausz, Gosavi, et al. in their report of calculations on the rearrangement of formylmethylene ($H\ddot{C}CHO$) to oxirene.[94] They show an amusing plot of the relative energies of these two species as a function of the "degree of sophistication" of the MO method. The plot resembles a Morse curve, with formylmethylene

30 kcal/mol more stable than oxirene by extended Hückel, oxirene more stable by 20 kcal/mol by MINDO/3 and NNDO, both equally stable by minimal basis set ab initio calculations, and formylmethylene becoming more stable again by extended basis calculations. Dewar might point out that MINDO/3 gives the minimum in the "Morse curve," but Strausz and co-workers suggest that the "dissociation limit" (formylmethylene ~18 kcal/mol more stable) will be the result of the most sophisticated calculations and experiment.

For the theoretical investigation of organic reactions at the present time, the safest (and economically most feasible) mode of operation would seem to be to thoroughly explore carefully chosen models of stable species, or small areas of a reaction surface by an ab initio technique, and to extract generalizations about the surface from such a study. Alternatively, the surface might be thoroughly explored by a technique such as MINDO/3, but important points should be spot-checked by an ab initio technique. As a rule of thumb, features of any type of calculation that can be explained in understandable terms (i.e., in terms of well-grounded principles, as opposed to "that's what the numbers say"), are the aspects of the calculation that should be believed.

XIX. REFERENCES

1.* J. F. Wolf, P. G. Harch, R. W. Taft, and W. G. Hehre, J. Am. Chem. Soc., 97, 2902 (1975).

2.* J. Weber, M. Yoshimine, and A. D. McLean, J. Chem. Phys., 64, 4159 (1976).

3.* J. M. McKelvey, S. Alexandratos, A. Streitweiser, Jr., J.-L. M. Abboud, and W. J. Hehre, J. Am. Chem. Soc., 98, 244 (1976).

4.* M. Taagepera, W. J. Hehre, R. D. Topsom, and R. W. Taft, J. Am. Chem. Soc., 98, 7438 (1976).

5. W. J. Hehre, R. T. McIver, Jr., J. A. Pople, and P. v. R. Schleyer, J. Am. Chem. Soc., 96, 7162 (1974).

6. The Gaussian 70 program used for these ab initio calculations is available from the Quantum Chemistry Program Exchange, Indiana University.

7. W. J. Hehre, Acc. Chem. Res., 8, 369 (1975).

8. W. J. Hehre, *Acc. Chem. Res.*, **9**, 399 (1976).

9.* A. Streitweiser, Jr., and J. E. Williams, Jr., *J. Am. Chem. Soc.*, **97**, 191 (1975).

10.* F. Bernardi, I. G. Csizmadia, A. Mangini, H. B. Schlegel, M.-H. Whangbo, and S. Wolfe, *J. Am. Chem. Soc.*, **97**, 2209 (1975).

11.* J.-M. Lehn and G. Wipff, *J. Am. Chem. Soc.*, **98**, 7498 (1976).

12.* N. D. Epiotis, R. L. Yates, F. Bernardi, and S. Wolfe, *J. Am. Chem. Soc.*, **98**, 5435 (1976).

13. R. Hoffmann, L. Radom, J. A. Pople, P. v. R. Schleyer, W. J. Hehre, and L. Salem, *J. Am. Chem. Soc.*, **94**, 6221 (1972).

14.* E. D. Jemmis, V. Buss, P. v. R. Schleyer, and L. C. Allen, *J. Am. Chem. Soc.*, **98**, 6483 (1976).

15. See, for example, B. M. Gimarc, *Acc. Chem. Res.*, **7**, 384 (1974).

16.* C. C. Levin, *J. Am. Chem. Soc.*, **97**, 5649 (1975); the same arguments have also been published by W. Cherry and N. Epiotis, *J. Am. Chem. Soc.*, **98**, 1135 (1976).

17.* J. H. Davis, W. A. Goddard III, and R. G. Bergman, *J. Am. Chem. Soc.*, **98**, 4015 (1976).

18.* W. T. Borden, *J. Am. Chem. Soc.*, **98**, 2695 (1976).

19.* W. J. Hehre, J. A. Pople, W. A. Lathan, L. Radom, E. Wasserman, and Z. R. Wasserman, *J. Am. Chem. Soc.*, **98**, 4378 (1976).

20.* P. Caramella and K. N. Houk, *J. Am. Chem. Soc.*, **98**, 6397 (1976). Optimized Geometries (4-31G) have since been obtained by Ruth Wells Gandour. The qualitative conclusions from STO-3G are reinforced by these more accurate calculations.

21.* A. J. Arduengo and E. M. Burgess, *J. Am. Chem. Soc.*, **98**, 5021 (1976).

22. (a) C. Y. Lin and A. Krantz, J. Chem. Soc., Chem. Commun., 1111 (1972); (b) O. L. Chapman, D. D. LaCruz, R. Roth, and J. Pacansky, J. Am. Chem. Soc., 95, 1337 (1973); (c) A. Krantz, C. Y. Lin, and M. D. Newton, J. Am. Chem. Soc., 95, 2744 (1973).

23. Reviewed by S. Masamune, Pure Appl. Chem., 44, 861 (1975).

24. Reviewed by G. Maier, Angew. Chem., Int. Ed. Engl., 13, 425 (1974).

25.* M. J. S. Dewar and H. Kollmar, J. Am. Chem. Soc., 97, 2933 (1975).

26.* W. T. Borden, J. Am. Chem. Soc., 97, 5968 (1975).

27. G. Maier, H.-G. Hartan, and T. Sayrac, Angew. Chem., Int. Ed. Engl., 15, 226 (1976).

28. W. G. Dauben, L. Salem, and N. J. Turro, Acc. Chem. Res., 8, 41 (1975).

29. L. Salem and C. Rowland, Angew. Chem., Int. Ed. Engl., 11, 92 (1972).

30. J. Michl, J. Mol. Photochem., 4, 243 (1972); Fortschr. Chem. Forsch., 46, 1 (1974).

31.* V. Bonačić-Koutecký, P. Bruckmann, P. Hiberty, J. Koutecký, C. Leforestier, and L. Salem, Angew. Chem., Int. Ed. Engl., 14, 575 (1975).

32.* P. Bruckmann and L. Salem, J. Am. Chem. Soc., 98, 5037 (1976).

33.* D. Grimbert, G. Segal, and A. Devaquet, J. Am. Chem. Soc., 97, 6629 (1975).

34.* L. Salem and W. D. Stohrer, J. Chem. Soc., Chem. Commun., 140 (1975).

35. L. Salem, C. Leforestier, G. Segal, and R. Wetmore, J. Am. Chem. Soc., 97, 479 (1975).

36. D. Grimbert and L. Salem, Chem. Phys. Lett., 43, 435 (1976).

37.* N. J. Turro, W. E. Farneth, and A. Devaquet, J. Am. Chem. Soc., 98, 7425 (1976).

38.* M. J. S. Dewar and W. Thiel, J. Am. Chem. Soc., 97, 3978 (1975).

39.* S. Inagaki and K. Fukui, J. Am. Chem. Soc., 97, 7480 (1975).

40.* W. R. Wadt and W. A. Goddard III, J. Am. Chem. Soc., 97, 3004 (1975).

41.* P. C. Hiberty, J. Am. Chem. Soc., 98, 6088 (1976).

42. S. P. Walch and W. A. Goddard III, J. Am. Chem. Soc., 97, 5319 (1975).

43. E. F. Hayes and A. K. Q. Siu, J. Am. Chem. Soc., 93, 2090 (1971).

44. K. N. Houk, J. Sims, R. E. Duke, Jr., R. W. Strozier, and J. K. George, J. Am. Chem. Soc., 95, 7287 (1973).

45.* P. Caramella, R. W. Gandour, C. G. Deville, J. A. Hall, and K. N. Houk, J. Am. Chem. Soc., 99, 385 (1977).

46. J. A. Berson, P. B. Dervan, R. Malherbe, and J. A. Jenkins, J. Am. Chem. Soc., 98, 5937 (1976).

47.* R. E. Townshend, G. Ramunni, G. Segal, W. J. Hehre, and L. Salem, J. Am. Chem. Soc., 98, 2190 (1976).

48.* M. J. S. Dewar, A. C. Griffin, and S. Kirschner, J. Am. Chem. Soc., 96, 6225 (1974).

49.* L. A. Burke, G. Leroy, and M. Sana, Theor. Chim. Acta, 40, 313 (1975). Professor Leroy has informed us that the "early" transition state reported in this reference becomes much like Salem's transition state[47] upon further geometry optimization.

50.* D. Poppinger, J. Am. Chem. Soc., 98, 7486 (1975); Aust. J. Chem., 29, 465 (1976).

51. M. J. S. Dewar, unpublished results reported at Louisiana State University, Baton Rouge, Louisiana, February 17, 1976.

52. P. Caramella, L. N. Domelsmith, and K. N. Houk, submitted for publication.

53. M. D. Gordon, T. Fukunaga, and H. E. Simmons, J. Am. Chem. Soc., 98, 8401 (1976).

54. G. Leroy and M. Sana, Tetrahedron, 31, 2091 (1975); 32, 709 (1976).

55. F. Keil and R. Ahlrichs, J. Am. Chem. Soc., 98, 4787 (1976).

56.* J. J. Dannenburg, J. Am. Chem. Soc., 98, 626 (1976).

57.* S. Scheiner, W. N. Lipscomb, and D. A. Kleier, J. Am. Chem. Soc., 98, 4770 (1976).

58. H.-B. Bürgi, J. M. Lehn, and G. Wipff, J. Am. Chem. Soc., 96, 1956 (1974).

59. H.-B. Bürgi, Angew. Chem., Int. Ed. Engl., 14, 460 (1975).

60.* J. E. Baldwin, J. Chem. Soc., Chem. Commun., 734 (1976).

61. H. Umeyama, A. Imamura, and C. Nagata, Chem. Pharm. Bull., 23, 3045 (1975).

62.* S. Scheiner, D. A. Kleier, and W. N. Lipscomb, Proc. Nat. Acad. Sci. U.S.A., 72, 2606 (1975).

63.* S. Scheiner and W. N. Lipscomb, Proc. Nat. Acad. Sci. U.S.A., 73, 432 (1976).

64. R. D. Gandour, G. M. Maggiora, and R. L. Schowen, J. Am. Chem. Soc., 96, 6967 (1974).

65. I. S. Wang and M. Karplus, J. Am. Chem. Soc., 95, 8160 (1973).

66.* X. Chapuisat and Y. Jean, J. Am. Chem. Soc., 97, 6325 (1975); Fortschr. Chem. Forsch., 68, 1 (1976).

67. R. Hoffmann, J. Am. Chem. Soc., 90, 1475 (1968).

68. G. A. Segal, J. Am. Chem. Soc., 96, 7892 (1974).

69.* K. Kitawa and K. Morokuma, Int. J. Quantum Chem., 10, 325 (1976).

70. W. A. Lathan, G. R. Pack, and K. Morokuma, J. Am. Chem. Soc., 97, 6624 (1975) and references cited therein.

71. R. S. Mulliken, J. Phys. Chem., 56, 295 (1952).

72.* N. D. Epiotis and R. L. Yates, J. Am. Chem. Soc., 98, 461 (1976).

73.* S. Inagaki, H. Fujimoto, and K. Fukui, J. Am. Chem. Soc., 97, 6109 (1975).

74. N. D. Epiotis, Angew. Chem., Int. Ed. Engl., 13, 751 (1974).

75. For a fuller description of the relationship between orbital interactions and configuration interactions, see K. N. Houk, in Pericyclic Reactions, R. E. Lehr and A. P. Marchand (eds.), Academic, New York, 1977, Vol. II.

76. K. N. Houk, Acc. Chem. Res., 8, 361 (1975).

77. L. Libit and R. Hoffmann, J. Am. Chem. Soc., 96, 1370 (1974); for applications to reactivity, see N. Fujimoto and R. Hoffmann, J. Phys. Chem., 78, 1874 (1974).

78.* A. Imamura and T. Hirano, J. Am. Chem. Soc., 97, 4192 (1975).

79.* S. Inagaki, H. Fujimoto, and K. Fukui, J. Am. Chem. Soc., 98, 4054 (1976).

80.* C. Santiago, K. N. Houk, R. A. Snow, and L. A. Paquette, J. Am. Chem. Soc., 98, 7443 (1976).

81.* H. Umeyama and K. Morokuma, J. Am. Chem. Soc., 98, 4400 (1976).

82. A similar method of calculation has been proposed by S.-Y. Chang, H. Weinstein, and D. Chou, Chem. Phys. Lett., 42, 145 (1976).

83. S.-I. Yamabe and K. Morokuma, J. Am. Chem. Soc., 97, 4458 (1975).

84. K. N. Houk and R. W. Strozier, J. Am. Chem. Soc., 95, 4094 (1973).

85. K. Fukui and S. Inagaki, J. Am. Chem. Soc., 97, 4445 (1975).

86. Quantum mechanicians often refer to the n^4 catastrophe (the number of integrals to be evaluated for n basis set members), but this catastrophe can also be applied to superscripts and subscripts used in equations to describe even simple concepts.

87. C. Santiago and K. N. Houk, J. Am. Chem. Soc., 98, 3380 (1976).

88. L. A. Paquette, D. M. Cottrell, R. A. Snow, K. B. Gifkins, and J. Clardy, J. Am. Chem. Soc., 97, 3275 (1975).

89. C. L. Liotta, Tetrahedron Lett., 519, 523 (1975).

90. N. T. Anh, O. Eisenstein, J.-M. Lefour, and M.-E. Trans Huu Dau, J. Am. Chem. Soc., 95, 6146 (1973).

91. J. Klein, Tetrahedron, 30, 3349 (1975); Tetrahedron Lett., 4307 (1973).

92. S. Inagaki and K. Fukui, Chem. Lett., 509 (1974).

93. P. Caramella and K. N. Houk, unpublished results.

94. O. P. Strausz, R. K. Gosavi, A. S. Denes, and I. G. Csizmadia, J. Am. Chem. Soc., 98, 4784 (1976).

INDEX